T0192071

THE FRONTIERS COLLECTION

THE FRONTIERS COLLECTION

The books in this collection are devoted to challenging and open problems at the forefront of modern science, including related philosophical debates. In contrast to typical research monographs, however, they strive to present their topics in a manner accessible also to scientifically literate non-specialists wishing to gain insight into the deeper implications and fascinating questions involved. Taken as a whole, the series reflects the need for a fundamental and interdisciplinary approach to modern science. Furthermore, it is intended to encourage active scientists in all areas to ponder over important and perhaps controversial issues beyond their own speciality. Extending from quantum physics and relativity to entropy, consciousness and complex systems—the Frontiers Collection will inspire readers to push back the frontiers of their own knowledge.

More information about this series at http://www.springer.com/series/5342

For a full list of published titles, please see back of book or springer.com/series/5342

Bernard d'Espagnat · Hervé Zwirn
Editors

The Quantum World

Philosophical Debates on Quantum Physics

Editors
Bernard d'Espagnat
(22 August 1921–1 August 2015)
University of Orsay
Paris
France

Hervé Zwirn
CNRS
Paris
France

ISSN 1612-3018 ISSN 2197-6619 (electronic)
THE FRONTIERS COLLECTION
ISBN 978-3-319-85655-1 ISBN 978-3-319-55420-4 (eBook)
DOI 10.1007/978-3-319-55420-4

Original French edition "Le monde quantique; Les débats philosophiques de la physique quantique" published by Editions Matériologiques, France, 2014

Printed on acid-free paper

This Springer imprint is published by Springer Nature
The registered company is Springer International Publishing AG
The registered company address is: Gewerbestrasse 11, 6330 Cham, Switzerland

Acknowledgements

This book relates the discussions that were hosted in Paris (France) between 2010 and 2013 under the *Académie des Sciences Morales et Politiques* and the *Collège de Physique et de Philosophie.* A French version of this book has been published in 2014 at Editions Matériologiques: Le monde quantique; Les débats philosophiques de la physique quantique.

The Collège de Physique et de Philosophie

His research in theoretical physics, primarily dedicated to the conceptual foundations of quantum mechanics, and his participation in debates surrounding this question, gave Bernard d'Espagnat a very simple idea: that of creating a society whose aim would be "the in-depth study of the contributions of contemporary physics to the theory of knowledge, especially with regards to the degree of plausibility of different conceptions of reality (and of our relationship with it) that have been or could be considered". Thus the *Collège de Physique et de Philosophie* was created in 2010, whose founding members, alongside Bernard d'Espagnat, were Michel Bitbol, Jean Petitot and Hervé Zwirn. The aim of the *Collège de Physique et de Philosophie* is twofold: (1) to organize meetings where physicists and philosophers can explore the new ideas arising from the latest research, and (2) to inform the public through conferences and other suitable means of the ideas that are stimulating debate.

The Académie des Sciences Morales et Politiques

The *Académie des Sciences Morales et Politiques* is one of the five academies of the *Institut de France*. Its actions include the organization of temporary think tanks bringing together periodically, over a few months or a few years, a small number of its members and external personalities for the study of a particular problem within its sphere of competence. Philosophy is one of them, which is why, aware of the advantages that exchanges between philosophers and physicists may bring to both parties, a think tank was set up to debate the possible contributions of recent research. Organized by the *Collège de Physique et de Philosophie*, these sessions were held at the Institute at irregular intervals between the end of 2010 and the start of 2013. The reader will find in this volume the transcript of these sessions.

Contents

About the Editors

Bernard d'Espagnat after spending time at the *École Polytechnique* and a few years at the French National Centre for Scientific Research (CNRS) with stints abroad, contributed to the creation of the theoretical physics research group at CERN (Geneva) and conducted research in elementary particles physics. Professor at the Paris-Orsay University, he continued his research in this field while developing an interest in the conceptual foundations of quantum mechanics. He organized workshops on this theme in 1970 in Varenna (Italy) and in 1976 in Erice (Sicily) (the latter in collaboration with John Bell) focusing mainly on quantum entanglement and non-locality, which was then in the process of being verified experimentally, and which he subsequently investigated further. His main works are: *Conceptual Foundations of Quantum Mechanics* (W.A. Benjamin 1971), *À la recherche du réel. Le regard d'un physicien* (Gauthier-Villars 1979), *Un atome de sagesse* (Seuil 1982), *Le Réel voilé. Analyse des concepts quantiques* (Fayard 1994), *Traité de physique et de philosophie* (Fayard 2002). Member of the Institute (*Académie des sciences morales et politiques*).

Hervé Zwirn is a physicist and epistemologist, and a senior researcher at the CNRS (the French National Centre for Scientific Research). He is a visiting researcher at the Institute of History and Philosophy of Sciences and Technology (IHPST, Paris1, CNRS & ENS), at the Research Centre for Applied Maths (CMLA, CNRS & ENS Cachan) and at the Paris Interdisciplinary Energy Research Institute (LIED, CNRS & Paris 7). His research focuses mainly on the interpretations of quantum mechanics and their philosophical consequences, on the axiomatic formalization of inference modes and on the behaviour of complex systems. He also proposed a model of the theoretical decision-making process using the mathematical framework of quantum mechanics. He is executive director of the *Consortium de Valorisation Thématique de l'Alliance Athéna* (UMS 3599, CNRS & Paris4), responsible for the commercial development and dissemination of the research output in social sciences of French laboratories. His main books are: *Les Limites de la connaissance* (Odile Jacob 2000), *Les Systèmes complexes* (Odile Jacob 2006), *Philosophie de la mécanique quantique* (with Jean Bricmont, Vuibert 2009), *Qu'appelle-t-on aujourd'hui les sciences de la complexité?* (with Gérard Weisbuch, Vuibert 2010).

List of Participants

Alain Aspect, an experimental physicist in quantum physics, is a professor and researcher at the *Institut d'Optique*, a professor at the *École Polytechnique*, and a member of the *Académie des Sciences*. With his collaborators, he has conducted experiments on fundamental quantum mechanics: (1) with entangled photon pairs to test Bell's inequalities, showing that theories of local hidden variables must be abandoned, and (2) with single photons in Wheeler's delayed choice experiment. His other work has focused on laser-induced atom cooling with Claude Cohen-Tannoudji, on atomic quantum optics and on quantum simulators with ultracold atoms.

Michel Bitbol is a researcher at the CNRS at the Husserl Archives (ENS), Paris. He has worked on Erwin Schrödinger's philosophy of physics, and has presented a neo-Kantian interpretation of quantum mechanics. He became interested in the link between the philosophy of quantum mechanics and the philosophy of the mind, and developed a conception of consciousness inspired by the epistemology of first-person knowledge.

Oliver Darigol is a senior researcher at the CNRS, and a specialist in the history of science. He works in the Sphere Laboratory (CNRS/ Paris-Diderot University).

Jean-Pierre Gazeau is an emeritus professor at the Paris-Diderot University. He is interested in questions of quantification and the relationship between quantum and information processing formalisms.

Alexei Grinbaum is a researcher at the Larsim Laboratory of the CEA in Saclay and member of the Paris Center for Quantum Computing. He is a specialist in fundamental quantum mechanics. Since 2003, he has worked in the field of quantum computer science, in particular on axiomatic approaches.

Michel Le Bellac is an emeritus professor at Nice University. He has published many books including the textbook *Mécanique quantique* (EDP-Sciences/Éditions du CNRS, third edition, 2013).

Catherine Pépin is a theoretical physicist at the *Institut de Physique Théorique* (IPhT) of the CEA. She works on the effects of quantum correlations, with a particular interest in superconductors at critical high temperatures.

Jean Petitot is a director of studies at the *École des Hautes Études en Sciences Sociales* (EHESS). He is interested in the epistemology of physical and mathematical models. He is a member of the International Academy for the Philosophy of Science.

Oliver Rey is a researcher in philosophy at the Institute for History and Philosophy of Sciences and Technology (IHPST) and teaches at the Paris 1 University.

Stéphanie Ruphy is a professor in philosophy at the Pierre Mendès-France University, Grenoble.

Alexis de Saint-Ours is a researcher in philosophy at the Sphere Laboratory (Paris-Diderot University).

Bertrand Saint-Sernin is a professor of philosophy and member of the *Académie des Sciences Morales et Politiques*.

Léna Soler is a philosopher of sciences, member of the Philosophy and History of Sciences Laboratory (Archives Henri-Poincaré, Nancy) and researcher associated with the IHPST.

Introduction

A Journey through the Quantum World

In the past, science and philosophy were inseparable. Aristotle was at the same time a physicist, a logician, and a philosopher. Closer to us, Descartes, Pascal and Leibniz were as famous for their philosophical contributions as for their mathematical discoveries. Even more recently, Henri Poincaré was equally a mathematician, a physicist and a philosopher. However, the links between science and philosophy have become largely distended over the course of the twentieth century and the gap between scientists and philosophers has increased to the point that we can say that a certain wariness, even hostility, has developed between the two classes of intellectuals. It is regrettable for two reasons that are symmetrical in the sense that one deals with the genesis of scientific questions and the other with the answers that are provided. On the one hand, it is important to remember that the questions that scientists ask themselves are quite often derived from fundamental philosophical questions about the Universe. On the other hand, uncovering the profound meaning of the results obtained using scientific theories often requires that a philosophical light be shown on them. The dialogue between scientists and philosophers must be restored for the benefit of knowledge in the broadest sense of the term. In 1984, a pioneering initiative was carried out by the *Académie des sciences*. Under the direction of Jean Hamburger, the *Académie* hosted a series of talks on the philosophy of science, each followed by a discussion between scientists and philosophers[1]. It is with the same ideal in mind that the *Collège de physique et de philosophie*, under the umbrella of the *Académie des sciences morales et politiques*, decided to organize a series of sessions bringing together physicist and philosophers and dedicated in its vast majority to recent advances exclusively published in specialized scientific journals. The present volume is none other than the transcript of these sessions.

It has become apparent that modern physics, and quantum physics in particular, can shed new light on profound philosophical questions and anyone who wants to seriously consider fundamental questions about realism, determinism, causality or locality cannot ignore the contribution of physics. It is not so much that quantum physics provides definitive answers to these questions but it eliminates certain philosophical positions that are no longer tenable today. Conversely, the formalisms used by physicists raise difficult questions of interpretation that physicists themselves are incapable of solving without an in-depth philosophical reflection. The dialogue between physicists and philosophers benefits both parties. Philosophers must take into account the teachings of physics so as not to support positions refuted by current research and physicists can rely on philosophers to enrich their reflection regarding the very foundation of their discipline.

Underneath all the themes covered during our sessions, there was the question of knowing whether independent reality existed "*per se*". In the volume cited previously, Jean Hamburger wrote: "Scientific exploration of the world is limitless, but it is also without the hope of attaining a reality free from the observer, its methods and its observational scale". The themes of these sessions were also linked to the problem of causality and that posed by the notion of information. The positions of physicists and philosophers are far from being the same, but some of these positions are now inadmissible in the light of the recent results of contemporary physics. What are the coherent concepts at this time? What refinements must we bring to the notion of realism in order for it to survive? Must we expand the concept of causality to take into account the fact that a putative independent reality may have appeared first, before space-time? Must we consider that nature has chosen a behaviour that is indeterministic by its essence?

It goes without saying that in such a field, we were not expecting definitive and clear-cut conclusions. The aim was primarily to allow each participant, through these discussions, to expand his personal reflection relative to a field undergoing rapid changes. The purpose of this volume is, of course, to inspire its readers to do likewise. Admittedly, quantum mechanics has raised from its inception questions of a philosophical order. But the rediscovery in the 1960s of quantum non-separability and, more generally, of entanglement at a distance (notions which were demonstrated by Erwin Schrödinger as early as 1935, but which strangely enough remained unnoticed for 30 years) led to new discoveries, and at the same time shed new light on what we knew before. Thus it stimulated pure research (and even applied research: e.g. inviolability in cryptography, perspectives in quantum calculations, etc.), where notions of relationship, linked to the state of knowing (*software*), take increasing precedence over notions of atomicity, classically linked to the state of being (*hardware*).

This current state of things surely justified a collective reassessment—taking recent advances into account—of the different ways of conceiving the very notion of reality, as well as that of knowledge and the relationship between the two, as much those already conceived by philosophers (ancient as well as contemporary) as new ones which can now be considered. Admittedly, with this approach, we had to allow technicality to have its place but thankfully, as both philosophers and physicists were present, we naturally avoided the pitfalls of using technicality and erudition for their own sakes.

One remarkable thing in this field is the variety of expectations that physicists have regarding the information they acquire. It has not always been this way. It seems that in the era of so-called "classical" physics, everyone expected physics to lift the veil on appearances, in other words provide an ever-increasing knowledge of physical reality "as it really is in itself". A point that clearly emerged from our discussions was that nowadays, only a small minority of established physicists believe that this is actually feasible. Considering the prominent role gained by quantum mechanics with its achievements in predicting observations, and the difficulties in interpreting it as a description of reality that is radically independent of human beings, some take the extreme opposite view. They ask nothing more of their field than to help predict what will be detected following such and such procedure. Others do not abandon the idea of finding a descriptive component, but consider that it is exclusively centred on communicable human experience. Others only expect that a theory gives them "food for thought", ideas for new experiments, and assess its validity on that basis. There are also those who, while denying the reality *per se* of objects, persist in upholding the notion of reality, some linking it only to structures, i.e. to relations and not to what is being linked, whereas others consider it a hypothesis that is necessary but not experimentally verifiable for solving a contradiction: that inherent to the idea of a "universally relative" reality.

In the following pages, we will never see a specific interpretation presented at the outset by the speaker. That is because they, physicists for the most part, generally provide evidence for these implicitly and even reluctantly, with the feeling that giving them too much thought would force them to overstep the boundaries of their own field. What they present are consequently purely scientific theories, with supporting experimental evidence. But also—and how can it be otherwise?—with the surprises these bring that compel us to reconsider matters. The discussions that followed these presentations highlight the different ways in which each tries to overcome these surprises. Perhaps the main feature of this volume is to capture the way some fundamental conceptual problems spring, so to speak, from a physics not at all designed for this purpose by its practitioners. Thus, unsurprisingly, this collection of presentations and discussions does not present or justify a specific philosophically preconceived manner of interpreting the knowledge provided by contemporary physics.

We will not be surprised, consequently, by the diversity of opinions expressed implicitly or explicitly on this matter, or generally by the fact that no attempt has been made to classify by topic what was broached in the various chapters. This diversity is seen as something valuable to take advantage of. Consequently, this volume essentially presents the transcripts of the sessions during which such and such a theory or such and such an experiment was studied and discussed. Édouard Brézin, member and former president of the *Académie des sciences*, presented the inaugural session. Entitled "The inescapable strangeness of the quantum world", he reminded us that it is decidedly not by using our sole "clear and distinct" (Descartes) ideas that we can interpret our experiments in this field. And furthermore, we must abandon considering a number of these ideas as having universal validity. The following session (session II) consisted of the extensive debate, which arose from this reminder, between physicists and philosophers concerning the notion of reality. We can already see during this session, from the frank and direct interventions of Michel Bitbol, Carlo Rovelli and others, some of the crucial problems mentioned previously. Sessions III and IV were dedicated to the important notion of decoherence, which appeared in the 1970's and accounts for the fact that macroscopic objects can never appear to us in a quantum superposition (that a cat will never been seen alive and dead, to reprise Schrödinger's famous example). The former (session III) consisted mainly of a presentation by the physicist Jean-Michel Raimond of an experiment conducted by Serge Haroche, Michel Brune, himself and other members of the Kastler-Brossel Laboratory (*École Normale Supérieure*) during the 1990's; a milestone in this field, it was the first to show that decoherence takes place over a very short but finite period of time. The latter (session IV) gave an account of the very rich discussion inspired by this experiment, covering various theoretical aspects of decoherence, including all those aspects that, evidently, touch upon questions of general philosophy (and once again without eluding the questions relative to the notion of reality).

We know that in parallel to so-called "orthodox" or "standard" quantum mechanics—the one exclusively taught in all the world's universities—there is another theory, still thriving and seen as more satisfactory by some renowned physicists, called the "pilot-wave" theory, devised in 1927 by Louis de Broglie and developed from 1952 by David Bohm, which provides the same observational predictions as the standard theory while being based on radically different ideas. Examination of this theory was the central focus of sessions V and VI. The former (session V) was centred on a presentation—given by Franck Laloë, also of the Kastler-Brossel Laboratory—on the main principles of this theory and its advantages, the main one probably being that it provides a simple, not to say trivial, explanation of what happens during a quantum measurement: an effectively remarkable feature considering that the explanations given by standard theory of this process are still matter for debate. The latter (IV) gave an account of the discussion on this topic, as well as an addendum by Franck Laloë where he detailed the serious reservations he has regarding this theory.

Sessions VII and VIII covered two distinct topics, non-locality and the relational interpretation of quantum mechanics, topics associated here for circumstantial reasons: the recent publication of an article by Carlo Rovelli and Matteo Smerlak which re-examines the former in the light of the latter. Session VII consisted of a presentation by the second author, firstly on the main principles of the relational interpretation of quantum mechanics initially conceived by the first author, and secondly of its impact on so-called non-locality. The subsequent discussion was fed in part by contributions by Carlo Rovelli himself, who attended this session. As for session VIII, it consisted of two short presentations, one by Michel Bitbol on non-locality and Bell's theorem, and one by Alexei Grinbaum on the notion of the observer in Rovelli's approach, followed by a more general debate where other questions relative to this relational approach were raised.

Session IX was dedicated to a presentation by Roger Balian on the methods for resolving the measurement problem using a statistical interpretation of quantum mechanics, through the detailed analysis of certain dynamic models. To finish, prospects for future research were highlighted in the tenth and final session of this volume by Carlo Rovelli, who rapidly covered quantum gravity theory which aims to unify quantum mechanics and general relativity, and therefore attempts to provide a unified framework for the whole of physical phenomena.

It so happens that in French, many idioms of the spoken language evoke the essence of ideas faster and better than their counterpart in the conventional written language. For this reason, the language transcribed here has been modified to match the written form only to the strict minimum required. Whenever possible, and therefore as often as possible, their spontaneous form has been voluntarily preserved. We therefore have voluntarily preserved this conversational style to capture the mood of these sessions; this sometimes leads to frank clashes of opposing views. The point was not to give an illusion of a unity of vision. From this point of view, this volume is fundamentally incomplete. It is perhaps in this incompleteness—this absence of conclusion—that we will find its relevance. At a time of important conceptual upheavals, this volume may, by its very incompleteness, suggest new lines of inquiry, which are varied and supported by new and practically indisputable facts. It highlights as much the positives as the difficulties encountered with each of these and aims to promote a comparative in-depth analysis of these ideas. A difficult task that requires knowledge, circumspection and an imaginative intelligence to a very high degree, but which is for that reason promising and consequently likely to attract the best minds.

1. *La Philosophie des sciences aujourd'hui*, sous la direction de Jean Hamburger, Gauthier-Villars, 1986.
2. These sessions took place from November 22, 2010 to March 25, 2013.

Bernard d'Espagnat
Hervé Zwirn

[illegible] relativity, and therefore attempt to pro- [illegible] of physical phenomena.

[illegible] of the spoken language [illegible] evoke the [illegible] conventional written [illegible] language transcribed [illegible] modified through [illegible] required. Whenever possible, and [illegible] passages [illegible] preserve [illegible] conversational [illegible] capture the [illegible]. The [illegible] point of view, the [illegible] this [illegible]

[illegible]

1. [illegible] Heidegger, [illegible]
2. [illegible]

Bernard d'Espagnat
Hervé Zwirn

Chapter 1
The Inescapable Strangeness of the Quantum World

Édouard Brézin

Bernard d'Espagnat. I am very happy to welcome you here, to this new working group entitled "Contributions of modern physics to the theory of knowledge" which you have all kindly agreed to participate in.

We will presently listen to Édouard Brézin, who needs no introduction especially here at the Institute as he presided over the Academy of Sciences for two years. I should nonetheless mention, very briefly, that his theoretical research has shed considerable light on the behaviour of condensed matter near critical points. However, not to worry, Mr. Brézin has no intention of lecturing you on this topic this evening. These are extremely cutting edge questions, and I think we will not reach such a high level of theoretical precision here, at least not during the first session! Mr Brézin is himself aware of this and has, on the contrary, kindly accepted to give an introductory presentation, which will be primarily an introduction to the questions we will discuss together, namely the highly conceptual questions regarding the relationship between physics and philosophy, in other words those touching upon the conception we may have of reality and what we mean by that. This is reflected in the title of his presentation "The inescapable strangeness of the quantum world".

1.1 Lecture by Édouard Brézin

First of all, I would very much like to thank Bernard d'Espagnat. It so happens that I attended his lectures on quantum mechanics as part of the Master of Advanced Studies in theoretical physics at Orsay, and it is therefore somewhat embarrassing

É. Brézin (✉)
École Normale Supérieure, Paris, France
e-mail: edouard.brezin@lpt.ens.fr

B. d'Espagnat and H. Zwirn (eds.), *The Quantum World*,
The Frontiers Collection, DOI 10.1007/978-3-319-55420-4_1

for me to talk about quantum mechanics in front of you. And that for two reasons in fact, as, like all modern physicists, I am only a practitioner of quantum physics, and one that has never seen it wanting. However, I have no particular opinion on quantum mechanics. I simply went on to teach the subject, before handing over this role to Jean-Michel Raimond. What I have to say on quantum mechanics only pertains to common practice: I will try to highlight its complexity, not from a technical point of view, but through the intellectual questions it raises. Is there a need to recall Richard Feynman's famous saying on this matter? Feynman is a great hero of quantum mechanics as the formulation he introduced in terms of path integrals, without calling into question the ideas of the Copenhagen School, has nonetheless completely changed our view of quantum mechanics in the modern period. On the first page of his lecture on quantum mechanics, in volume 3 of his famous lectures [1], he wrote: "*Because atomic behavior is so unlike ordinary experience, it is very difficult to get used to it, and it appears peculiar and mysterious to everyone—both to the novice and the experienced physicist. Even the experts do not understand it the way they would like to, and it is perfectly reasonable that they should not, because all of direct human experience and human intuition applies to large objects*".

Quantum mechanics has led us into a strange world that is nevertheless our own. I will not cover the history of quantum mechanics. This would require lengthy presentations and there are people more qualified than me to do this. I would like to remind you that from the last quarter of the 19th century until 1926, the inability of classical physics to describe numerous phenomena was apparent everywhere. Let us mention for example the study of the emission and absorption of atomic spectral rays, a type of atomic music similar to the music of vibrating strings. There is an accumulation, an absolutely extraordinary "numerology" for classifying these rays. It was, as you surely know, one of Niels Bohr's greatest triumphs to completely explain this "numerology" for the hydrogen atom, for which he received the Nobel Prize in 1992. The official "birth" of quantum mechanics took place in 1900 with Max Planck's article on black-body radiation, i.e. the radiation of a heated body, which in classical physics is totally incomprehensible in many respects. Classical reasoning leads in effect to radiation continually increasing as the frequency increases, a divergence that is of course inadmissible. In addition, the experiments of Otto Lummer and Ernst Pringsheim (1894) clearly show that there is a complete contradiction between what we observe and classical ideas. Planck tried to resolve this by formulating a very complex hypothesis of "quantified energy levels" and "radiation quanta" that he spent a lot of time trying to justify, and which was only completely validated many years later by Satyendranath, Bose and Albert Einstein.

It seems to me, however, that the most serious crisis followed the discovery of the atomic nucleus by Ernest Rutherford in 1909. The atom became completely incomprehensible and its existence opposed itself violently to classical physics. In 1909, Rutherford discovered that there was at the centre of the atom a hard nucleus that makes up practically all of its mass. He calculated the size of this nucleus to be approximately 10^{-15} m. It is therefore an extremely small nucleus, if we recall that the size of electronic orbitals is around 100,000 times greater. Let us imagine what 100,000 times greater means. If we image that the nucleus takes up the entire room we are in, which is 10 m in length, then the electrons would be 100,000 times 10 m, i.e. 1000 km away. Between this room and the 1000 km where the electrons orbit, there is nothing. This means that our matter is empty. Why is it empty? (Note: it is empty in ordinary matter, but astrophysics describes very different types of objects, such as stars only a few kilometres in diameter with a mass as large as that of the Sun—neutron stars). However, the matter that surrounds us is nearly empty. Why is that?

In other words, why do electrons not lose potential energy by moving closer to the nucleus? Normally, in classical physics, a charged particle that orbits around a centre is endowed with non-zero acceleration, since only rectilinear motion is not accelerated. Yet all accelerated charged particles emit radiation and thus lose energy. This is what happens in accelerators designed to produce synchrotron radiation: the latter is produced by electrons that are maintained in circular trajectories contained within a ring. Why do electrons within atoms not radiate while losing energy and falling on the nucleus? This was completely baffling and Bohr's reasoning that allowed the calculation of the size of electronic orbitals and postulated this absence of radiation did not really provide any qualitative explanation. This came with Werner Heisenberg and Erwin Schrödinger.

Before I try to give an intuitive explanation, I will describe quantum physics in a bit more in depth. We must start with the notion of *state*. States should be seen as vectors, only they are not vectors in ordinary space. They are vectors in abstract space. The vectors in question here are objects that can be added to each other and can be multiplied by complex numbers. Let $|\Psi_1\rangle$ and $|\Psi_2\rangle$ be vectors; we obtain another vector by adding vectors $|\Psi_1\rangle + |\Psi_2\rangle$, or more generally $\lambda_1|\Psi_1\rangle + \lambda_2|\Psi_2\rangle$, where λ_1 and λ_2 are complex numbers. Quantum physics is formulated in this way: there is a linear structure in state space, which is not three-dimensional space, but an abstract space. Under these conditions, we can superimpose states and consider state $|\Psi_1\rangle + |\Psi_2\rangle$ or state $|\Psi_1\rangle - |\Psi_2\rangle$ in the same way we can speak of sums or subtractions of two vectors.

We must realize that straight away this hypothesis implies a representation that is extremely different from our classical view. To illustrate this, I will briefly describe how chemical bonds present themselves in quantum physics. Imagine that we have two positively charged nuclei and one electron. If these two nuclei are far apart, it is possible to have a left state where the electron is bound to nucleus 1, and also a right state where the electron follows the right nucleus. However, if the two nuclei are close to each other, there is in quantum physics a phenomenon that is not allowed classically, which involves crossing potential barriers, which we call *quantum tunnelling*. Let us consider the potential energy "seen" by an electron: in the vicinity of nucleus 1, it sees an attractive potential well, and likewise in the vicinity of nucleus 2; there is a barrier between these two wells. Classically, if the electron was located in the vicinity of nucleus 1, this potential barrier would prevent it from approaching nucleus 2. But, because quantum mechanics, as a result of its wave-like nature—I will of course return to this point—allows a wave to travel beyond the space where classical particles are confined, known in optics as an evanescent wave, the electron can transit from one nucleus to another. A basic calculation shows that the possibility to cross the barrier implies that it is the state $|\text{left}\rangle + |\text{right}\rangle$ that has the lowest energy. This is to say that the two nuclei are bound by this electron: the energy is lower when an electron is in the state $|\text{left}\rangle + |\text{right}\rangle$ than when it accompanies one of the two nuclei. This is the origin of the bond since energy would have to be supplied to return to the situation where the electron would accompany only one nucleus.

The two nuclei are bound by the electron, and they are bound by this impossibility, in the state $|\text{left}\rangle + |\text{right}\rangle$, to say whether the electron is left or right. It is completely delocalized, and it is that which underlies the electronic bond. It is that which acts in molecules where multiple nuclei bind. It is therefore not an esoteric phenomenon since it operates in all atomic and molecular chemistry. (The covalent bond only reiterates the same phenomenon for two electrons rather than one, with opposite spins.)

The evolution of states is as deterministic in quantum mechanics as it is in classical mechanics, meaning that if we know the forces present, then the state at time t can be deduced from the state at time 0 by a perfectly defined algorithm. The algorithm is called Schrödinger's equation in non-relativistic instances, and becomes Dirac's equation when speeds approach that of light. When there are phenomena of particle creation and annihilation, we must enter what we call field theory. However, the evolution remains perfectly defined there.

I will introduce straight away Feynman's point of view, which is essential for understanding the difference between classical and quantum mechanics. The evolution is characterized by Feynman by a sum of the *histories*. In classical mechanics, if we know that at time 0 a particle is located at the initial position x_{in} and that at time T it will be at the final point x_f, than its trajectory is defined: it is defined by a principle that we call the principle of least action, or in certain instances Fermat's or Maupertuis' principle, which allows the determination of a classical trajectory from an initial point to a final point. Action minimization is at work for example in the Snell-Descartes law of refraction. To give you an illustration, we can imagine a lifeguard on a beach who sees someone drowning. What will the lifeguard do to save the victim? He will take the trajectory that requires the least amount of time to get to the victim. However, he obviously runs faster than he can swim: the straight line from the lifeguard to the victim is not the one that minimizes the length of time. Indeed, it would be preferable to travel a bit further along the beach than along the straight line (towards the victim) where the lifeguard would have to swim more. The result for light is the Snell-Descartes law of refraction. In reality, this representation of classical physics shows that it is rather strange. Indeed, the lifeguard chooses his trajectory because he knows exactly where the victim is located. It is under these conditions that he can determine the trajectory that will take the least amount of time. However, light that travels from air to water, for example, does not know where the "victim" is. In a way we can say that, strangely, classical physics is not causal: how can light determine its trajectory if it does not know where to go? It so happens that the quantum point of view explains, and enlightens, this paradox. We are here within a framework where it is classical physics that is paradoxical and quantum physics that resolves this paradox [2].

Indeed, in Feynman's vision, we must not imagine that there is only one trajectory, the one that minimizes action. In quantum physics, all trajectories are possible: any *path* that goes from x_{in} to x_f over a time T is realized. And to find the probability amplitude (I will come back to what we call probability amplitude), the amplitude that allows us to characterize in a quantum manner the passage from an initial point to a final point over time T, we must add up all the histories. We associate a complex number with each history (forgive me for introducing equations). This complex number is the exponential of the action divided by a constant, which is the Planck constant. And to find the quantum result, we must imagine that the particle going from the initial point to the final point uses all these trajectories, and that, for each, we add this factor, this complex number, called "amplitude" $e^{iS/\hbar}$.

In the end we have the sum:

$$\sum_{\text{all trajectories}} e^{iS/\hbar}$$

In all macroscopic situations, this sum is dominated by classical trajectories. Allow me to explain this a little bit more intuitively: if I throw a stone in the water, this stone will create a wave that will travel. If I throw two stones in the water, then it is a bit more complicated. We can see there will be points where there will be crests because two arriving waves will coincide and there will be points where the two waves are opposed, one is in a crest whereas the other is in a trough, and there will be a flat area. This results in interferences, which are very easy to visualize with two stones. But imagine that you throw a billion stones, or rather a billion billion stones in the water without stopping. This will produce a lapping of waves that is so dispersed that in the end barely anything will move. This is what happens in Feynman's sum. When we are in a situation where the quantum effects are negligible, the interferences produced by adding these diverse trajectories balance each other out nearly everywhere. Only the "dominant" trajectory of this sum, which is the classical trajectory, will remain. Therefore, this non-causal view of classical mechanics, this view where we know in advance where we need to go if we apply the principle of least action, is only an appearance that results from the interference of non-classical paths. Causality is a lot more present in quantum mechanics than in classical mechanics.

I will now refer back to the well-known experiment in the quantum world of Young's double-slit experiment. The experiment consisted, as you know, of the following set-up: a first plate pierced by two slits is placed between a source emitting a coherent light beam and a screen. Thomas Young in 1801 used a light source, and by observing the interferences on the final screen, deduced the wave-like nature of light. However, it has been repeated with many other types of particle beams. Clinton Davisson and Lester Germer discovered in 1927 that, like light, electrons could diffract, confirming Louis de Broglie's prediction in 1924 (which they had no knowledge of!) of the wave-particle duality associated with all particles. Today, electronic microscopes routinely use this property of electrons of being equally waves and particles.

So, let us return to Young's quantum slit experiment: if we first block slit number 2 on the plate, we have a blot on the final screen centred on the geometric image of slit number 1. If slit 1 is blocked, we have a blot around the image of slit number 2. These are the two possible classical trajectories. But as you know, if no slit is blocked, the Feynman sum over these two histories results in an illuminance that is not the sum of the anterior illuminances (*please note that the Feynman sum is a sum of probability amplitudes; the probabilities, in this case the illuminations, are the squared modules of the amplitudes: the square of a sum is not the sum of all squares*). The resulting illuminance law can only be explained because the electron passes through the two slits. This interference experiment has been much commented on, particularly by Feynman. Nowadays it is carried out routinely. We can now send electrons one by one and observe the successive impacts dispersed randomly on the end screen. When we compile the results, we uncover an illuminance pattern with dense impacts zones and zones with no impact, an image none other than the interference fringes of Young's experiment. Feynman has analysed this experiment at length. In particular, he imagined putting a lamp behind the pierced screen between slits 1 and 2 in order to determine, since electrons interact with light, whether the electron passed through slit 1 or slit 2. However, as soon as we establish through which slit the electron went, the interference fringes disappear. We find ourselves in the same situation as when we throw classical marbles (i.e. under conditions where the wave-like nature does not manifest itself due to the smallness of the associated de Broglie wavelength). Feynman wondered if the disappearance of interferences was provoked by the radiation of the lamp, placed there to determine through which slit the electron passes, that *perturbs* the electronic wave by this interaction. He imagined that we try to minimize this perturbation; to do this, we need to illuminate the slits with a radiation with the longest possible wavelength: in this way the corresponding photons, which have an impulse that is inversely proportional to their wavelength, are extremely "nice" to the electrons and do not perturb them too much. Unfortunately, the moment we do this and when the wavelength of the lamp's radiation becomes in the same order of magnitude as the distance between the two slits, we can no longer distinguish the two slits and therefore we cannot determine if the electron went through slit 1 or slit 2. We must therefore conclude that the classical view that an electron passes through a single slit is wrong: the electron passes through the two slits. As before, the electron was left and right and we could not say whether it was left or right. This is the reality of quantum mechanics.

Heisenberg has allowed us to qualitatively understand many quantum phenomena. There are indeed two main origins of quantum mechanics: one that, following de Broglie, is implemented by Schrödinger who looked for the equation that determined the propagation of the de Broglie wave. Heisenberg followed an independent path by studying the rules of transition between atomic levels and he came up with an a priori unrelated *matrix mechanics*. Following on from Schrödinger, Pascual Jordan showed that the two points of view were identical. In 1927 Heisenberg discovered that within this mechanics, observables can be *incompatible*. He used as an example the position and momentum of a particle (the momentum, for low speeds compared to c, is the speed of the particle multiplied by its mass). In the same way—although the spin had not yet been discovered in Heisenberg's days—I will use the spin as an example: with each particle, for example an electron, is associated a type of internal vector that we call its spin, which like any good vector of three-dimensional space has three components: S_x, S_y, and S_z. These are the observables: we can theoretically measure each one of them: S_x or S_y or S_z. Yet these observables are incompatible, meaning that if we measure S_x, we must abandon knowing S_y or S_z, and vice versa. In the same way, if we accurately measure the position of a particle, Heisenberg explains why we must abandon knowing accurately the momentum of the particle. To explain this, he devised a thought experiment which we call nowadays *Heisenberg's microscope*: in order to know where a particle is, we need to "see" it and therefore interact with it. We can for example send a light beam on this particle and deduce its position from the effect of the particle on the beam. To localize it with an uncertainty of δx, the radiation wavelength must be smaller than δx; indeed, if the wavelength is greater than δx, we will only observe a diffuse image that cannot be "resolved" below this wavelength. De Broglie's relationship tells us that within this light beam, the photons that make up wavelength l have a momentum $p = h/\lambda$, the inverse of λ, to within a constant which is the Planck constant. If λ is too small to localize the particle, this means that the photons have a large momentum. (Note: this is why we build accelerators: to have good microscopes, we need short wavelengths; to have short wavelengths, we need large momenta. Large momenta, requiring much energy, cannot only be achieved by accelerators. Accelerators are, in a way, only gigantic microscopes. The LHC [3] is a microscope that allows us to go down to wavelengths in the region of 10^{-18} cm.)

Therefore, the photons that are used to localize the particle in question have a large momentum in the order of $h/\delta x$. Under these conditions, because their momentum is large, the collision between photon and particle will be very "hard", and that will result in a high uncertainty δp on the momentum of the particle that will be greater than $h/\delta x$. Thus the momentum of the electron, which we perhaps knew before measuring its position, for instance because it was at rest, becomes even less knowable the better we have measured its position. These two variables are therefore incompatible and the microscope imagined by Heisenberg provides a qualitative explanation for this. For a long time, we thought that the rather paradoxical nature of quantum mechanics, as seen in these incompatible observables that do not exist classically, was limited to Heisenberg's microscope. In fact, as we will show later on, the situation is far more complex than that.

Heisenberg's reasoning nevertheless explains why a measurement can change the state of the system. Furthermore, it allows us to understand qualitatively the previously mentioned mystery surrounding the size of atoms. Indeed, if the size of atoms decreases, if the radius of gyration of an electron around a nucleus decreases, this means, according to Heisenberg's reasoning, that it is better localized. The closer it gets (by thought experiment) to the nucleus, the better localized it is. However, if we improve our knowledge of its position, we decrease, as Heisenberg demonstrated, our knowledge of its momentum; kinetic energy, proportional to the square of the momentum, can then take on high values. Thus, thinking we are reducing potential energy, we are actually increasing kinetic energy. However, in the *fundamental* state, the electron actually minimizes the total energy: the sum of kinetic energy and potential energy. More precisely, potential energy $V = -\alpha/r$ effectively decreases when the electron gets closer to the nucleus. However, according to Heisenberg's reasoning, kinetic energy takes on typical values that are inversely proportional to the square of the localization length, which is in the same order of magnitude as the distance r from the nucleus:

$$T = p^2/2m = h^2/2mr^2$$

Therefore, he adds potential energy to kinetic energy, the former increasing as r decreases. Minimizing the whole provides an optimal length: the minimal r^* of the sum $T + V$ is not in the order of 10^{-15} m, i.e. the size of the nucleus, but 10^{-10} m, i.e. the size of the atom, since this sum reaches its minimum at:

$$r^* = h^2/me^2 \sim 10^{-10}\,\mathrm{m}$$

Therefore, the gigantic size of atoms, the astounding emptiness of our matter, results from the incompatibility of the observable position and momentum as illustrated by Heisenberg's reasoning.

The greatest mysteries of quantum mechanics stem from *measurement*. In quantum mechanics, the outcome of a measurement is random, and this randomness seems irreducible. We are no longer in the usual framework of using probability calculations as in classical mechanics. Indeed, classically, we often resort to probabilities because we are in the presence of extremely complex systems. If we want to describe the slightest gram of matter, the gigantism of Avogadro's number, in the order of 10^{23} particles in any grain of matter, prevents us from following the evolution of all the degrees of freedom. Consequently, we frequently use statistical or probabilistic calculations. However, this is only a theoretically convenient way to bypass enormous calculations. Besides, nowadays, the computing power of computers allows us to follow the movement of thousands of particles without the need for any calculations of a statistical or probabilistic nature. Nevertheless, the use of probabilities, of *statistical mechanics*, is in the end the best method for understanding the behaviour of matter. But let us repeat that the use of probabilities in classical mechanics is solely for the sake of convenience.

For a mathematician, the notion of probability does not pose any particular problem: a probability is a measurement of an ensemble. Thus for an alternative with 50% for one term, and 50% for the other, we know how to calculate the probability of the various possible outcomes without any problem. If we state that a coin has one in two chances of landing on tails, the probability calculation to land on tails 100 times for 200 throws poses no problem for a mathematician. For a physicist, it's completely different: the coin is a physical object and the probability assigned to it is just an estimate that relies on our knowledge of this coin. If we have no reason to believe the coin is biased, we will assign a priori a probability of ½, allowing ourselves to change our point of view if we notice when tossing that 50% does not fit. Note that the probability is not a characteristic of the coin; it is only a convenient way to bypass lengthy calculations. By tossing a coin, we could imagine modelling a priori the way it is thrown, the way it spins in the air, the way it falls on the table, and without doing any probabilistic calculation, we could try to deduce which side it will land on depending on experimental conditions. This is possible theoretically in classical mechanics, even if it is very complicated. In quantum mechanics, however, the notion of probability is of a different nature: it is irreducible; there is no way of avoiding the introduction of random variables. This will be illustrated below.

Let us come back to the electron spin, and the vector that is associated with it. The outcome of the measurement of this spin component along the z axis is either + or − within the units associated with the Planck constant. Thus, there are two possible values $\pm\hbar/2$, designated either by $\pm$ or by an arrow pointing up or down; two possible values and nothing more. This is already very different from classical mechanics, since if we measure the component of a vector along an axis in classical mechanics, the outcome, which depends on the angle of the vector with respect to the axis, can be any value between the length of the vector and its opposite. It is not like this in quantum mechanics. We must accept that this is the measurement outcome. There are more complicated things: if the state of the spin prior to measurement is state $|\Psi\rangle$, then after a measurement producing for example the outcome + along the z axis (+ means $+\hbar/2$), the state becomes an upwards state of the spin. The measurement has changed the state of the system. This does not seem extremely paradoxical in itself, if we recall Heisenberg's microscope: a measurement has perturbed the spin, but even so this immediately raises the question: what is a measurement? First of all, we find ourselves with two evolution principles. The first is the one we mentioned above; knowing the initial state, we can deduce the state at time t, for example by using Schrödinger's equation: $|\Psi(0)\rangle \to |\Psi(t)\rangle$. However, the measurement introduces a second evolution principle, called the *reduction postulate*, since if the state prior to measurement is $|\Psi\rangle$, after knowledge of the outcome, this state changes to $|\Psi\rangle \to |\Phi\rangle$. This raises the question of knowing when we should take a measurement, and what a measurement is.

Could we do without this second law of evolution which is specific to the measurement process? For this, we could try to describe the measuring apparatus as a complementary element of the physical world in question. We would thus consider the ensemble made up of (i) the system, which is initially in state $|\Psi\rangle$, (ii) the measuring apparatus, which is in state $|X\rangle$; we could try to describe the evolution of the state of the ensemble $|\Psi, X\rangle$, i.e. the electron/measuring apparatus ensemble. Is the evolution $|\Psi, X > (0) \rightarrow |\Psi, X > (t)$ of the ensemble, studied by a gigantic Schrödinger equation in which we include the evolution of the measuring apparatus and its coupling with the studied system, likely to explain the second principle, namely the reduction of the initial wave from the evolution equation? As we will see later on, the answer is probably no: the measurement process is not a stage that we can deal with in the same way as the studied system, using ordinary quantum dynamics. Is the random nature of measurement limited to Heisenberg's microscope, i.e. to the inevitable perturbation of the studied system by the measuring apparatus? Here as well, the answer is no, as we will see shortly.

Let me come back to the paradoxes that have not ceased to haunt quantum mechanics. Among the most famous are "Schrödinger's cat" and "Wigner's friend", to use their now popular names. What is Schrödinger's cat? The well-known image was derived from discussions in which Bohr and Schrödinger distinguished themselves. The question was to know whether the strange nature of the quantum world could be transposed to the everyday macroscopic world. In order to do so, they imagined a radioactive particle; it is in an excited state prior to emitting its radiation, and de-excites after emission. However, it can be in one state or the other. For a given particle, it is well-known that it can be in one of the two states, or in any superposition of these two states. Schrödinger said: now imagine that this radioactive particle is in a box where there is cat. The box is sealed, we do not look inside; if the particle has disintegrated before the box is opened, than the radiation has killed the cat; if it has not, then the cat is still alive. Therefore the cat, like the particle, can be in two states: one state $|\text{live cat}\rangle$ or one state $|\text{dead cat}\rangle$, or still, as long as we have not looked inside the box, we can have superposition of states for the macroscopic object "cat" itself. Before opening the box to see the state of the cat, it can be in a superposition $|\text{live cat}\rangle + |\text{dead cat}\rangle$. This seems absurd, much more so than a superposition of states for a particle. What can we make of this? I hope that Jean-Michel Raimond, who has carried numerous experiments on photon cavities, will tell us about certain explicit experiments that show how we go from a microscopic object to a macroscopic state. That is, here, the first paradox.

The second paradox is called "Wigner's friend": when we open the box with the cat inside, we reduce the state $|\text{alive}\rangle + |\text{dead}\rangle$ to the state $|\text{live cat}\rangle$ or the state $|\text{dead cat}\rangle$ depending on the state in which we find the animal. However, if Wigner's friend looks inside the box, and tells him the result only much later, at what moment must Wigner reduce the state of the cat? We are faced with the astounding conclusion that it is consciousness that produces the reduction. In this case, we have to introduce the observer's consciousness to carry out state reductions. This is very troubling, even more so when we apply quantum mechanics to the study of the primitive universe prior to the existence of any observer. What does consciousness mean in this case?

There is a way to resolve this, but I must tread carefully. Our colleague d'Espagnat has thought about this more than anyone else. There is no paradox, so it seems to me, if the state of the system, as part of the description of the system by the observer, corresponds to the observer's *own* subjective knowledge of the system.

In other words, as long as I have not carried out any observation, describing the cat in a state $|\text{live cat}\rangle + |\text{dead cat}\rangle$ is not surprising or paradoxical: it is my knowledge of the system, I describe it thus in the absence of an ulterior measurement that would lead me to change my description. (Allow me to mention the conversations with Rudolf Peierls in which he stated that there was no paradox if we considered that the state attributed to the system by the observer was only a reflection of his own knowledge of the system.) However, from this perspective, we do not really know where the physical reality is in this subjective description of the world. Does a physical reality exist independently of the observer? We would much prefer to be able to say that the reduction postulate related to measurement is a convenience that takes the place of a description of a complicated measuring apparatus. How satisfying it would be to think that if we included the measuring apparatus in the equations, we could avoid the problem of measurement and its subjectivity. However, as you will see, that is probably not possible.

Indeed, in 1935 things became considerably more complicated with a famous article [4], which at the time gained much interest from Schrödinger and Bohr who discussed it at length with Einstein, but which afterwards remained forgotten for a long time. When I was a young physicist, we generally thought that there was no difficulty, that quantum physics worked perfectly, that Einstein, Podolsky and Rosen (EPR) had dwelled on questions of no interest. It was only after the publication of John Bell's article in 1964 [5] that things changed and that the *EPR paradox* became once again the centre of much attention. EPR imagined a particle, for instance a positronium, which disintegrates into an electron and a positron; in a laboratory system where the positronium is at rest, the electron and the positron drift apart in opposite directions. Their spins, that of the electron and that of the positron, are in an *entangled* state, which we must describe. The emitted electron and positron both have spin; the conservation of the angular momentum (a consequence of invariance by rotation) implies that their spins, measured along any axis, are opposite. We can easily see what state $|+, -\rangle$ would be, where the electron spin is positive and that of the positron is negative; and equally for the state of opposite spins $|-, +\rangle$. However, quantum mechanics requires that these particles are emitted in state $|+, -\rangle + |-, +\rangle$, where we can no longer say what the spins are either for the electron or for the positron. Let us imagine that under these conditions a first investigator (nowadays conventionally called Alice) measures the electron spin and finds a positive outcome. This implies that the initial state of spin $|+, -\rangle + |-, +\rangle$ has been reduced through measurement to state $|+, -\rangle$. Consequently, after this measurement, which produced a positive outcome for the electron, we know with certainty that a second investigator (Bob) placed at an arbitrary distance from

Alice, and who has decided to measure the positron spin, will find a negative outcome. This appeared to EPR as extraordinarily surprising: Alice, by measuring, can predict from her measurement the outcome that Bob will find. There is no violation of relativistic *causality*, which postulates that no information can be transmitted at superluminal speed. When the experiment is over, the two investigators meet up. The list of outcomes obtained by Alice, as the experiment is repeated many times, is, let us say $\{+, -, -, +, -, -, -\}$; she knows that Bob will produce the opposite list: $\{-, +, +, -, +, +, +\}$. She does not need to look at the second list to know it. However, Alice can tell Bob afterwards what he will find only by ordinary means of information transmission.

For a long time, we considered that it was not that problematic. I will try to show you to what extent it is really surprising. (We are at the heart of the problem concerning knowledge.) In a certain way, we can superficially consider that the EPR situation is no more surprising than the following common experience: I arrive at my desk, I search my pockets, I find only a single glove, the right hand one; I therefore know the one I left at home is the left hand glove. Nothing paradoxical or mysterious: I know it straight away and if I want to transmit the information to someone at home, I can phone to say that the glove on the cabinet is the left hand one. However the following experiment is more complex: imagine that, before leaving their respective laboratories where they will analyse emitted twin particles, Alice and Bob agree that they will both measure the spin component along the x axis. Here, we know that if one finds +, the other will find—with certainty. But imagine that Alice, playing a trick on her colleague, measures the y component instead of S_x without telling him. Therefore she knows that Bob, who is still measuring S_x, instead of finding a well-determined outcome, will find with 50% probability $S_x = +$ or $S_x = -$. In other words, the effective outcome that observer 2 will find depends on what observer 1 is doing. It is a bit as if I said: if I am a magician and I can transform my right hand glove into a left hand glove, then I have also transformed the one at home from a left hand to a right hand one [6].

This implies that measurement is a non-local process. Once again it is not a violation of relativistic causality, it is a violation of measurement locality. Let us come back to what we were saying before: if I imagine that the measurement of the electron spin is an interaction between it and the measuring apparatus in Alice's laboratory, it will be localized to the ensemble "measuring apparatus/studied system". Now in the EPR situation, although the positron has travelled far, the measurement of the electron spin has also measured the positron spin. Therefore the measurement is not limited to a local interaction with a macroscopic apparatus. This tells us that, presumably, we cannot include the measuring apparatus in the studied system. If we included the measuring apparatus in the evolution equation, we would only put local interactions between the studied system and the apparatus. This non-locality of the measurement process shows that, presumably, it is not possible to leave out the second postulate of wave reduction. Nowadays, this experiment is carried out routinely with twin photons. (We even sell cryptography systems that rely on these twin photons; in optical fibres, the twin photons move kilometres apart. This thought experiment by EPR has now become a practical object that is used to know if communication between Alice and Bob has or has not been intercepted. This is known as *quantum cryptography*. If a spy has intercepted the communication, the absolute correlation between the polarization states of the twin photons is lost.)

One last point: are hidden variables a substitute for quantum mechanics?

Let us examine how quantum mechanics differs from classical mechanics. For example, we saw that the measurement of the spin along Ox and the measurement of the spin along Oy are incompatible variables. If we measure S_x, and the outcome is positive, the measurement of S_y is completely random and produces either + or − with equal probabilities. This is what quantum mechanics tells us. We could interpret this result thinking that it is the measuring apparatus that has modified the particle for which we have measured the spin. This modification would be such that, once we know S_x, then S_y becomes indeterminate, in the same way that once a position is measured, the momentum becomes indeterminate. Therefore, in this "naive" version of Heisenberg's vision, it is the measurement of S_x that prevents us from knowing S_y. However, we will see that the situation is even more complicated.

One way to "represent" this is to imagine that in reality, the electron population is made up, for example, of four types of particles: one where S_x and S_y are positive (+, +), one where S_x is positive and S_y is negative (+, −), one type (−, +), and one type (−, −). Four population types I call N_1, N_2, N_3, N_4. Thus the measurement of S_x will produce a positive outcome + with a probability that is the ratio of the number of favourable cases $N_1 + N_2$, the number of particles with a positive S_x, divided by the total number: $(N_1 + N_2)/(N_1 + N_2 + N_3 + N_4)$. After measuring S_x, I imagine that I abandon knowing S_y, because the measurement of S_x has perturbed the system too much. Let us note that we have stepped outside of the quantum framework by imagining these four population types, as this framework does not say that particles of the type S_x positive, S_y positive, etc., exist. It simply says that after having measured S_x the outcome for S_y is random, with equal probabilities. This limitation is not called into question by the principle of hidden variables, i.e. the replacement of the quantum description by the introduction of these four particle types which are not accessible to our measurements but provide a representation of the system. This seemed to be an alternative representation of the same physical facts.

For a long time, this interpretation went against quantum mechanics as the latter does not allow us to consider one of these types, such as S_x positive, S_y negative, since we cannot measure them simultaneously. Quantum mechanics insists on the fact that we can only conceive of what is measurable. It rejects the idea of a finer underlying reality that remains hidden because our measuring apparatus are too crude and violently perturb the system. Quantum mechanics claims that there is no finer reality. For a long time, quantum mechanics and hidden variables appeared as two points of view which were opposite but in a way equivalent in their practical predictions of measurement outcomes. We spoke of another interpretation of quantum mechanics. It seemed to be an alternative framework that was better suited for providing a more intuitive view of what was going on.

This ended a few years after the important contribution of an Irish theoretician, you (Bernard d'Espagnat), must have known him well at CERN, namely John Bell.

He showed in 1964 [7] (it was not what he was trying to show at the time) that no physical theory that replaced quantum mechanics with a local theory of hidden variables was compatible with it. We cannot reproduce quantum mechanics with such a theory. The two points of view are not equivalent. Bell demonstrated that they were opposed by simple inequality. I will try to give you an intuitive version, without calculations, of what Bell's inequality is. To do so, I will consider a simple model of hidden variables, devised by Eugene Wigner, which extends the above discussion on the four types of spins. Wigner considered an EPR-type situation, therefore with two emitted particles with opposite spins. Two investigators can measure the spin values, S_1 and S_2, along three axes Oa, Ob and Oc. In a non-quantum interpretation, either particle can be in the following states: the first particle can be for example of the type: along Oa = +, along Ob = +, along Oc = +, that is (+, +, +), etc. There are eight possibilities ranging from (+, +, +) to (−, −, −): {(+, +, +), (+, +, −), (+, −, +), (+, −, −), (−, +, +), (−, +, −), (−, −, +), (−, −, −)}.

So much for the first particle. If, for example, it is in the state (+, +, −) and if I chose to measure Sa, I would find +, and if I measured Sc, I would find −. In the EPR situation, if particle 1 is in the state (+, +, −), then we know that particle 2 is in the opposite state (−, −, +), and this applies to all eight types.

Let us describe the EPR experiment from the two points of view: (i) hidden variables with eight possibilities for each particle; (ii) quantum mechanics.

1° *Hidden variables*. Each particle is in one of the eight possible states, the spins of the two particles are opposite, meaning first particle (+, +, +), second particle (−, −, −); first particle (+, +, −), second particle (−, −, +), etc. In total eight possibilities for which we know the populations:

	1	2
N_1	$(a+, b+, c+)$	$(a-, b-, c-)$
N_2	$(a+, b+, c-)$	$(a-, b-, c+)$
N_3	$(a+, b-, c+)$	$(a-, b+, c-)$
N_4	$(a+, b-, c-)$	$(a-, b+, c+)$
N_5	$(a-, b+, c+)$	$(a+, b-, c-)$
N_6	$(a-, b+, c-)$	$(a+, b-, c+)$
N_7	$(a-, b-, c+)$	$(a+, b+, c-)$
N_8	$(a-, b-, c-)$	$(a+, b+, c+)$

We imagine that we repeat the EPR experiment many times. Suppose that Alice measures S_a and Bob measures S_b. What is the probability of Alice finding + along Oa and Bob finding + along Ob? For Alice to find + along Oa, I have the first four possibilities but only the third and fourth ones are compatible with the fact the Alice finds $a+$, and Bob finds $b+$. Therefore the probability $P(a+, b+)$ of $(a+, b+)$ is $N_3 + N_4/N$ (where $N = N_1 + \cdots + N_8$ is the total number of possible outcomes). If Alice measures S_a and Bob S_c, I find $P(a+, c+) = N_2 + N_4/N$. If Alice measures S_c and Bob S_b, the probability is $P(c+, b+) = N_3 + N_7/N$, etc. You can see that if I add

$P(a+, c+)$ and $P(c+, b+)$, I find $N_2 + N_4 + N_3 + N_7/N$ which is greater than $P(a+, b+) = N_3 + N_4/N$ since there are two additional terms:

$$P(a+,c+)+P(c+,b+)=N_2+N_4+N_3+N_7/N \geq N_3+N_4/N=P(a+,b+)$$

Here is Bell's inequality: the probability of $(a+, c+)$ added to the probability of $(c+, b+)$ is, in the hidden variables model, greater than the probability of $(a+, b+)$. This is what theory of hidden variables says, in a very naive way.

2° *Quantum mechanics*. We know how to calculate the probability of $(a+, b+)$; I will use the result that is found in all quantum mechanics textbooks,

$$p(a+,b+)_{QM}= \tfrac{1}{2}\sin^2(\vartheta/2)$$

where θ is the angle between a and b.

For the first time—this is what Bell found—there is a hidden variables prediction and there is a quantum mechanics calculation. Are they compatible or not?

Consider a particular configuration: let there be three coplanar vectors a, b, c: a and b are perpendicular, and c is the bisector of angle (a, b). For $P(a+, c+)$, quantum mechanics tells us that it is half the sine squared of 45° divided by 2, which is approximately 0.0732: $P(a+, c+) = \tfrac{1}{2}(\sin^2(45°/2)) \sim 0.0732$. Same thing for $P(b+, c+)$.

For $(a+, b+)$ however, the angle between a and b is 90°, the $\sin^2$ of 45° is ½ thus $P(a+, b+)$ is undoubtable $\tfrac{1}{2}(\sin^2(90°/2)) = \tfrac{1}{2}$. ½ = 0.25. 0.25 is greater than 0.0732 + 0.0732: $P(a+, b+)$ is greater than $P(a+, c+) + P(c+, b+)$.

This quantum mechanics result violates Bell's inequality which, relying on hidden variables, led to an inequality in the opposite direction. An experiment was designed [8] and put in place, with further refinements by Alain Aspect and his team [9] to eliminate all causal interactions between "Alice and Bob".

The experiment demonstrated that the quantum mechanics result was indisputable: Bell's inequality is violated and massively so. There are even now experimental processes where it is completely violated [10]. Classical impossibilities are realized 100% of the time in quantum situations.

There is no doubt that quantum mechanics and local hidden variables are incompatible. This condemns us to adopt this strange quantum mechanics, strange because it is not intuitive. We do not really know what a measurement means, and in any case, its non-local nature is something astounding. Nevertheless it is an experimental reality.

Nowadays? Nowadays, quantum mechanics describes all atomic and subatomic phenomena known to this day. This was Richard Feynman's, Julian Schwinger's and Sin-Itiro Tomonaga's great achievement to understand, in the post-war years, the compatibility of quantum mechanics and electromagnetic interactions [11]. It was the great achievement of the 1970s to understand the compatibility of quantum mechanics and weak and strong nuclear forces. What we call nowadays the standard model. However, attempts to make quantum mechanics compatible with Einstein's theory of the last force that was not included above, namely gravitation, are currently only pure speculations with no experimental counterparts.

Various theoreticians, in particular Gerard 't Hooft, have highlighted the problem of using quantum theory when considering black holes. You know that black holes are now a reality. We even discovered less than ten years ago that at the centre of our galaxy, there is a huge black hole of some millions of solar masses. Therefore, this is not speculation: black holes exist. 't Hooft was the first to show that if we had a black hole, what we call unitarity, meaning the conservation of information, is a priori violated. A pure state will transform itself in what we call a mixed state, a density matrix, which is a loss of unitarity. This is somewhat technical: it should be a lecture in itself. However, after Stephen Hawking showed that a black hole "evaporates" in a quantum manner, physicists wondered whether this radiation produced a restoration of information compatible with quantum mechanics. This is one of the claims of the superstring theory, and I believe it is one of its great achievements, even though for now, there is no experimental evidence [12].

In any case, the length scale introduced by Planck, for dimensional reasons, which combines the Newton constant, the speed of light and the Planck constant ($l_P = (Gh/2pc^3)^{1/2} \sim 1.6 \times 10^{-35}$ m) is an extraordinarily small length scale, much smaller than what we are capable of measuring at this time (at CERN, we can achieve 10^{-12} m). At that scale no one knows what happens. For now, the validity of quantum mechanics at the Planck length remains an open question. Perhaps one day, at that scale, we will have to modify quantum mechanics. Even if it was modified at that scale, at the scale where we carry out our experiments in the laboratory, there is no substitute for quantum mechanics and there never will be, in the same way that the theory of relativity has not called into question the application of Newton's theory for all that concerns the macroscopic physics of planetary movement. Therefore, there is no doubt that the validity of quantum mechanics will never be called into question, even if at the Planck length there could be conceptual modifications which are at present completely unknown. This is where we are at: there is, as Feynman used to say, for the physicist as for the novice, much difficulty in having an intuitive view of what a measurement could be.

1.2 Discussion

Bernard d'Espagnat. A big "thank you", dear Édouard Brézin, for this magnificent introduction to our field. With superposition, measurement theory, problems of non-locality and so on, we have covered a range of questions as formidable as they are fundamental. We have our work cut out for us in the future!

I suppose there are questions. Léna Soler.

Léna Soler. I would like to begin by putting in perspective that which impacts the way our work is going. I believe our working group is called "Contributions of modern physics to the theory of knowledge". We need to place ourselves from the start within a large perspective, which does not presuppose that at the start physics needs to be interpreted in a realistic sense, in the broad sense; that is to say there is no way of escaping the fact that our physical theories and the way we interpret them nowadays tells us something of the real world. I think we should take this questioning seriously, that maybe, maybe… I know the positions and leanings of, at least, some of those present today regarding this question…

We should be able to take the possibility of, let us say, instrumentalist interpretations of scientific theories seriously; where it would be a tool that allows us to do things, to answer certain questions, but which would not necessarily be a literal interpretation. I believe as well that we should take into account the possibility of multiple interpretations. Here, in the case quantum physics, we have an alternative interpretation. However, you must be able to take it seriously. It seems to me that what you have explained today, in a completely intuitive and very clear manner, is the counter-intuitive nature of a certain interpretation of this formalism, which is the interpretation that is primarily taught nowadays. However, I am not sure that this is inevitable insofar as there is an interpretation, known as Bohm's, which is associated with a very different scenario, and with a very different measurement theory. I would really like to know your position regarding this other possibility, before going into any detail.

Édouard Brézin. Quantum mechanics involves a certain number of rules that allow us, when faced with a process, to try to describe, if we know the forces present, what we will find when we take a measurement, when we question the system. Quantum mechanics is well-defined. Beyond that, what is the representation: can we have an image of the rules we have applied and what underlies them? The most intuitive interpretation, proposed by de Broglie, and I believe favoured by Einstein, and which John Bell had in mind when he came up with this inequality, is that of hidden variables. I believe that even if Bell had expected the experiment to tip the balance in favour of local hidden variables, he admitted that the experiment was decisive: "*It now seems that the non-locality is deeply rooted in quantum mechanics itself and will persist in any completion.*" Nevertheless, he remained attached to the hope of an objective *observer-free* formalism of quantum mechanics. Let us add that if the hidden variables are non-local, then we cannot really see in what way these are hidden variables, because they are no longer just linked to the system but also to completely separate entities: the situation is therefore even stranger than that of quantum mechanics. Therefore, I do not believe it is correct to say that there is one interpretation that is quantum mechanics and others that are opposed to it. No, there are rules in quantum mechanics that no experiment has disproved, and there is the interpretation, the intuitive view that we can try to have of it. It is true that there is another view, which I do not know well, that was developed following Louis de Broglie by David Bohm. However, this does not call into question quantum mechanics. How, then, do we represent it to ourselves? If Bohm's interpretation helps you and is not contradictory, that is fine. As for me, I do not know of any interpretation that helps me form an intuitive representation [13].

I am confronted with this reality, with its experiments, like that of Aspect et al., which show me that all I could have that was naive or intuitive for understanding what a measurement is disappears. This is all I can tell you. Whether there are other interpretations… I am a not a theoretician of knowledge.

Bell's breakthrough was to say: there are not just "interpretations"; there are two ways of looking at things, quantum mechanics and local hidden variables, and these do not necessarily lead to the same results. Therefore, what would be interesting would be to be able to show that Bohm's formalism leads to something measurable that is different from quantum mechanics. In which case, that would interest me greatly. Or is this an interpretation we have of it. If it is an interpretation, then it does not help me much personally.

Léna Soler. I do not know this very well personally. However, for example, they say they have resolved the problem of measurement. Hervé [Zwirn] probably knows this better than I do. For them, there is no longer a measurement problem with this interpretation, for example.

Édouard Brézin. Yes, perhaps. Does Jean-Michel [Raimond] want to add anything? Have you thought about this?

Jean-Michel Raimond. I know very little about Bohm's formalism. As far as I know, it is an extension of the formalism of guided waves, of a particle guided by a probability wave. This formalism describes the movement of particles very well. I do not know whether it can describe all that has been described by standard quantum mechanics or whether it has been successfully applied to all situations, including massive entanglement, solid physics, etc. that are described by standard quantum mechanics. On the other hand, what Bell's and Aspect's experiments tell us is that this formalism necessarily obeys non-local evolution equations, which are therefore, from a conceptual point of view, as surprising as quantum mechanics. In short, I completely agree with what Édouard [Brézin] was saying a minute ago. What is surprising in quantum mechanics is not so much our interpretation as the experimental results. And that, in a way, if indeed we trust experiments—but if not then we should stop doing physics—is what is counter-intuitive. Beyond that, if we put this counter-intuition on the standard Copenhagen approach with which I agree with completely, or if we put it in equations of probability wave propagation which are non-local and terribly complicated, is a matter of choice. But we must know whether the alternative scenarios are equivalent or not. We could discuss these things technically with real specialists in these alternatives. Are they equivalent or not and do they have the same predictive power?

Léna Soler. Normally yes.

Jean-Michel Raimond. Can they address all the issues?

Léna Soler and Stéphanie Ruphy. They are empirically equivalent. We have shown that they are empirically equivalent but associated with different scenarios.

Édouard Brézin. Including in EPR situations?

Stéphanie Ruphy. To my knowledge, yes.

Olivier Darrigol. The equivalence between standard quantum mechanics and Bohmian mechanics is now sufficiently completely demonstrated. There are even very recent relativistic extensions that cover the results of field theory. The problem is to know whether Bohm theory is more intuitive or not. The non-locality of hidden variables in this theory remains counter-intuitive. The advantage for supporters of this theory is that the observable spatial values are defined at all times (in particular, a Schrödinger cat is either dead or alive). Therefore, it all depends on what intuition criterion we have. One big flaw of pilot-wave theories, which has already been pointed out by Wolfgang Pauli, is the large redundancy in the representation. In fact, there is not just one but multiple possible Bohm theories, depending on whether we place ourselves in p space, q space, etc. There are actually an infinite number of Bohm theories. If we apply the intuition criterion where we must have the closest correspondence between the theoretical representation and the experimental possibilities in a physical theory, then from this point of view, the traditional Copenhagen interpretation represents a considerable economy. This economy is lost in Bohm theory. In a way, I do not think we will ever have a definitive answer. Since, underlying these choices of interpretation that you [Jean-Michel Raimond] mentioned, there are different intuition criteria that vary from person to person and with the general philosophy of knowledge they favour.

Hervé Zwirn. I agree with all you have just said. I believe that if we want to try to learn from this at a rather global level, the first thing to remember is that initial attempts to find an interpretation that is compatible with intuition, which was in fact the goal of the first hidden variable theories which were local, were bound to fail as we already know. Non-local hidden variable theories or quantum mechanics in its traditional form work well in terms of empirical predictions. However, it is not possible in either case to find an interpretation of the world that is compatible with our vision of the macroscopic world. This lesson has not necessarily been learned by everyone. There are still physicists, not necessarily interested in these topics (because you can be a physicist without being at all interested in the philosophical problems of quantum mechanics), who think they can say that these problems, in fact, are "foundation" problems but that in reality we can save realism in a quote "naive" way. Meaning we could find interpretations of quantum mechanics or Bohm theories (which are often used as examples in these discussions) where the properties of the systems have localized values or defined values prior to measurement. We now know that this is no longer possible. We can already draw one lesson from this, and there will be no need to come back to it; it is that the belief that systems have locally defined values prior to measurement does not work. It is contrary to experience. That is a really powerful lesson, because it crushes any hope to think that the world is really as it is intuitively perceived at the macroscopic level.

It is true that in theories like that of Bohm, it is possible to consider that particles have a defined position at all times, the trajectory of particles being determined by pilot-waves. Nevertheless, this interpretation has its limits. As Michel Bitbol pointed out during one of his interventions, which has been transcribed in volume 2 of *Implications philosophiques de la science contemporaine* [14] where he cites a recent interferometric experiment on neutrons, the experimental results lead us to admit, even in Bohm theory, that the mass of neutrons is not localized at their position but spread throughout the volume of the interferometer. Under these conditions, it is clear that the gain we obtain with intuition is very small, especially if we think furthermore that these theories must be non-local (due to Bell's inequalities) and contextual (due to the Kochen-Specker theorem). Therefore, the assessment of these theories seems to me to be rather negative, considering the fact that on the one hand their formalism is far more complex than that of quantum mechanics, and that there are many hurdles, despite some progress, to extend them at the relativistic level, and on the other hand, the intuitive benefit we gain is extremely small. In addition, I think there is in certain cases divergence between certain predictions of quantum mechanics and Bohm theories during the relaxation time of the system. But for the time being, these divergences are beyond the reach of experience.

Olivier Darrigol. That is to say that the equivalence presupposes a pre-established harmony in the initial state of the Universe between the statistical distribution of particles and the probability amplitude of waves. If we suppose this harmony did not exist in ancient times on a cosmological scale, then we can have violation of the equivalence. Certain supporters of Bohm theory look for tests of that kind. In short, we have a rigorous equivalence only under a certain hypothesis that is not necessarily satisfied.

Édouard Brézin. Allow me to insist: I was not placing myself here in an interpretative framework. I was trying, as Jean-Michel [Raimond] said so well, to describe what experience tells us. From there, everyone is free to find an interpretative framework. As you said very well, what was eliminated by Bell and the experiments that followed was the most naive interpretative framework. It is no longer valid. Now, there may be other interpretations. But whatever happens, there is no simple and intuitive interpretative framework that corresponds to those astounding experiences which are our physical reality.

Bernard d'Espagnat. Yes exactly, I think that, as you have just said, and as Hervé Zwirn pointed out, this is already a considerable achievement if we think about all that we have been taught. I do not know what teaching is like now, however when I and my daughters were students, it was implied, tacitly, without openly saying it, and with considerable power of persuasion (we did not demonstrate it, it seemed so self-evident it was never even *spoken of*) that everything presents itself in accordance with Cartesian "common sense". Physics was by definition the study of a reality that exists outside us, and to increase our understanding of this reality independently of us, we would use without transgression fundamental concepts of position in space, of force, of movement, of trajectory, etc., which we all have at our disposal (like the "clear and distinct" ideas of Descartes), with possible additions borrowed from mathematics. It was self-evident that these concepts, and only these concepts, were suitable for the study in question. Nowadays we know this is not the case. Or at least, we know this with certainty for the atomic world you spoke about. Now there are very competent people who still hope that the macroscopic world is not affected by this surprising conclusion. I remember Jacques Merleau-Ponty for example, an excellent philosopher of science [15], in his reply to one of my articles, posited that the Sun really exists, completely independently of our existence; or, in any case, there exists, independently of our possibility of knowledge, a natural being to which the word "Sun" refers. So, is it or is it not the case? Those who have worked on or thought about decoherence theory do not react the same way to this matter. Some consider that decoherence cannot be used to justify such theories, labelled by them as "metaphysical", of a Sun existing independently of all human knowledge. Others, at the beginning of their research on this topic, seemed to think that it could and should be interpreted this way. We will have in the near future a presentation by Jean-Michel Raimond on decoherence, which will provide an opportunity to discuss this question, which from a philosophical point of view seems to me to be of great importance.

Jean-Michel Raimond. I am not sure I know how to answer the question of knowing whether the Sun really exists, particularly at this time of year. I will place myself from the "experimenter's" point of view to show what decoherence is. I do not think it is the only solution to the interpretative problems of quantum mechanics. It tells us certain things regarding what happens during a measurement, but not everything. The standard interpretation of decoherence itself relies largely on the postulate of projection and the notion of partial trace which is technically a direct emanation of it.

Bertrand Saint-Sernin. I would like to ask the following question: when we consider physics from the 19th century, and I will take the particular example of Antoine-Augustin Cournot (1801–1877), who came first at the École Normale Supérieure, etc. He asked the question of whether we can be a realist in physics. What seemed crucial was the fact that synthetic chemistry was starting to develop at the end of the 1820s and at the time Marcelin Berthelot gave his great series of lectures at the Collège de France in 1866, or 1864, I cannot remember. Berthelot announced that there could be hundreds of thousands of compound chemical substances not created by nature, and that chemistry, he said, created its object. Then Cournot said, if that is true, that if we master the building of chemical substances, then being a realist simply means that we are capable of reproducing natural substances and introducing within nature substances what we have not spontaneously discovered there. Is this mode of reasoning still valid or has quantum mechanics created a completely different situation, so that we can no longer speak of realism in the sense meant by Cournot. This is important because this need for realism exists for instance in medicine. We want to be absolutely sure that chemical molecules are real. If someone is diabetic, it is very important for him to know that the insulin molecule that has been created by genetically modified yeast has exactly the same properties as human insulin. In other words, the concern with realism exists from a practical point of view. However, can physicists tell us that it is simply a belief, a type of comfort, etc., but does it also have any scientific meaning?

Édouard Brézin. I am not sure I can answer to the depth of your question. I think there are not really, on this point, any differences with 19th century physics. Nowadays we prepare systems, even very small systems. We build carbon nanotubes, as an example, and there is now engineering at the molecular level that, for example, Jean-Marie Lehn has carried out in the most extraordinary manner, which continues to be carried out and where quantum phenomena are sometimes strongly at play but which nevertheless does not change this view of a world that we can build from a combination of atoms. Our combinations have become more complex since beyond atoms, we have nuclei, and beyond nuclei we have constituents which are neutrons and protons, which are themselves made up of sub-nuclear objects we call quarks. This is where we may encounter a difference with what you mentioned because quarks are effectively the ultimate components of matter, well ultimate as far as we know; perhaps they also have a structure. However they are greatly different in that we know they exist but we cannot generate them. Therefore, this may represent the first step away from the simple realism of the 19th century: they are there, everything happens as if they were there, but we know that at present we cannot free ourselves from the attraction that binds them to the interior of our particles, whereas in the first moments of the universe when the temperature was considerable, they were deconfined. They no longer are and never will be again, at least not until temperatures reach once again billions of degrees, or even more.

Bertrand Saint-Sernin. Thank you very much.

Bernard d'Espagnat. It seems to me that Bertrand Saint-Sernin's question raises the important but tricky question of what we mean by "real". There is, let us say, the "abstract" definition which refers to the notion of existence considered as primordial and posits that what is real is what exists independently of us. It is the one that many of us (myself included) have in mind. There is, at the other extreme, that which is implicit in the common expression "we have to be realistic", which is a reflection on ourselves since it means that we must take into account data and phenomena as they present themselves and act accordingly. I think I understand, but I could be wrong, that Cournot adopted a more operationalist definition than the so-called "abstract" definition. A similar definition, in fact, would be to consider that what is real is what we can act on; seen from this angle, this definition by Cournot comes closer, I believe, to that corresponding to the common expression regarding realism, considering both are centred on human action.

Bertrand Saint-Sernin. No, he precisely ruled out common sense situations, and positioned himself for defining realism... He had a general theory that said that the only important crises of knowledge are scientific crises. He does give a definition of realism when he said: the criterion that seems to me to be the most solid of my time, namely the 1860s and 1870s, is to say that in chemistry we can reproduce existing substances by creating new ones. However, he said that this may not work elsewhere, or may not be extendable to all realms of reality. This is the reasoning behind my question to Mr. Brézin.

Édouard Brézin. I believe the current limit is life, to reproduce in a test tube a system which… Venter [16] has claimed to have succeeded in building, piece by piece, synthetic replicating DNA, the holy grail of our times of physical chemistry applied to biology. But in this physical world, this combination, i.e. Cournot's realism, has not ceased to multiply itself with recent discoveries.

Jean-Michel Raimond. We need to say that you, Édouard, described very well the quote "mysteries" of quantum physics. It is true they are counter-intuitive, it is true that they resist any form of simple interpretation, whichever interpretation we seek. But at the same time it is extraordinarily predictive. Never has a physical theory been so predictive. Quantum physics takes the calculation of an exotic parameter of the electron which we call $g - 2$ (the difference between the real gyromagnetic ratio of the electron and the theoretical value 2 given by the Dirac equation for free electrons); we push this calculation to 12 digits and we can also measure these 12 digits: we find they are the same. Never has a physical theory been able to predict a physical quantity at the 10^{-12} degree of precision and confirm this through experiments. Carnot's thermodynamics was in the region of 30%. Therefore, we are in a situation that is, despite the difficulties of conceptual interpretation, extraordinarily predictive. We can image quantum systems, combine them, then realize them and witness that they function effectively as predicted by quantum mechanics. Like the rules of assembly of quantum bricks, it works perfectly well and has great predictive power. Therefore, it is mysterious, it is counter-intuitive, but it works very well. We need to keep this in mind.

Bernard d'Espagnat. Is the question to know whether physics is reduced to a system of prediction of observations or is there more to it in the teaching of physics? I think that is the question, or at least one of main questions, that we would like to address during our meetings. And I would say in this matter I sense a considerable shift in the way our fellow physicists think. When I was young it went without saying that, even for physicists dealing with quantum mechanics, as Édouard Brézin hinted at earlier on, there was no real problem. Physics described reality as it was, end of story. It was so self-evident that it would have been pointless to be interested in the arguments between Bohr and Einstein, outmoded and misleading… that anyway Bohr had resolved "with certainty" (so they thought!) in a manner consistent with realistic evidence. I have the impression that this mentality is being replaced by a very different, practically opposite, state of mind which Jean-Michel Raimond has just described, which consists of considering that in physics the only assertions that matter, because they are the only ones that have a real value of truth, are those of a purely operational nature. They are of the type "if we do this, then this will happen", as well as very general rules of calculation (the axioms of quantum mechanics are nothing more than that) from which such assertions are deduced. However, on the one hand this way of thinking has not, or has not yet, become widespread, and on the other hand we are all a little bit schizophrenic: we think in this manner when at work or when we consider the technical aspects of our work, but nevertheless for many of us there is a strong sense of incompleteness linked to the idea that human thought is not the only thing that exists. We must also tell ourselves that going from one extreme to another is seldom the right solution. It is all probably not that simple. Therefore we continue to think.

Alexis de Saint-Ours. I would like to come back to a term that has come up a number of times and whose ambiguity was pointed out by Olivier Darrigol, which is in fact the term intuition. I believe the question is, and it is fundamental, whether intuition is an immediate element, an indisputable, primary element, of our representation of the world or whether we can build intuition. This would mean we could speak of multiple intuitions for example. I would like to cite two examples: firstly, what happens in set theory; secondly what happens in special relativity. We can see what happens when we teach set theory to students. What happens in set theory? We describe Cantorian constructions. What does Cantor show? That there is a scale of infinities and that some infinities are bigger than others. From there, two choices present themselves to the student, either he refuses this construction, saying it is not possible to have certain infinities bigger than others, or alternatively, he realizes that our first intuition, which is naive when it comes to what sets and infinity are, needs to be reassessed regarding this construction. It seems that something relatively analogous is happening in special relativity regarding relativity of simultaneity. Special relativity can appear counter-intuitive, because we believe in the absolute nature of simultaneity. When we finally get our hands dirty, we realize that, here again, our primary intuition, namely the belief in the absolute nature of simultaneity, does not hold up to scrutiny. In other words, our primary intuition often relies on a lack of analysis, in a certain way. I would like to know to what extent it is possible to have something analogous in quantum mechanics, and to what extent it is possible to build what we could call a quantum intuition of the world.

Édouard Brézin. I find it hard to say. It is clear that we get used to it progressively. The first time we hear about relativity, it seems absolutely astounding, and we struggle to accept it. The moment we know that there is only one finite speed of information transmission, we understand immediately that simultaneity is the cause. The moment we cannot exceed a certain speed of information transmission, we can see that two observers in motion will not have the same definition of simultaneity. Thus, we end up having an intuitive vision where we integrate the fact that there is a finite speed of light. In quantum mechanics, it is true that practitioners, those who do quantum chemistry every day, have a very intuitive vision of orbitals, of the way they combine them, of the way orbitals will link a given system that contains another molecular system. Therefore, there is a broader intuition in quantum mechanics, it is true. Things really become more complex in those EPR situations. Up to EPR situations, we can modify our intuition in such a way as to include quantum mechanics. For now, these situations of system entanglement are still very difficult to assimilate intuitively. Perhaps those, like Jean-Michel Raimond, who are immersed every single day in entanglement…

Jean-Michel Raimond. Being immersed in entanglement every day, 10 h a day for the past 35 years does inevitably lead you to adapt the functioning of your neuronal circuits to this type of situation. We remain more or less normal in other areas. So it seems we can develop a quantum intuition, which allows us to say without any calculation that if we do this then this will happen and the state of the system will be like this. I have the feeling I am starting to get there, in a very modest way. Some manage much better than I do, and they are indeed people who manipulate entanglement and non-locality: those who do quantum informatics for example, those who work on quantum algorithms for quantum computers (only a theoretical fantasy at present), those who develop a real intuition for doing calculations by handling quantum states. The problem with this intuition is that it can easily be misleading, it can take us up the wrong path, and it is fragile as it is an acquired intuition. It is not the intuition of common sense that says: if I drop this stone, it will fall, which is really anchored in our everyday experience. It is not an intuition that is anchored in everyday experience and therefore it is more fragile and more prone to error.

Olivier Darrigol. On this subject I would like to mention—it is kind of amusing—that Heisenberg's article on uncertainty relations is called "On the intuitive content of quantum mechanics". Thus he was hoping through his arguments, through his thought experiments, to establish a new intuition. Basically, the aim was to show that the concrete possibilities we have of manipulating measuring devices correspond exactly to the possibilities of formal definitions within the theory. Thus there is a sort of harmony in quantum mechanics, strange as it may be, between the possibilities of measurement and the theoretical dispersion of magnitudes. This is quite a remarkable thing. Obviously it does not solve all problems, as you pointed out today. There are the fundamental problems of measurement, of non-locality, etc. It is remarkable that physicists like Jean-Michel [Raimond] can develop, in the everyday context of their laboratory, a new intuition that integrates all these quantum incongruities. But I suppose you [Jean-Michel Raimond] must still ask yourself the question of the compatibility of this intuition with macroscopic physics. Quantum mechanics, applied to the interaction between a quantum object and an ideal measuring device generally does not lead to a defined value of the measurement outcome. Decoherence aims to remove this stumbling block. However—and this is a point that has been raised by Bernard d'Espagnat regarding certain naive interpretations of decoherence—the question remains: why, at the end of the day, is a macroscopic system always seen by us in *one* well-defined state? It goes against the prediction of the formalism, even in decoherence theory. Bohm theory provides an answer to this question, but that can be challenged for the philosophical and practical reasons mentioned previously. As for Everett's theory, mentioned by Léna Soler, it addresses this problem in a very extraordinary way compared to our usual way of thinking. In this case, we really need to switch to a new, a very new form of intuition. There are advocates for this. For instance Thibault Damour is a strong supporter of this theory. So there are very serious people, cosmologists, who are interested in this theory.

Jean-Michel Raimond. When I spoke of intuition of the quantum world, it is clearly an intuition that takes place within the framework of quantum physics sensu stricto. It remains impossible, I think, to have a visceral intuition of what a particle localized in two places at once may be. That, I think is… However, an intuition of the outcome that would be obtained if we played with the formalism, which is then confirmed by the application of this formalism, can be developed.

Hervé Zwirn. On this notion of intuition, why do we ask ourselves the question of the importance of intuition? If we come back to the problem of realism, which you mentioned at the start, in fact intuition is important because it is in relation to it that we define the macroscopic world, and thus realism. The definition of what we call "realism" must be clarified: we can define it as the fact that something exists that is independent of the observers who are there to observe it. That is the Sun conundrum: does the Sun exist or does the moon shine if no one is there to observe it? I we accept this as the definition of realism, then intuition is important because quantum physics goes against our macroscopic intuition that is developed throughout life, from childhood to adulthood, through our experience of the macroscopic world that leads us to believe that effectively the world exists in itself, independently of us, and that ultimately we are neutral observers. What quantum physics shows is that this position is no longer tenable. We can no longer think this. I believe that Bernard d'Espagnat has shown this clearly in many of his works, the idea that we can describe quantum physics while completely excluding the observer does not work. We can do this in classical physics, no problem. We can have an objective description of classical physics and we can exclude the notion of observer, thus we can be realist. However, quantum physics does not allow us to do this and the observer must be mentioned at some point. There had been some hope that with decoherence we could think that ultimately referring to an observer was not mandatory, but in fact a precise analysis of the way decoherence works—and I hope we will have the opportunity to discuss this—shows that, as Olivier [Darrigol] was saying, it does not work; that is to say at that the final stage of decoherence, through this partial trace operation that allows us to free ourselves from some degrees of observation of the observer and end up with a determined state, the observer is still there! Otherwise, everything remains superimposed. Thus, the problem of realism as presented earlier is not solved by the investigators' intuition. I am convinced that investigators who perform experiments on a daily basis can have an intuition of the quantum world, in the same way that we can acquire with practice an intuition of a completely random game. The game follows certain rules and we do not necessarily try to understand them, but with practice we develop a certain knowledge of what will happen. However, this intuition is not sufficient to allow us to equate the description of the world with a realist description, since for that to be the case, this intuition would need to be like a macroscopic intuition in order to free ourselves from the observer. And that, which will be the topic of further discussions, seems no longer possible nowadays. The traditional realist position—of which there are many different descriptions—no longer seems to work.

Bertrand Saint-Sernin. Anyway, that is not the position of Cournot.

Hervé Zwirn. It is not that of Cournot.

Olivier Rey. Just a brief comment. I was struck by what Poincaré said: "For a perfectly immobile being, there would be no geometry". This means that the space we are in is not a space that is independent of us, but a space that is built by our actions. It seems to me that there are many paradoxical things in quantum physics that lose their paradoxical nature from the moment we consider science not as a science of the world as it is, but of the way we interact with it. From this point of view, the theory becomes perfectly coherent. The question that comes up next is to know to what extent we can equate the space of our actions with an objective space, which is independent of our actions. Secondly, to what extent can we devise a unitary theory which would cover all our possible interactions with the world? Paul Veyne described in his book *Did the Greeks Believe in their Myths*? that during a single day, we incessantly jump from one system of truth to another depending on the context we are involved in. When we are in a laboratory, or with our family, etc. we do not move inside the same systems of truth. What physics would like to have at its disposal is a single system of truth. But it is not at all clear that this could be achievable. By analogy: in mathematics, differential varieties are topological spaces with the property to be locally homeomorphic with $\mathbf{R}^n$ openings. From this point of view, $\mathbf{R}^n$ is obviously a variety, since the identity application makes it homeomorphic with itself. We could say that $\mathbf{R}^n$ is just a special case of these new more general spaces that are differential varieties; except that we need $\mathbf{R}^n$ to define these varieties. It seems to me that there is something similar between classical physics and quantum physics, an association of a different nature from what happens between relativity and classical mechanics. Within relativity theory we find elements of classical mechanics when we make c tend towards infinity. In the case of quantum physics, the situation is more complicated, insofar as we cannot simply consider classical mechanics as a limit of quantum mechanics. Indeed, we necessarily need classical reality to exist to even be able to formulate quantum physics. From there arises the question of whether we can conceive in a unitary fashion this association between classical physics and quantum physics.

Édouard Brézin. We would all have wished for it to be so. In particular, that was what I was trying to say, a joint quantum distribution of the measuring apparatus and the system: the perhaps naive hope was that this would allow us to free ourselves from these overcomplicated things.

Olivier Rey. Precisely, experience shows us that these naive things cannot work.

Édouard Brézin. In any case, it does not work in a simple manner. That said, we apply quantum mechanics every day, to situations such as the inflationist models of the beginning of the Universe that explain the fluctuations in cosmological radiation, which have now been observed in great detail. We apply it to a range of situations where there was no observer, with total reliability and confidence in its application. It is true that there are now only a few situations where this is problematic. I can name two: first of all there are EPR-type situations, where I find we have a really hard time understanding what is going on whichever interpretation we use. And then, at the Planck length, there is no convincing quantum theory at present and it is more than likely that it will imply radical departures from quantum mechanics as we know it. Perhaps the quantum mechanics we know is like thermodynamics compared to an underlying statistical corpuscular physics. Statistical physics works very well, but there is something else which allows us to understand where it comes from. Perhaps that is what quantum truth is. Until we have a quantum theory of gravitation that also works in those situations, in these problems of black holes, which are themselves properties of quantum evaporation, we cannot say "we did it, we have a definitive theory!", if that can ever make any sense. I do not know if a definitive theory can exist, this is a matter for philosophy, but for the time being we are not there yet.

Jean-Michel Raimond. There are no experiments at those scales, we cannot even conceive of them, except perhaps one day at the cosmological scale.

Olivier Rey. The idea behind the analogy with differential varieties, for what it is worth, was precisely that physics should perhaps admit the fact that there is no global theory, and that depending on which scale we place ourselves, depending on the type of phenomena we study, we should each time use a particular formalism in agreement with the experimental possibilities at our disposal regarding the type of reality under examination.

Édouard Brézin. Everything is going the other way, actually. If we look at the history of physics, nothing suggested that we could achieve a unified vision of electricity and magnetism, yet this was Maxwell's great achievement. Nothing allowed us to understand it, and Maxwell did not try to say that he was going to come up with a unified theory. Nothing suggested that one day—and Mr. d'Espagnat has worked considerably on weak interactions—we would have a unified theory of electromagnetism and weak interactions. And this did not arise from a desire to unify, it arose from quantum understanding itself. In other words, everything went in the direction of a total unification of all the concepts at our disposal, without it stemming from a desire to unify such different forces. But as long as there is no formalism that shows experimentally its ability to reconcile Einstein's gravitation with quantum mechanics, the problem remains open.

Jean Petitot. I would like to comment, coming back to what you said: "We cannot have an intuition of a particle that is in two places at once". This crisis concerning intuition has been very well formulated by certain philosophers at the start of the 20th century. I am thinking for example of a text by Husserl, who was quite removed from physics, where he said that the fundamental problem posed by quantum mechanics is that spatial localization is no longer a principle of individuation, when it is for us in our intuition derived from the macroscopic world. This is truly something critical. I do not wish to come back to Kant, who is one of my masters in philosophy, but the intuition we speak of is essentially a spatio-temporal intuition and there is a conflict between non-commutability in the quantum mechanics sense and spatial intuition as a geometric framework. We encounter here a fundamental theoretical problem that has remained open for a very long time, and which is now just beginning to be overcome. Throughout the entire history of quantum mechanics, and up to relatively recent theories, there has been this conflict between the geometric framework of space-time and the descriptions in terms of non-commutative algebra of operators on Hilbert space, etc.

Personally, I really like the way Alain Connes has tackled this problem, when he said: non-commutability needs to be placed at a more fundamental level, we need to take it down a level from the non-commutability of observables to a geometric intuition that needs to be completely rethought from the a priori that is non-commutability. From there, we can rebuild spatial intuition. This idea that we need to build a spectral geometry from, for example, the basic experiments you have mentioned on spectral rays, which are completely incompatible not only with classical physics but also with the geometrical framework of classical physics, this idea to rebuild geometry itself is, I believe, a step which was lacking until now and which is very significant. Connes has done this but there are other possibilities, like Edward Witten for instance. I hope that, even if it is technical, and without going into formulas, we will have the opportunity to talk about the way certain mathematical physicists, like Carlo Rovelli who has worked extensively on this, try to transform the most fundamental a priori of our intuition using the bases of quantum mechanics, of the spectral and of non-commutability.

Bernard d'Espagnat. If anyone manages to do this without going into technical details, and finds a way to present them, then yes of course. However, this is a difficult task.

Jean Petitot. Yes, it is difficult to geometrize non-commutability.

Édouard Brézin. It is true that we are at time of crisis, or real difficulty. I believe that Alain Connes was greatly inspired, he says so himself, by the mechanics of Heisenberg matrices. This was his starting point. He introduced a non-commutative geometry, which is an a priori view of space-time. Physicists have followed a different approach. And in the current vision of a part of physics, in particular string theory, space and time, in the visions we now call "branes" [17], appear as non-commutative variables. Space and time, as physicists now say, are emergent variables. They are emergent concepts, not a priori concepts in which we place objects as in the classical view. Space is no longer indefinitely divisible. The concepts of time and space, of infinity, to come back to what you were saying, are completely different in this view of things. I believe we should expect, as we progress in our knowledge, and there is still some way to go, many changes to our vision, I do not dare say intuition, of what could be space, time and the physics that takes place there, which is still undergoing radical change and debate. This may well take a long time, but there are many people working on it.

Jean Petitot. I believe it is a really essential philosophical step.

References

1. Richard P. Feyman, Robert B. Leighton & Matthew Sands, *The Feynman Lectures on Physics*, 1964, volume III: *Quantum Mechanics*.
2. See Feyman's exceptional book entitled *QED* (i.e. Quantum Electrodynamics); it is not technical and contains no equations. *QED: The Strange Theory of Light and Matter*, 1985.
3. Large Hadron Collider, at CERN.
4. Albert Einstein, Boris Podolsky & Nathan Rosen, "Can Quantum Mechanical Description of Physical Reality Be Considered Complete?", *Phys. Rev.*, vol. 47, 1935, p. 777–780.
5. "On the Einstein-Podolsky-Rosen Paradox", *Physics*, 1(3), 1964, p. 195–200.
6. *Comment added following an exchange with Bernard d'Espagnat and Franck Laloë*: This situation is nonetheless not sufficient to explain the "strangeness" of quantum mechanics: we could qualitatively reproduce by thought-experiment the same strangeness in another representation mentioned further on where spins are described by hidden variables. Only qualitative considerations derived from Bell's inequality enable the EPR experiment to differentiate between quantum mechanics and local theories of hidden variables (to be precise, let us call thus any hidden variable theory, unlike that of David Bohm, that does not subject the variables in question to influences having effects at a distance).
7. See 5.
8. John F. Clauser *et al.*, "Proposed experiment to test local hidden-variable theories". *Phys. Rev. Lett.*, 23, 1969, p. 880–884.
9. Alain Aspect *et al.*, *Phys. Rev. Lett.*, 47, 460 (1981); 49, 91 (1982); 49, 1804 (1982).
10. Initial idea by Greenberger, Horne, Zeilinger, 1989, realization in 1998 by Bouwmesteer et al.
11. These three physicists are the founders of quantum electrodynamics (QED).
12. On this theme, see Leonard Susskind's book, *The Black Hole War: My Battle with Stephen Hawking to Make the World Safe for Quantum Mechanics*, Little, Brown and Company, 2008.
13. *Comment added during proofreading*: To my mind, Bohm's mechanics does not provide an "intuitive" framework: it is completely non-local. Furthermore, as a minute example, when a wave function is real, the particle piloted by the wave is immobile: therefore, an electron in the fundamental state of a hydrogen atom is immobile – which, in my opinion, is a strange representation of the electronic orbital.
14. Bernard d'Espagnat (ed.), *Implications philosophiques de la science contemporaine*, tome 2, PUF, 2002.
15. Specialist in the philosophy of cosmology.
16. Craig Venter is a pioneer in synthetic biology.
17. Extended object of string theory.

Author Biography

Édouard Brézin is an emeritus professor of physics at the *École Normale Supérieure* and member of the *Académie des Sciences*. His work focuses on quantum field theory, especially on the applications of the renormalization group to phase transitions and their critical fluctuations. He has also worked on the topological developments and applications of string theory.

Chapter 2
Quantum Physics, Appearance and Reality
Round Table

Bernard d'Espagnat. You have all received the report [1] of the previous session. I will now ask those who, after having read through it, would like to comment to put their hands up. Very well, here we go! We shall start with Michel Bitbol, as he was not here last time.

Michel Bitbol. Thank you very much, Mr. d'Espagnat. Indeed, I was not here last time, and therefore I read with much interest and attention what was said during the last session. In particular, I found the debate following the excellent presentation to be very rich and very revealing. What struck me was to see, now more than ever before, a greater consensus emerging around what I would call an antirealist interpretation of quantum physics; the idea that quantum theory is not a direct and unequivocal representation of the world. It is true that certain symbols of this theory have been given names that suggest a representation, such as "vector state", which is meant to represent the state of a system; meant to represent, if we take seriously the semantics of the word "state", something that is *intrinsic* to the system. In Édouard Brézin's presentation [2], I noted a number of important points that directly call into question this standard conception of vector state. For example, Édouard Brézin asked himself whether we could not say that a quantum state is the expression of our knowledge rather than the expression of an intrinsic characteristic of the system. He cited on this matter Rudolf Peierls who would also have made such a proposal, presenting it as a solution to the problem of measurement. As soon as I read this statement in Édouard Brézin's presentation, I said to myself that it suggested a conception that was totally compatible with Carlo Rovelli's relational interpretation. According to Rovelli, the quantum state is not a characteristic of physical systems but a relational characteristic that depends on the position of the observer during the process of knowledge acquisition. This potential convergence really struck me. All the more so that the significance of this consensus was very well expressed by Olivier Rey, who pointed out that the paradoxes of quantum

B. d'Espagnat and H. Zwirn (eds.), *The Quantum World*,
The Frontiers Collection, DOI 10.1007/978-3-319-55420-4_2

theory disappear if we accept that it is not "a science of the world as it is, but of the way we interact with it". This statement recalls the famous formulations of Heisenberg and Bohr. Consequently, this strengthened my conviction that a new consensus was emerging around a similar interpretation to that of Bohr and Heisenberg, after being out of favour for so long.

Under these conditions, it seems that our working group could devise a programme that could go even further than the debate opposing (1) strong scientific realism and (2) a conception combining scientific antirealism with "open realism", according to your definition, Mr. d'Espagnat. Our working group could consider that this debate is now more or less closed, and that it would be interesting to go as far as we can with the antirealist interpretation of quantum mechanics by trying to draw out its implications and test its validity.

We could first of all show how such an interpretation resolves or dispels the most confounding paradoxes of quantum physics. Some participants have pointed out that this interpretation could defuse certain paradoxes (such as the measurement paradox). As for me, I posit the hypothesis that a fully antirealist interpretation is capable of resolving *all* the paradoxes; it would be a good project to try to test this hypothesis point by point.

Likewise, a second project for our working group could consist of investigating why a theory that, as is now accepted, does not represent a reality that is completely external to us, completely independent of what we do in it, is nonetheless so effective. Why this effectiveness? The famous miracle argument of scientific realists claims that if a theory does not represent the world, we could not understand why it works so well. Would there be a way, going against scientific realists, to understand why a theory is effective when it does not represent the world faithfully, point by point?

Finally, the third project I see for some of us could consist of trying to understand why we have resisted this other vision of scientific theory, and in particular quantum theory, for so long; a vision where quantum mechanics is not at all a representation *of* the world, but an inventory of the ordered, coherent multitude of ways we can position ourselves *in* the world. Is there a cultural factor that makes us balk at the full acceptance of such a conception?

There is one last point, which I cannot fail to mention as it is the one that is of greatest interest to Mr. d'Espagnat, and rightly so. Since we can no longer consider our current quantum theory, which is the theory-framework for most physics theories, as a more or less faithful copy of the world, the question that needs to be asked is the following: how should we conceive of the reality we are in so that it is resistant in this way to all attempts at representation?

Bernard d'Espagnat. Yes, indeed, this would be an entire area that would be interesting to explore. However, I have one or two comments or questions to ask you.

The first comment is that you alluded to the notion I have introduced of "open realism". I believe I have always made clear that this notion is no way contradictory with the way you present physics. On the contrary, I consider like you that physics is not capable of describing reality per se. I am even tempted to consider quantum physics as an essentially operationalist theory, in other words reducing everything, directly or indirectly, to the notion of prediction of experimental outcomes, which obviously implies that it always relates to us, what we see or feel. Is this how you consider the theory for which you have just drafted a programme? Or do you see something more than just operationalism?

Michel Bitbol. Thank you for this question, Mr. d'Espagnat. Yes, I see more than operationalism in quantum theory, but not "more" in the sense of "more representation of reality". I see more in the sense of "more justifications of the remarkable capacity of this theory to predict the outcome of our experimental operations". This additional justification of course calls on transcendental philosophy, meaning a philosophy derived from that of Kant. Although work remains to be done in this direction, I have good reasons to believe that the effectiveness of this theory can be understood if we think of it, somewhat in the manner of Kant, as a compilation of the conditions of possibility of a certain modality of knowledge. Of course not the same modality of knowledge that Kant reflected on; not the same modality of knowledge that the historical Kant, with only Newtonian mechanics at his disposal, examined. But the identification of the structure of knowledge as a product of knowledge can be done in the same spirit as him in any case, including when we depart considerably from Newtonian mechanics.

Working in the spirit of Kant simply consists here of trying to see in what way a modern physics theory like quantum mechanics is not just some sort of random heap of recipes, but a totally coherent structure of operative prescriptions that express the general presuppositions of our own interventions in the world around us. A good example of the relationship between theoretical structure and conditions of possibility of action was given by Olivier Rey in the report of the previous session: namely Henri Poincaré's conception of space. For Poincaré, space is built as an inventory of our possibilities of motion, and Euclidian geometry ensues. According to this conception, the theory of space that is Euclidian geometry is neither a passive representation of an external reality nor a compilation of recipes: it expresses in an optimal manner the coherence of the motions accessible to a being both sensitive and capable of motion.

Bernard d'Espagnat. That is right. Ultimately, you relate motion, the concept of motion itself, to the concept of conscious being.

Michel Bitbol. I have not yet spoken about consciousness.

Bernard d'Espagnat. We are heading in that direction.

Michel Bitbol. Let us say that up to now, we can avoid talking about consciousness. We can avoid doing so up to the point that consists of collecting the conditions of possibility of a certain coordinated action in the world. Obviously, we could also decide to talk about consciousness, and that would be very interesting. But perhaps this would lead us too far from our topic of discussion.

Nevertheless, perhaps I could briefly introduce the theme of consciousness by saying how I apprehend the notion of independent reality. You often speak of independent reality, Mr. d'Espagnat, meaning a reality independent of the mind. At the same time, a reality independent of the mind is only thinkable by the mind. When we mention independent reality, we perform an operation of self-abstraction, we subtract ourselves from the reality we would like to think of as independent of ourselves. However, we should not forget that this operation of abstraction is itself only an act of the mind! Thus, we are forced to recognize that all reality facing us presents itself only as an object (positive or negative) of a conscious act. We must not lose sight of this absolutely fundamental fact, which is not a scientific fact but an everyday, immediate, fact. Each time I describe something, I do this through an act of consciousness; each time I think of something, it is an act of consciousness; each time I think of something my consciousness cannot grasp in its field, that again is an act of consciousness (the consciousness of something that escapes my consciousness). All in all, reality is accessible only as a direct, or indirect, correlate of consciousness.

Bernard d'Espagnat. You exclude the act of faith, which consists of saying that there is a reality outside of ourselves, in other words, that we are not the only existing beings. But what seems to me to be problematic in your approach is precisely that, ultimately, we have the impression that it amounts to saying "we are the only existing beings". So, to counter that, you could, obviously, reject the notion of existence, and say the word "existence" has no meaning.

Michel Bitbol. No, obviously, I agree with you, we are not the only existing beings, but we need to examine carefully how this certainty manifests itself in us, how, deep down, we *know* we are not the only existing beings. How can we be sure? I think we know we are not the only existing beings when we become aware of our own limits. It is ultimately the limited nature, understood as limited, of our own existence that tells us that we are not the only ones to exist. We are not able to do everything, we cannot know everything at once, we cannot will the motion of our counterparts, and therefore there is something else other than ourselves. Ultimately, even the notion of otherness must be analysed from our own experience. Such is the lesson of a well-known branch of philosophy called Husserlian phenomenology. If we wanted to express this lesson in the more abstract terms of phenomenology, we would say: *even transcendence is only apprehended in immanence*.

Bernard d'Espagnat. Thank you for this answer which gives me much food for thought. For now, others would like to join in, in particular Jean Petitot.

Jean Petitot. I would like to comment on what Michel (Bitbol) said in his second to last intervention. What is important is the term "correlate", the fact that any reality is a correlate of consciousness. The way you criticized it was to say that a correlate of consciousness is a form of inclusion within consciousness. This brings us immediately to a modern version of solipsism. The great achievement of Husserl's phenomenology was indeed to understand how transcendence of the outside world can be based in the immanence of consciousness all the while remaining an external transcendence, so that solipsism is avoided. This problem began with Kant. Transcendental idealism is totally compatible with empirical realism. It is not in any way—solipsism, but rather a problem of the constitution of objectivity—in particular the objectivity of the outside world—through acts of consciousness. As we maintain an objective transcendence, we do not fall into psychological solipsism. We need to investigate further the concept of correlation.

It is a very complicated concept, but it so happens that in another field—which takes our reasoning outside the realm of physics, which Husserl investigated a lot, and which is the subject of much current cutting-edge research—namely that of perception, we encounter a similar problem. The objects of the outside world that we perceive as external to us in three-dimensional space are built from the treatment of information that is internal to and immanent of the retina, and we are starting to understand better how this exteriority develops, this conviction that objects are on the outside and transcend us, that is to say how perception frees itself from solipsism. Therefore, in the case of perception, we are starting to understand what the correlation between subject and object is. In physics, we would need to do something similar for objectivity.

Bernard d'Espagnat. We would need to think in depth about the way that Husserl, you or others actually manage to free yourselves, not so much from solipsism but from the idea that there is only a single being, which is of the nature of consciousness. We would need to understand this.

Others would like to contribute to the discussion.

I will conclude by saying that the lines of thought you propose, Mr. Bitbol, would likely be of interest to scientists outside of physics. In any case, I would be very pleased if we were able to move the discussion forward here.

Hervé Zwirn. Allow me to return to the report of the previous session, as this was our starting point.

During that session, we had a discussion initiated by Bertrand Saint-Sernin on the problem of realism, in particular regarding realism sensu Cournot. This required that we define in a more precise manner the term "realism", in order to know exactly what we meant during the discussion by the different types of realism we attempted to describe. We are already at the heart of the matter, however it seems to me that there are different levels here. What we have been talking about is essentially what we could call metaphysical realism, i.e., knowing whether, yes or no, there exists a reality that is independent of the mind of any given observer. This is the first problem.

The second problem, which is sometimes confused with the first and this introduces a number of ambiguities, is to know whether this reality is intelligible and whether current physics, as it is, adequately describes this reality.

There are, I would say, three problems of a different nature but which are obviously linked; we sometimes have a tendency to mix them up and this can lead to approximate reasoning. Is there a reality that is independent of the mind? Is this reality, if we accept it exists, intelligible? Is physics the right tool to account for it? These are three topics that were perhaps grouped together during the previous session without sufficiently differentiating them.

On the topic of metaphysical realism, there are a number of important historical and philosophical arguments—they are not new. It is true that new light is shed on these discussions by quantum mechanics.

The most important arguments, the most famous ones, what are they?

There is, first of all, what Hume called the relation of cause and effect. Why do we posit the existence of an external reality? For instance, why is it that, we when hear someone speak in the room next door, we will tend to infer that someone is there? Well, because each time we have heard someone, there was someone next door when we checked. Hume questions this as it is an inductive inference, and you are all aware of Hume's critique of induction. This type of inference is not valid.

Along the same lines, when we see a shape in front of us, a table top with four legs, we have a tendency to infer that there is a table that exists outside of us which accounts for this perception. Hume is critical of this as well, by saying that the hypothesis according to which there really exists a table is nothing more than a convenient arrangement of our perceptions, and I see in this argument the beginnings of phenomenology. Husserl went much further, and it would be very interesting to conclude on this point.

And then there is the famous argument that was also highlighted by Bernard d'Espagnat, according to which our experiments do not always produce outcomes that are congruent with our scientific theories. As he said, there is something that says "no". Therefore reality resists, and this is often taken as an argument in favour of the fact that there is something other than the mind. This argument can be criticized. The simplest example we can give of a criticism is that, in our dreams, we do not do as we please and yet our dreams are nothing but a pure invention of our mind. Our dreams resist us. Therefore the very fact that a certain number of things resist us does not prove that they are external to us.

There is another argument that can be put forward, which I defend but which takes us further, which is that we have to make two mental constructs correspond: on the one hand, we have a mental construct of external reality in the sense we have indicated previously. How do we build in our mind the fact that there are external objects? As Jean (Petitot) said, it is not because we have the impression that we are building something external that it is a sufficient reason to legitimize the fact that there is something external. On the other hand, we have a construct that we will say is mathematical, which is that of our scientific theories. We achieve these two constructs through completely different mental modes. The perception of external reality is developed from childhood and is intuitive in a certain way; the construct of scientific theories is progressive, and is based on different notions. So the fact that the two sometimes do not perfectly match up is ultimately not that surprising. I think on the contrary that the idea that something resists us is an argument for saying that these are two things we have built independently, and making them match up is not that simple. The coherence of a mathematical theory is a complex problem and the coherence of two independent constructs, the fact that they are congruent with each other, and the ability to demonstrate this congruence, is something extremely hypothetical. Obviously, these are analogies, since a mental construct of reality is not a mathematical theory; however, these analogies tend in that direction.

Lastly, as a final point, there is the argument put forward primarily by Hilary Putman that states that there are no miracles. That argument seems very weak to me. It consists of saying that our theories work simply because there is a reality to which they correspond. And if that was not the case, it would be a miracle if our theories worked… However, the large number of past theories that did work and were empirically adequate, i.e., that saved phenomena as Bas van Fraassen said, and yet which turned out to be false shows that it is not because a theory works that it corresponds to an external reality that it describes precisely.

This is an introduction to the theme of today's session which is structural realism. What seems to remain ultimately, if we really want to keep something, is not the ontology of theories, for which we can easily see that it was put into question and dismantled as scientific theories evolved, but seemingly the structure, in a sense that undoubtedly needs to be defined, of these theories, which leads us effectively to structural realism. Structural realism has had various, more abstract, developments, which seem to save a type of realism which is not an ontological realism in the sense of naive scientific realism.

Bernard d'Espagnat. In your first book and also in a recent article [3], you very clearly introduce the notion of a reality external to us. You consider, when you define your version of empirical reality—we were talking about this earlier in private—that there is something that is not us, which prevents us from thinking just anything, that there is something that resists us. I do not clearly see the link between this and what you have just said, as from what you have said you seem to conclude that, ultimately these things are built by us.

Hervé Zwirn. The link is the following: I am not a solipsist. Anyway, solipsism is a position that we know is irrefutable. I am not a solipsist because to reduce all that exists to something internal to us (or that is internal to me) is a position that seems to me to be both sterile and inappropriate, in a certain way.

Olivier Rey. An impregnable fortress defended by a madman!

Hervé Zwirn. The position that claims that nothing exists other than what I think or conceptualize does not satisfy me.

Bernard d'Espagnat. I was under the impression that it satisfied Michel Bitbol!

Michel Bitbol. Perhaps I can explain myself a bit more to dispel this idea. I am not the madman of the impregnable fortress!

Olivier Rey. I can confirm that, I share his office!

Michel Bitbol. The question is to know what the meaning of this exteriority we speak of is. What does a reality external to us mean? Given that we have an immediate intuition of external space when we speak of external reality, we have a tendency to represent this exteriority in the image of the distance between my body and all other surrounding bodies. However, I believe that the true exteriority that we should address here, the exteriority of external reality, is not of this nature. It is a non-spatial exteriority, in the sense of going beyond what I can do or grasp; it goes beyond my own finiteness. A certain number of non-spatial characterizations of the "exteriority" of the thing itself have been proposed over the course of the history of post-Kantian philosophies, and these may help us formulate an adequate, non-metaphorical, definition of its break from us. I think for instance of Schopenhauer who, instead of characterizing the thing in itself as an actual *thin*g, as a thing that would be external to us in a spatial sense, characterized it as a "will", meaning a sort of completely obscure drive that we feel interiorly and that leads to us carry out actions without really knowing why. The obscurity itself of our motivation to act shows that we are overwhelmed, that we are not alone, nor are we the sole master of our lives and of our knowledge; it shows that there is something much greater than us that acts by us, within us and through us. Surpassing and freeing the "Will" sensu Schopenhauer could be an alternative characterization of the exteriority of the thing per se.

Bernard d'Espagnat. Do you accept this?

Michel Bitbol. I think it is not bad at all. In any case, it gives us a plausible alternative view of exteriority. Besides, Kant had anticipated Schopenhauer when he wrote a beautiful sentence, full of perplexity, in his *Critique of Pure Reason*. This sentence, which I wrote down with great passion and which I analysed in my book *De l'intérieur du monde* [4] is as follows: "[…] regarding this noumenon, we do not know whether it is within us or outside of us […]" [5]. Of course, Kant thought the "noumenal" thing itself (in the sense of it being only thought of as a limiting concept) could be external to us in the traditional, primarily spatial, sense; but it could also be a simple extension of ourselves, which surpasses us in every way and which shows by this excess that we are not alone, that we do not hold all that exists within the boundaries of our consciousness. Therefore, we do not fall into solipsism, but we do not either posit more exteriority in the most ordinary sense of the term nor a complete separation of what there is in relation to us.

Bernard d'Espagnat. We could attempt to bring together what you have said with the Neo-Platonic stance, in particular that of the Christian disciples of Neo-Platonism. Even Saint-Augustine, with his "interior master", makes me think of what you are saying.

Michel Bitbol. Indeed, I completely agree with you. I think it would make a wonderful topic for future discussion; but perhaps we should not address it today, as we have other topics to discuss.

Hervé Zwirn. To go in the same direction regarding the rejection of solipsism, I would say that what was said earlier could effectively suggest that I did not accept that there are things that exist that do not come from us. However, these arguments, along with the rejection of solipsism, form the basis from which we can build what exists. That is to say: rejection of solipsism, therefore everything does not stem from us, but at the same time rejection of the arguments mentioned previously i.e., to think a reality in the traditional manner in which it can be thought, i.e., as Michel (Bitbol) said, in an exterior manner, etc.

I quite like the metaphor you borrowed from Schopenhauer. In a certain way, it is all that is constrained, while being in part external to us since it does not stem from us. However, it is at the same time linked to us, as it is all that constrains the way in which we apprehend the world, all the constraints that concern our perceptions. Earlier we spoke of the difference between empirical reality and phenomenal reality. This is what I called empirical reality, which is not spatially separated from us, which is not made up of objects, things, entities, fields, vectors, whatever we want, but which is in fact a mix between something external and something that stems from our categories, of the way we can then apprehend phenomena, or create phenomena, because I think that in a certain way we create what we perceive.

Carlo Rovelli. Some thoughts as they come in reaction to what you have said and in support of what Michel (Bitbol) has said. What I would like to say concerns the certainty and the development of knowledge, particularly the certainty. It seems to me that the discussion here revolves around the possibility of basing with certainty a belief in reality, or a lack of belief in reality, or an interpretation, a reading of what knowledge is, how knowledge works. The aspect I would like to bring to the centre of the discussion, on the contrary, is that of the lack of knowledge, the lack of certainty in our knowledge.

Michel, you spoke of this fundamental intuition which is that the elements of reality are always perceived within our consciousness, and are therefore correlates of consciousness; a fundamental, and I would say very ancient intuition: these are the Upanishads [6]. It is true, without a doubt. At the same time, it is also true that this consciousness develops slowly as we learn to think about the world. Developmental sciences teach us that, at first reality is without a self, then we add a self to reality, and later do we begin to distinguish reality by the self, by developing an idea of a reality that is independent of the self. So what comes first in our development? First question.

Second question: in all this, we spoke of the mind. You have all spoken of the mind, of us, our position. Each time I hear this, I get somewhat confused. I felt this confusion when I was at school, when I was told about Hegel. Who is the mind? Is it I, Carlo? Is it you, Michel? Do we speak of all of us, as a group? The entire community that speaks together? Who is this mind? Who is consciousness? My mind? Our mind? Where must I put it? I believe there is something to be learned here. That is to say: the constitution of reality is of course as *per* the critique of classical empiricism, because there are regularities in perception, etc. There is this well-known process, but there is also something else. I have an idea of myself, I have an idea of my consciousness, but I also have friends. I have had friends from an early age: I speak with you and I have an idea of you. In my idea of you, I am starting to realize that you are very similar to me, and that when I see a shadow, you also see a shadow. You tell me "I see a shadow" and I hear that it is the same thing as when I tell you I see a shadow. Thus I have a mental construct in which there is not simply reality, but where there are also others who perceive reality, and there is a surprising similarity between my perception of reality and what others say they perceive of reality. Therefore, I place something completely new in this reality: it is I, as one of many around me who are similar to me. So within reality, something has now formed within my construct which is my consciousness—my own—perceived as an element of external reality, only a copy of what I see others say and do.

Obviously, we all go through this, humanity has been through this and we ask ourselves the question: if I want to establish knowledge that is reliable, where do I start? This is kind of the question we ask ourselves, but is it the right question? Is this what we are interested in?

You have fragmented the problem of realism. First of all, is there an independent reality? Secondly, is it intelligible? Thirdly, can it be described by physics? We have learned much since the end of classical physics. Our current theories are probably false in the sense that classical mechanics is false… Therefore we know our knowledge is limited. What does that mean? It means we accept the idea that we can think of the world with a degree of uncertainty, and despite that, live in and interact with the world. I think this is the crucial lesson of the end of the last century, that is to say that we have learned that we can think of reality without asking ourselves what the foundation of our knowledge is, while still using this knowledge. So, what is reality? It is the reality described by physics theories, with the strong awareness that there are obvious limitations: in the same way that Newton's classical mechanics is false, these are probably also false. What we have in physics is a better way of thinking about the world today, now, and a better way of conceptualizing the world now: no more, no less.

Therefore, in this theory that apparently works well, i.e., quantum mechanics, in this discovery of the world that is not personal but stems from a community of scientists who have developed science, we find strong limitations to the idea of a precise and realistic description, in a stronger sense than in classical mechanics.

It seems to me that we realize—this is how many people understand this—that quantum mechanics tells us that a certain vision of this reality, as formed for example (I am simplifying) in classical atomism or the atomism of Democritus, as the space where particles can travel, does not work. The best description we can make takes into account the fact that there are systems that interact with one another and that is more a description of these systems, and there Michel (Bitbol), as you know, I agree with you 100%.

We have learned that this complex reality we have developed is more complicated that what we thought, but we have just learned that. Is it really necessary to know the precise role of the notion of reality to use the notion of reality? I think not. This is a recent discovery of science.

I now have a more precise question. You (Michel Bitbol) speak of Kant, of Kantism. You ask us to go back to it. Obviously, I agree with you, but I also have a question. I find two things in what I know of Kant. One is a strong reminder that we must take into account the necessary conditions for knowledge. I cannot describe reality per se, I must remind myself that what I describe is a reality perceived by me and that there are conditions for perceiving it. Science itself gives us this information as—I now come back to our starting point—I can reinterpret myself not as the holder of everything but also as the object described by science, where very incomplete information about the world arrives, which generates mental images, mental constructs that concern the exterior but which are totally separate from the exterior. I think Kant himself had this idea of the unknowability of reality, due to concrete conditions. That is Kant I, however Kant II says: despite this, there is a possibility to reason on this, which gives me a priori information about the world, about the possible structure of reality as I perceive it, which I can arrive at individually through knowledge. It seems to me that the development itself of quantum mechanics leads us away from that. We realized that there were limits within realism, not by thinking about the conditions of knowledge but by doing experiments. The world is not what I thought it was; particles… they are more complicated than that. Perhaps to think it, I need to rebuild a description of the reality where I belong, thus with the construction of a relationship that includes me, and I must take into account the knowledge I could not have had a priori. In this sense, we are distancing ourselves from classical Kantism, of the Kant of *Critique of Pure Reason*.

Last point, before wrapping up. If we take the stance where we say we do not know the starting point, what we have learned of the world is what science is telling us of the world, knowing full well that there are limits to this and that there will be changes. The conceptualization we carry out of the world itself changes, therefore it is not that necessary to cling onto a central starting point of a certainty regarding what does and what does not exist. Of course there is reality! There is no need to question that. Everyone agrees there is an external reality and of course it is within my own perception. This is sufficient to start understanding how things work.

If I take this point of view, I find that structural realism on the one hand clearly tells us: "look at these structures, they are really interesting; perhaps we could conceptualize these objects in term of structures", and this interests me. However, it interests me a lot less if it is suggesting that "we have finally found the ultimate thing that does not change". No, we have not found the ultimate thing that does not change. I was not able, in all the texts I have read, to understand what this structure that does not change during the course of the history of science could be. I cannot find, during the course of the history of science, a structure that does not change. On the contrary, I find that all structures have changed, but there are objects that have changed very little: the Sun is still the Sun, we still speak of the Sun, from Ptolemy to our current solar theory, and the Sun is still the Sun. There is a permanence of objects, a permanence of structures, and also a weakness of objects and a weakness of even the strongest structures. There are laws, it is true, like the Fresnel law, which have passed from classical mechanics to quantum mechanics without suffering too much damage, but there are laws that were good before and that were not so good afterwards. Therefore I cannot see structure as being a possible reliable anchor.

Michel Bitbol. The problem is that you have raised too many important points during your intervention! I will concentrate on one or two. Let us take your last argument, that of the constancy of certain structures throughout the course of the history of science. A number of contemporary philosophers of science, including Bas van Fraassen, have pointed out that the only structures that persist in an *absolutely* permanent manner are purely empirical structures, for example the apparent trajectory of Mars, or the calculations that are directly related to the apparent trajectory of Mars. Yes, this remains constant. There are also certain predictive formal cores that remain practically unchanged from one theory to another as of course we must find the effective predictive structures of the old theory in the new theory. Thus the constancy of certain structures is not as surprising as is sometimes said, because all it does is express the durability of certain regions of effectiveness that were attained in the past since they limited themselves to structures that were as close as possible to the empirical. There may be some permanent structures in scientific theory that are much more removed from the empirical, such as the laws of conservation or the optimization principles, however, others will speak of these better than I can.

A second very important point regarding Kant and quantum physics: what is the difference between Kant's original approach and the Kantian approach we may take regarding quantum physics? Well, one of the major differences—I think Jean (Petitot) would see many others—is this: according to Kant, the constitution of objectivity that is suitable for the macroscopic universe, the universe described by classical physics, is sufficiently effective that in the face of this physics we can act *as if* it was describing a reality completely independent of us. It is the famous Kantian "as if" (*als ob*). By using this expression, Kant insisted implicitly that it is precisely just an *as if*: the object of classical description behaves as if it was an independent reality, however it is not, and we can easily convince ourselves of this through an epistemological argument. Indeed, if we want to explain the effectiveness of science other than through the well-known device of the highest correspondence with reality, which Kant considered highly problematic, we must absolutely consider that objects are *not* things in themselves, but are *built* from the processes we need to gain knowledge, which is valid from practically any point of view, for everyone, at any time and place. It is the famous intersubjectivity clause. However, this "as if" does not work in quantum physics. There are a number of structures of quantum predictions, a number of predicted phenomena corroborated by experiments that are not congruent with the remarkably structured way of classical physics. In the absence of these structures, we can no longer act "as if" we were dealing with objects that are completely independent of us. The main difference between these two types of physics is this possibility in one case and the near-impossibility in the other case to act "as if" what we are studying is radically independent of us. The basic epistemological idea is still the same; in classical physics as in quantum physics, the structures of knowledge follow on from the conditions of possibility to form knowledge that is shared and can be communicated. There are cases where we can forget this construct and act exactly as if all things were independent of our ability to build, and other cases where we cannot. Quantum physics is that: the scientific situation where Kant is mandatory, whereas he was only optional in classical physics.

Bernard d'Espagnat. I would tend to qualify the difference you speak of by saying that Kant believed in a descriptive physics, describing phenomena "as if" they really existed, whereas quantum physics seems to me to be essentially predictive.

Michel Bitbol. Absolutely.

Bernard d'Espagnat. Indeed. I believe that the essence of quantum physics, ultimately, is to be predictive, that is how it works. It works every time, when we see it from this angle.

Michel Bitbol. Absolutely.

Jean Petitot. However, classical mechanics is very predictive.

Bernard d'Espagnat. Yes, but it is also descriptive! It is both.

Jean Petitot. This is why it is so paradigmatic. It is predictive in an extraordinary way.

Michel Bitbol. Exactly. But classical mechanics is predictive because it is descriptive. The prediction follows on here from a description. By contrast, in quantum physics, the prediction is put forward in a sort of naked state. There is no real description of a spatio-temporal process that underlies the predictive capacity of a theory. The real reason why this is so: quantum phenomena depend entirely on the experimental conditions of their appearance; this is not the case in classical physics. These are phenomena in the most obvious sense of the term, i.e., in the sense of events that take place in the laboratory for all to see; but not in the sense used by physicists who, in my opinion, often confuse in their vocabulary process and phenomenon. A process is what we describe as happening in the world; a phenomenon is what takes place in the laboratory, or before our eyes.

Jean Petitot. There would be much to say on this matter, because, as you know, the retina is a very nice quantum device. It is an admirable photon detector.

Jean-Michel Raimond. If I may say so, the principle of quantum superposition plays absolutely no role there. The difficulties of quantum physics play no role in the way biological organisms function.

Jean Petitot. Yes, but I meant it was a very nice measuring device.

Jean-Michel Raimond. Yes, but the fact of seeing unique photons, or almost, does not imply that the way the retina functions relies on everything that makes quantum mechanics conceptually difficult.

Jean-Pierre Gazeau. I have a question on this topic: what do you actually call "descriptive"? What does this word mean? What limitations do you impose on this notion? Why do we claim that quantum mechanics is not a descriptive science? I have no background in philosophy but I am very interested in this type of discussion.

Michel Bitbol. To keep things short, we consider there is something external (to the intrinsic properties of the object) for which we give a detailed account. We say we have made a description when we present a symbolic copy of the properties and processes that exist independently of the experiments we carry out to gain knowledge of them. Suppose there is nothing that can be considered external to us, that there are no intrinsic properties, that there are only experimental phenomena in the sense of apparent manifestations generated by the workings of a device. In the latter case, we have nothing left to describe because there is nothing we can make a symbolic copy of. It is true we use symbols in quantum physics, but these symbols are simply predictors of phenomena.

Carlo Rovelli. Are we not going too far here?

Hervé Zwirn. Many of us at this table agree with the fact that this is not descriptive, however we agree because we all share a point of view that is not that obvious. It is not shared by everyone, notably by many physicists who do not ask such questions. What you have just said Michel (Bitbol), presumes that we agree to say that quantum mechanics is not descriptive, that we have already adopted the point of view that scientific realism is a not a tenable position. When we dismiss scientific realism which assumes that there is effectively a reality external to the observer, which scientific theory is meant to describe, then we say that "quantum mechanics is not descriptive". However to say this implies that we abandon the idea of an external reality.

Bernard d'Espagnat. We abandon the idea of a describable external reality but we do so for good reasons.

Hervé Zwirn. We have of course good reasons to do this, however this is an a posteriori not an a priori position, where all attempts to maintain the idea of an independent reality external to the observer appears incompatible with a number of things that pertain to quantum mechanics. Or rather, to the quantum world in the broad sense. We can include theories other than quantum mechanics. Experiments of non-locality, of contextuality, etc., do not necessarily assume that we are within the framework of traditional quantum mechanics. These experiments show that trying to maintain a realist, we could say "classical", stance does not work. We are thus compelled after analysis to abandon this view, and if we abandon it then effectively scientific explanations are no longer descriptive, since there is nothing to describe, or at least, what there is to describe is not external to us.

Matteo Smerlak. Like Jean-Pierre Gazeau, I am troubled by this claim that quantum mechanics is not "descriptive". I have a feeling that what underlies this opinion is the idea that phenomena are generated by the intervention of an apparatus. My question is therefore as follows: we know that neutron stars are stable as a result of the quantum repulsion of fermions; what do we mean in this case by the intervention of an apparatus? Answering this question would perhaps allow us to define the sense of the term intervention in quantum mechanics.

Carlo Rovelli. It seems to me that quantum mechanics tell us that there are limits to the possibility of describing what is going on, in the sense where classical mechanics was capable of describing what was going on. But to say that there is no description at all… seems to me to be too much of a leap.

Realism is this idea that there is a world independent of us. If realism means the belief, in the limited sense of certainty, that there is a world external to us, then I think we are realists. There is a world outside of us. Personally, I do not think that the only thing that exists in the world is Carlo Rovelli. Do you think that the only thing that exists in the world is Carlo Rovelli? I think that outside of Carlo Rovelli there are other things, which I call the world, even if it is a projection of my dreams. See the dream of the butterfly: the Chinese philosopher asking himself upon waking whether he is not the butterfly's dream. Even that does not really change the fact that, in my dream, I interact with the world, I am confronted with the world. Therefore it seems to me the notion of reality is essential for our work. The fact that we set limits to this reality, and to what we can say about it does not mean that it is useful to abandon the notion of reality. Between realism in classical mechanics, or Democritus and Leucippe, and complete irrealism, where we do not need to think of a reality external to us, there is an enormous gap. Therefore it is not because we are compelled, empirically, to abandon a precise, strong descriptiveness that we must take unnecessary precautions, so it seems to me. For example: neutron stars.

More than that, it seems to me that quantum mechanics—to speak bluntly—does not tell us at all that my observations of a phenomenon are the only things I can speak of. Quantum mechanics does not only speak of what happens in the laboratory. It speaks of what happened at the other end of the universe, of what happens in astrophysics, of what happens inside atoms when we do not observe them, of the colour of light, constrained by the Schrödinger equation, even when we are not looking at it. I am convinced of it. It speaks of interactions, of what happens within a system when it interacts with another.

We may think about quantum mechanics without necessarily invoking the investigator's consciousness when we speak of neutron stars, for example, while rejecting the descriptiveness of reality. Thus we abandon the possibility of describing reality as precisely as in classical mechanics, but we still speak of a reality where there are interactions, where there are things I can describe.

Hervé Zwirn. I believe that in this debate on realism, Bas van Fraassen provides a good description, which is deliberately exaggerated, of what scientific realism is. He said: we must take the theory literally. What does literally mean? I means, when we have a theory like quantum mechanics which says that there are electrons, there are atoms, there are particles, there are forces, we must understand it as "there are electrons, there are atoms, there are particles, there are forces" in the same way as when we say "there is book on the table, or this is a microphone". That is scientific realism in the strong sense of the term, as described by van Fraassen, and he insists on this point: it is not a metaphor, it must be taken literally.

If we adopt this stance, well, it does not work. When we say for example that quantum mechanics speaks of things other than laboratory experiments, for example the Big Bang or the beginning of the universe, things like that, it is only a story it reconstructs to correctly and coherently organize the observations we make; quantum mechanics has no means of speaking directly of the Big Bang. All that it does is conceive of a number of things like the Big Bang, which are coherent within a formal framework, which then allows us to account for our perceptions in the laboratory, but it does not speak directly of the Big Bang. It simply speaks of what we perceive.

When it speaks of neutron stars, it speaks of the observations we make, by different means, with instruments, of what we perceive and that we interpret in a coherent manner by the algebraic formalisms of quantum mechanics which reconstructs all that we need to account for our observations. There again, what quantum mechanics does is nothing more than order our observations. And to postulate there are really neutron starts, or that the Big Bang really took place, or that there really are quarks, is itself an additional postulate of scientific realism, which raises a number of problems if I take it far enough. This is what we mean when we say that in fact external reality, in the strong sense of the term, cannot be thought of like this. This does not mean there is no reality.

Carlo Rovelli. The same thing can be said from a classical mechanics viewpoint.

Hervé Zwirn. Yes, of course.

Michel Bitbol. Yes, except that in classical mechanics, we could still act "as if"…

Carlo Rovelli. Yes, I agree, in classical mechanics we could act "as if" the world was exactly like that and in quantum mechanics we cannot act "as if" but without necessarily rejecting the notion of reality. If we see a veiled reality, a reality we cannot see completely, we can say certain things but not others, but we do not need to throw the baby out with the bathwater.

Hervé Zwirn. I did not throw it out completely since, precisely, what we are trying to do is reconstruct reality in a form that is different from the usual scientific realism, by saying that it is not built in a spatially external manner with objects that are taken literally, that are the ontology of a theory. It is something else. It is an attempt at reconstructing something external, otherwise we are solipsists, and effectively we say that there is only Carlo Rovelli, or Hervé Zwirn. As we reject this, we try to reconstruct what is compatible with the minimal postulate according to which "there is something else besides my own mind". This is the minimal postulate, and from there what can we reconstruct that is compatible with quantum mechanics or, in the broad sense because I think this is broader than that, with everything that touches upon the quantum framework or that extends beyond quantum mechanics? Non-locality or contextuality experiments do not presuppose traditional quantum mechanics. They are, at a push, independent of quantum mechanics. What these experiments teach us is that realism, let us say the usual scientific realism, raises a problem when we want to take it literally, as van Fraassen said. Therefore we need to find something else, and if want there to be something else to avoid postulating that we are alone in the world, then what is coherent with that? It is the attempts that seek to construct the minimal thing that is compatible with both the quantum world and the fact that we cannot be solipsistic.

Bernard d'Espagnat. Yes, but when we try to do this, to construct a description of this "other thing" which is not us, it seems to me that ultimately we fail. For this reason, I would say that quantum mechanics is essentially predictive, since if you try to replace particles, small dots, etc., by wave functions, to say that it is the wave function that is real, then it does not work. It does not work because if you say that what is real within a particle is the wave function, then as soon as you have a collision between two particles you no longer have a wave function for each of the two particles, you simply have a global wave function. And as soon as a third collision occurs, there is only the wave function of the three particles. And—since this has been happening for a very long time!—ultimately you only have as reality the wave function of the universe: an idea that is somewhat hard to swallow. You could try changing tack and say: "no, what really exists is not the wave function but the density matrix!". Effectively, after a collision you could associate a density matrix with each particle. But if you say that the reality of particles is described only by their density matrices, after a collision phenomenon between two particles the only reality, in your eyes, of this particle pair can only be made up of the density matrix of each one. However, data from the two matrices do not show at all the correlation resulting from the impact, even though it is there. As a result, a description by density matrices cannot be a complete description of reality, since it suggests an absence of correlation when the experiment clearly shows that there are correlations.

In other words, the density matrix is not a satisfactory description of physical reality. Ultimately, when we try to make a description, we come up against this type of problem, whereas when we decide once and for all that it will be solely predictive, predicting observations, then everything works. We only need to do calculations and experiments, nothing else. If we carry out an experiment, this is verified each time.

Jean Petitot. In this type of discussion, I always tend to reintroduce the question of mathematics. The difficulties we are addressing here seem to me to be practically intractable on a conceptual level and we are not taking into account what is, in my opinion, a fundamental characteristic of physics theories, not only quantum mechanics but all current physics theories, namely the possibility to do what some call a computational synthesis of phenomena, a mathematical reconstruct of phenomenal reality.

We have often had this discussion; however I feel it is not sufficient to say "we do calculations". I think there is much philosophy within calculations and I have a tendency to consider that the true philosophy of physics has, precisely, more to do with calculations. Why do we absolutely want to reintroduce, in addition to calculations, a conceptual, classical philosophy where there would be good old substances, like in the past, and metaphysical entities like objects and relations, including within structural realism? Old Kant, of which we spoke earlier, said there was "strictly speaking" no science unless there was mathematics. If not, there is a classificatory, taxonomic ordering like in botany, etc., coupled with conceptual analyses. All sciences perform science in that sense. They collect a substantial amount of empirical data and use our cognitive abilities to put these in order, to classify, categorize, introduce cause and effect relationships, perform conceptual analyses, find concepts that are more fundamental than others, etc. We formulate in this way sociological theories, anthropological theories, etc.

Physics does a lot more: it does calculations and uses the extraordinary generative power of mathematics to reconstruct a tremendous empirical diversity from a few principles. This generativity has a very precise meaning. It is, for example the difference between a differential equation and its solutions. Differential equations express only general principles, however the solutions nonetheless match up with the details of the empirical data. There is something spectacular about this: the principles provide the infinitesimal generators and the models of reality are derived from the associated integral.

I believe that certain debates you have had with your colleague Roland Omnès centred on the role of mathematics and on the question of whether we can shift reality towards the fundamental equations of physics. For my part, I would tend to agree with this idea that reality in physics is formulated by laws and complex equations.

Bernard d'Espagnat. Do you believe in Platonic reality? We are the ones doing the calculations.

Jean Petitot. Effectively, I have shifted the problem of physical realism towards the problem of mathematical realism. Is mathematics simply a cognitive creation? Or is there something else, within mathematics, other than our cerebral activity? This is a very delicate problem, since mathematics is based on calculations, and we must take into account what certain logicians and computer scientists are saying regarding what a calculation is. I believe we can really support the theory that there are fundamental objective aspects within calculations. I cited Alain Connes [7] during our last meeting. He is not at all a Platonist, but at the same time, he did strongly insist during his exchanges with Jean-Pierre Changeux [8] on the fact that all objectivity criteria are united by mathematics.

Michel Bitbol. Mathematics is objective *in the Kantian sense*…

Jean Petitot. Yes, all objectivity criteria, in the strictest sense, are satisfied by mathematics. And yet mathematics is not ontological. It is the proof that there is a non-ontological objectivity. Many philosophers, not only Wittgenstein, have tried to understand why mathematics expresses transcendence *par excellence* while being of purely human origin. This is why I think it is interesting to shift the paradoxes of physics towards the paradoxes of mathematics, using the fact that mathematics is fundamental in physics.

Bernard d'Espagnat. Jean-Michel Raimond has something to say.

Jean-Michel Raimond. Yes, I apologize in advance for doing a *non sequitur* with what has been previously said, but I have from the start a problem with this idea of realism, which comes back to the comments of my young colleague (Matteo Smerlak). I understand to what extent the predictions of quantum mechanics are linked to the measuring apparatus, etc. Nevertheless, when we carry out, for example, a collision experiment, as you were just saying, the outcome of this prediction is highly dependent on knowing whether the particles that meet have an integer spin or a half-integer spin, which seems to me to be the archetype of the intrinsic property and of the intrinsic reality of a particle. I understand the problem of mental conception, etc., I understand completely that we could say that "there is no realism; anyway, everything exists within the field of consciousness and within the consensus that stems from our consciousness"—I must add I know nothing about philosophy—, but I find it hard to understand how we can completely reject all objective reality, independent of the measuring apparatus, for a number of properties of objects that are manipulated by theory. I need to know the charge, the mass, the spin of an electron to do anything in quantum physics. Is this not an element of reality?

Bernard d'Espagnat. Personally, I would not systematically eliminate things like the charge of an electron or the Planck constant, things like that, from the collection of what could constitute the elements of what we call independent reality: *mind-independent reality*. Therefore I agree with you in this sense, but I think that, firstly, it is conjectural and, secondly even if it was true, even if these elements were really the constitutive blocks of independent reality, their knowledge in itself would not be sufficient to reconstruct independent reality. This is my small contribution, which far from answers the question.

Jean-Pierre Gazeau. I do not know whether I may follow on from this contribution, but I was thinking of classical physics, of the evolution of what we call a physical quantity. A physical quantity, according to the definition of the

International Bureau of Weights and Measurements is something that can be characterized qualitatively and measured quantitatively.

If we take the example of instantaneous speed, it is a concept that was impossible certainly before the 18th century. It has become with mathematics a physical quantity in its own right that is listed by the International Bureau of Weights and Measurements. This physical quantity can cross the quantum barrier, i.e., we will find the same name for the physical quantity that we now analyse within a quantum framework, such as energy, such as quantities previously identified in the classical framework that will continue to be identified in another framework. We will describe them, once again, in a quantum framework. I do not know, personally, whether we could have described them better in a classical framework. What is the length of this table? Are there natural boundaries? Is there an impossibility to define concretely the length of an object? Simply compared with units, etc., at one point or another there are limits that are impossible to overcome.

I do not know, I find it hard to discriminate between something that is descriptive within the framework of one theory and that is no longer descriptive within the framework of another theory. I think mathematics is precisely there to help us.

It seems to me that in quantum physics we speak a lot about spectrum: the spectrum of a physical quantity. It is through the spectrum that we will identify a physical quantity. We also find a spectrum within the framework of classical physics. Simply, the spectrum is different, even infinitesimal, much more idealized in the classical framework than in the quantum framework. Within the quantum framework, there are many quantities, ultimately, for which identification relies on a discrete spectrum, more easily attainable than the elements of the "classical" continuous spectrum, as it happens in classical physics with the concept of length. You see.

Bernard d'Espagnat. We can try to give a partial answer to what you are saying. You speak of speed. We can also speak of momentum, it is the same thing. If we try to construct a theory in which physical quantities such as position or momentum are not inventions, ways of describing our experience, but correspond to reality as it is, it means that we seek a theory that is ontologically interpretable, and the model for a theory that is ontologically interpretable is the theory of David Bohm. Besides, we should not say "Bohm theory", it is in fact a theory (or one of many theories) of Louis de Broglie, which Bohm developed further [9]: it is a great injustice to speak of "Bohm theory", but never mind.

We therefore need to consider Bohm theory which reproduces all the predictions of quantum mechanics and which at the same time allows us to interpret the position of an electron effectively as a reality in itself. So if we consider momentum, Bohm theory effectively defines the momentum with a very precise formula; however this momentum cannot be measured. The momentum we measure is something else. We can measure a quantity which, in textbooks, is called momentum and which corresponds to what we call momentum in classical mechanics, but this quantity is not a momentum according to Bohm theory. The momentum predicted by Bohm theory cannot be measured: when you measure a momentum, you automatically measure something else. We are therefore faced with real difficulties that were not there in classical physics but are here now.

Hervé Zwirn. I would like to try to counter the two objections, which are not exactly the same but which go in the same direction.

First of all, I would like to go back to the concept of length found in classical mechanics that is also found in quantum mechanics, as in the concept of speed. Regarding the notion of description, it is not because we have observables, properties, etc. that are comparable or similar in quantum mechanics to what we have in classical mechanics, that we necessarily recover a description. The concept of length and the concept of speed you (Jean-Pierre Gazeau) were alluding to are properties that we effectively find again; we use the same words and they intuitively mean the same thing, but in order to say "we have recovered a description" we would need to assign these properties to an object. However, the problem we have in quantum mechanics is not that we cannot speak of the same quantities; we have the same observables in quantum mechanics and many of things are similar to what we already have in classical mechanics. The main difference is that we can no longer, for the most part, consider that the property is directly assigned to the particle, either in quantum mechanics or in de Broglie-Bohm theory. In classical mechanics we could consider that these quantities could have defined values for particles, whereas in quantum mechanics we no longer can, since the values are determined only after measurement. To suppose that the measurement is there, as in classical mechanics, only to observe their value does not work. Therefore, the moment we can no longer consider that we are describing something that exists, that has values and that we content ourselves with observing using instruments, the notion of description loses much of its meaning. The problem we have in quantum mechanics stems from there, it does not arise from the fact that we can no longer think of the same quantities. We think of them, but we no longer have the right to consider that these quantities can be applied, as in classical mechanics, to objects and that we will observe and describe what there is because we do not describe what there is: if we try to think like that, we fail.

Jean-Michel Raimond. There is still one thing that physics, including quantum physics, relies on. It is the capacity, the postulate that we are capable of, to properly define what a sub-system of the universe is, without which we come up against, as Mr. d'Espagnat was saying earlier, the wave function of the universe, and we might as well just give up and go to the madhouse.

To describe a sub-system means I am capable for all practical purposes—I agree it is not a philosophical point of view—of saying that "at this time, I am working with a system that is made up of three or four electrons, which are doing things, which interact in a local quantum manner". I am dealing with a situation with predictability issues, but I am nonetheless capable of saying that my reasoning, my theory, applies to a defined sub-system. I need to be able to say this, at one time or another. This sub-system is defined, it is made up of entities which themselves have, within the limits we all understand, I think, a certain number of established properties that I would tend to call, perhaps because I am completely ignorant, reality.

Carlo Rovelli. I agree with both. It seems to me that you are in fact, by saying two complementary things that are not in contradiction, focusing the problem. We are to some extent at the heart of the discussion. On the one hand, I hear "There are things in reality that are not defined until you observe them". On the other hand, I hear, "Yes, but still, there are things in reality that seem to be defined, in any case, we can make a list". It seems to me this is where we are at.

I would be tempted, to move this discussion forward, to look for a position that does not necessarily say "since a certain type of realism, where I can imagine an precise description, in a sense, of reality, is clearly blocked by quantum mechanics…". It has been blocked in an empirical manner. Given this, are we forced from then on to abandon the notion of reality? And the notion of description? Is there not a weaker way to speak of description? Is there not a weaker way to speak of reality? Which one?

It seems to me that in this discussion, and in quantum mechanics, there are things we can use, and not construct a philosophy out of nothing. In the image of the world provided by quantum mechanics, some realistic aspects still remain. We still use the notion of sub-system that seems to work very well and at the same time is the source of all problems, since the real problem of quantum mechanics is that of the limits of the notion of sub-system. If I have a particle, I can think there is a wave function, but at the same time if I consider a bigger system, I know that this wave function will not have all the required information regarding the correlations with the rest. I need the wave function of both.

It seems to me that if we want to find "neorealism", it is in the opposite direction. We are still saying "we measure this". What does that mean? It means that a device or someone interacts with the system, and the device itself is subjected to and is affected by this interaction. The outcome of this measurement is what is considered in Heisenberg-type quantum mechanics. What is predicted by quantum mechanics, written down on paper, is not the state but the effect of the interaction.

If I take the ensemble of these "measurement outcomes" as my notion of reality, I still have a problem since I realize that what is measured can become for me, for someone who considers me within a quantum system, only a fragment of my wave function, by opposition with another me in another branch of the wave function. Therefore, the difficulty is that I have another unresolved problem. I have not resolved the problem by simply following Heisenberg, by saying "I see the electron here, here and here and I do not ask myself where it is elsewhere but that is all I see". I have not resolved the problem; however it seems to me, if I remember that I want a weaker definition of reality, that quantum mechanics is telling me that these measurement outcomes must be attached to the affected system. Therefore if I imagine that any system of the universe, any sub-system of the universe, as it interacts with other systems, produces measurement outcomes, I can think of the world without necessarily invoking the mind, that is to say without… imagining a *mind-independent* reality, where there are interacting sub-systems and, *with respect to each system*, elements of reality that are real; however, only real regarding that system. I can understand this as a theory of knowledge, as a very partial description of the world.

Let me come back to what you have said. When Newton introduced the concept of force, the whole of Europe said "No, no, no, that is irrealism!" The most vehement criticism from France, Germany, Italy and Spain of the *Principia* was that it was too antirealist: realism should be strong, in the manner of Descartes, or like the atomism of Ancient Greece. Interactions should exist only for things that are in contact. The concept of force was part of irrealism. As time went on, we all became Newtonians and, for us now, the concept of force is part of realism. We have completely changed our conception of reality.

It seems to me that we must search for a new conception of reality, one that is weaker, where the key factor would be to focus on the measurement outcome and to realize that this outcome is relative to the observed system. I think we can do this without necessarily speaking of consciousness, without necessarily speaking of the *mind*, by staying within physics, yet still speaking of a description, but obviously a description in the weak sense, in quantum mechanics. It seems to me that it is our idea of reality that must evolve to adapt itself to quantum mechanics.

Bernard d'Espagnat. It seems to me that you base most of your arguments on the notion of sub-systems. I think that in fact the notion of sub-system is a creation of our way of seeing things. To say that we have a particle that disintegrates in two, therefore that we have two sub-systems, is to say that when you look you will see two sub-systems. But until you have looked, you cannot really say that you are dealing with two sub-systems, so it seems to me. That would be to interpret them as things in themselves; this does not seem possible to me.

Jean-Michel Raimond. I know that these two particles are correlated; however, I want to be able to speak of these two particles which are not each a sub-system independent of the rest of the universe.

Bernard d'Espagnat. You speak of them because you know that when you will observe them, you will observe two things in two different locations, but you speak of future observations, you do not speak of something that is totally independent of yourself.

Michel Bitbol. Just one point to add to what Mr. d'Espagnat said: a great physicist named Asher Peres wrote a book on quantum theory from an entirely empirical perspective [10]. According to him, what we call a physical system is only one way of translating a class of experimental set ups. When we say, for example, that there is a physical system made up of two particles, that simply means that we have prepared an experimental set up such that the value associated with the value of the observable "number" is 2. The usual vocabulary where we "prepare a system" is redundant so to speak; the word "system", according to Peres, is the name we give to partially controlled potentialities that become available through preparation.

Carlo Rovelli. What of Matteo's (Smerlak) neutron star?

Matteo Smerlak. We still have the same problem of the apparatus that would interfere with the system.

Michel Bitbol. However, I do not need the apparatus to intervene straight away. Quantum mechanics allows you to predict that you will never observe a neutron star breaking spontaneously. You can try to associate a vector state with a neutron star and it allows you to predict that, in future experiments, the probability of observing the spontaneous collapse of the system is close to zero.

Matteo Smerlak. How could I prepare a neutron star? You define a sub-system through a preparation process.

Michel Bitbol. Indeed, the problem here is that the preparation is not controlled; however we can obviously define a preparation that we call for example the Big Bang, or another process that plays exactly the same role. But I have to admit this is not yet the right answer.

Carlo Rovelli. As soon as I allow myself to do this, I can think of the world as something that was prepared and which I study. Therefore the notion of preparation is not that essential after all.

Michel Bitbol. Using the notion of preparation could simply mean that we circumscribe the conditions of our experiments, meaning that we put in place instruments that allow us to select a certain number of objects, for example neutron stars. Once we have done that, we can perhaps associate a wave function with these selected objects… The problem you (Matteo Smerlak) raise is the following: in general, we associate a vector state with a system using the knowledge we have of a previous preparation. For example, I can associate an electron with the vector state $|+\frac{1}{2}\rangle$, the specific vector of a spin component, and I can say "I associate this vector state with it because I made it pass through a Stern-Gerlach device which served as a preparation, and because I have selected the upper electron beam". In the case of neutron stars, the problem is that I do not have this capacity for control…

—Addendum by Michel Bitbol—

After this discussion I realized that my response to Matteo Smerlak's objection was incomplete and unsatisfactory. A more convincing conception needed to be formulated, but I only thought of it afterwards. I therefore sent a letter to Matteo to clarify what I meant by "preparation of a physical system". Once this clarification is accepted, nothing stops us from extending the notion of preparation and attributing an associated predictive symbol to systems that are not controlled in the laboratory such as neutron stars. Here is my letter:

"Dear Matteo,

I thought about your interesting objection to neutron stars last night (and unconsciously during the night). It seems to me that I now have the answer. In short, it rests on an *informational* definition of preparation rather than on the standard idea that considers preparation as an operation that is materially controlled in the laboratory. It is amusing that I did think of this straight away: it proves I still carry with me remnants of realism despite my best efforts!

1. In the standard conception (first established by Henry Margenau in 1937), a preparation is a filtering operation of physical systems controlled in the laboratory.
2. Following this operation, depending on the type of filter used, we associate a vector state with these systems. Some very classical examples: the emission of photons by a lightbulb and the passage through a vertical polarizer, the emission of silver atoms by an oven followed by a Stern-Gerlach (device) and the selection of one of the two trajectories, the emission of electrons by a heated filament, the acceleration by an electric field and the interposition of a screen with one hole. The associated vector states are respectively: two states that are specific to a spin component and one state that is specific to the observable spatial position.
3. What matters in this process of association of vector state and preparation is not the a priori instrumental control of the systems, but the *information* made available by this control. As proof of this: the "*delayed choice experiments*". It is particularly apparent in the first thought-experiment of this type, i.e., Heisenberg's microscope, reviewed and improved by (Carl) von Weiszacker. The vector state of an electron lit up by a photon does not depend on what was initially done to the electron, or on the momentum transmitted to it by the photon, but by the position of the photographic screen near the microscope's eyepiece. What changes with the position of the screen is only that the information obtained is optimal *either* for the position variable *or* for the momentum variable. If the information is optimal for position, we assign a vector state to the electron that is nearly identical to a specific vector of the observable "spatial position", and if the information is optimal regarding the quantity of motion, we assign a vector state that is nearly identical to a specific vector of the observable "quantity of motion".
4. By extrapolation, we can perfectly speak of the "preparation of a system" even if it is absolutely not controlled in a laboratory (as is the case with astronomic objects!). "*Preparation*" *in this case is nothing more than the discrimination at a distance between different classes of objects that allows us to obtain initial information about them.* Selecting a neutron star in the sky based on spectral and/or gravitational criteria amounts to making initial information about it available, i.e., to "prepare" it (in the broad sense of the term). From this initial information, we will be able to (at best) attribute it a vector state or (at worst) a density operator.
5. It is true we can also not attribute a vector state to a neutron star and use more global strategies to show its stability. This is the case, using the most well-known example, when we prove the stability of atoms using Heisenberg inequalities. This simply means that we consider an entire class of preparations

rather than a single preparation; however, even here the initial information is made available by what we could call a "generic preparation", and we are again in the standard situation.

I confirm that ultimately quantum mechanics is a formal process that allows us to derive an instrument of probabilistic prediction from the initial information made available by an operation called "preparation". This instrument of probabilistic prediction can be modified as new information is obtained (this is the "state reduction"!)".

The case of a neutron star does not call into question the easily interpretable general operational scheme of quantum mechanics.

—Remainder of the session (31st of January 2011)—

Matteo Smerlak. My example aimed to show the limits of the very Peresian idea, of preparation of a sub-system. This operationalism is relevant in the laboratory, were experiments are carried out on atoms, on systems that we can master in our immediate vicinity. However, quantum mechanics goes beyond that. It seems to me that this type of point of view is problematic in the context of astrophysics.

Hervé Zwirn. There are effectively problems with preparation within the context of astrophysics that must be dealt with differently.

Michel Bitbol. The problem is always this one: what are we doing to attribute a vector state to something? Effectively, when we attribute a vector state to an electron in a laboratory, we know what we are doing: we prepare, we pass the electron through a Stern-Gerlach (device), we select those which are at the top, for example, and not those which are at the bottom, therefore we can attribute a vector state. What are we doing in practice—it would be interesting if you (Matteo Smerlak) could tell us—to attribute a vector state to a neutron star?

Matteo Smerlak. What I had in mind was more related to statistical physics. The properties of a neutron star are those of a very large number of neutrons that we do not try to describe individually; this is an important difference. The quantum mechanics we speak of in this type of epistemological discussion is the quantum mechanics of prepared and measured individual systems.

Michel Bitbol. I now understand, you speak of a generic system and not of an individual system. Effectively, it is not prepared, however, you consider a class of systems, with a possible hypothetical preparation, and you say "with this class of systems I associate, let us say, such and such density operator r".

Hervé Zwirn. When we observe a neutron star, we are not in same situation as when we observe an individual neutron. We have macroscopic properties that emerge independently, for which there is no need for preparation because they are affected by a sort of global thermodynamic property. What emerges is statistics. It is exactly the same thing as when we look at a table.

Matteo Smerlak. Indeed. I would like to point out the fact that quantum mechanics is also critical in this framework, which is not that of atomic physics. I think that these questions can be asked slightly differently, if we keep in mind this other aspect of quantum mechanics, i.e., its statistical aspect.

Hervé Zwirn. Is it a counter-argument to what we were saying previously, or not? Since we could easily say that, independently of the fact that every neutron of a neutron star has its own existence, quantum mechanics shows that the collective properties of this ensemble of electrons are still the same macroscopically. Therefore it works. I had some difficulty with this argument earlier, but in fact it is not a counter-argument to the fact that we refute the individual existence of each neutron.

Matteo Smerlak. I agree completely.

Hervé Zwirn. Yes, completely, since it is macroscopic and therefore it emerges without the need for a measuring device.

I would like to ask Carlo (Rovelli) a question regarding what he said earlier. Why say, if I understood correctly, that when two sub-systems interact we do not need to have the intervention of an observer to say that a measurement has been done? The interaction of two systems does not produce a measurement!

Carlo Rovelli. I think it does. I say this rhetorically. Imagine you make measurements. Someone else has made them in Beijing. There is the same Stern-Gerlach (device) in both locations, but imagine the Chinese scientist stepped out of the room. The measurements were made without him being there. Were the same measurements made? I think so. What happened had nothing to do with the mind.

Hervé Zwirn. Yes, because you made the measurements. In that case, the two systems, as long as we have not made any measurement, are considered as one. They are entangled. We are in an EPR situation. It is the same system. What do you mean by an unread measurement?

Carlo Rovelli. No, no, it is not EPR!

Hervé Zwirn. So the Chinese scientist has gone out, and has not looked?

Carlo Rovelli. There is a white dot, but he has not looked at it. There is nothing in quantum physics that tells me I must discriminate between the two. Therefore I think it has nothing to do with the mind, it has to do with something else, with an interaction between the electron and the measuring device, however if it was not a measuring device, if it was a magnetic field and neutrons, I think something would happen in exactly the same way. I am personally convinced, through the study quantum mechanics, that there is nothing that speaks of the mind, that directly speaks of consciousness. On the contrary, there is something in quantum mechanics that strongly compels me to say that I cannot take the fact that the dot is there as a property of absolute reality, but I must relate it to something else, for instance my point of view, or someone, or someone else.

Michel Bitbol. I did not expect this from you Carlo. I have the impression that you are being unfaithful to your relational interpretation. You say there is no difference between the instance when the Chinese scientist has looked and when he has not. But of course there is a difference! In one case, *relative to* the Chinese scientist who has not looked, the ensemble of the system is still in a state of superposition, whereas, *relative to* the Chinese scientist who has looked…

Jean-Michel Raimond. No!

Michel Bitbol. According to Carlo it is. According to his interpretation, there is a difference between the description relative to the Chinese scientist who has looked and the one who has not! Carlo, relative to the one who has looked, you must say…

Carlo Rovelli. I think that quantum physics forces me to say that, relative to the screen, a measurement has been made.

Hervé Zwirn. No because the screen and the electron are in a superposed state and nothing in quantum theory allows us to say there is a dot at a given location.

Carlo Rovelli. We are at the heart…

Jean-Michel Raimond. We now come to the wave function of the universe!

Carlo Rovelli. I will try a different approach. You (Hervé Zwirn) make a measurement; I am outside. You are in a box, and you can see what is there; I am outside and I have very precise instruments that allow me to see very close quantum correlations. For me, no measurement has been made. In relation to you, something has been measured; with relation to me something has not yet been measured.

Michel Bitbol. *That* is typical Rovelli!

Carlo Rovelli. Therefore, relative to the screen, a measurement has been made. Relative to you, a measurement has been made. Relative to me, a measurement has not yet been made. If we accept this weakening of reality, we can speak of reality, we can still speak of the elements of reality. It is very strong and we do not need to invoke the mind, things happen relative to other physical systems not relative to consciousness.

Jean-Michel Raimond. Except that to detect subjective correlations that prove that your friend is not in a superposition of states, we would need in practice a device that is bigger than the Universe. These correlations are so inextricably hidden (like a needle in a haystack) that we need devices of unfeasible complexity to detect them. Therefore we would need to say, perhaps, at some point—it is not philosophically correct—"in practice, we do not care, a measurement has been made". One knows the outcome, the other does not, but the measurement has been made.

Carlo Rovelli. Yes but I think that if we keep the distinction, we can salvage certain suggestions for solving certain paradoxes we do not understand. I agree but if we keep in mind that it is only *in practice* [11] that we can do this, we still have a weakening of reality, but on the other hand we gain the fact that certain paradoxes are resolved, and we do not need to invoke human consciousness. Within the universe, human beings do nothing that is unique. Except that they are capable of gathering information on the world, and perhaps of other things like interacting.

Michel Bitbol. Something human beings are capable of is understanding that they do not do everything in the world, as you do in a very strong manner Carlo when you say: "human beings do nothing that is unique other than collect information". This ability is very important; in it resides our power to objectify, out power to define a domain of existence that we can consider as not entirely controlled by us. At the same time when you highlight the fact that human beings do nothing unique, it is a construct you are doing just now with your human intellectual powers! We had to start somewhere to define a domain which is not affected by humans, and this starting point is none other than our human point of view… Human beings also have this great capacity for defining the conditions in which they can consider that they do not intervene.

Carlo Rovelli. I follow you 100%.

Hervé Zwirn. I think everyone agrees, from a practical point of view, effectively, that the measurement has been made, since to see whether a superposition still exists would require measuring devices completely beyond our reach. From a practical point of view, it is "as if", we all agree.
To see whether in fact, "in reality", the system is still in a superposed state would require measuring devices completely beyond our reach. As human beings, we can consider that the measurement was made "for us". As physicists, we do not need to ask any more questions and we can carry on doing physics, but, as a philosopher, we differentiate between "for us" and "in itself".

Jean-Michel Raimond. This measurement is beyond the reach of the universe as a whole.

Hervé Zwirn. It is out of reach if we consider measuring devices as being created by human beings but, technically, as long as we have not reduced the density matrix by the partial trace operation linked to the fact that we will not look at certain degrees of freedom (and that is due to our own limitations), this density matrix remains non diagonal. Therefore if we want to reason not from the point of view of mankind's capacity, but on a philosophical level, technically, we cannot say that the density matrix is diagonal and thus that the measurement has not been done.

Jean-Michel Raimond. In that case, if we want to reason on a philosophical level, it is urgent that we stop reasoning on quantum physics, since it says that from the beginnings of the universe all particles are in an entangled state to which we belong. There is no possible reasoning on all this and we come back to the wave function of the universe.

Either there is a "for all practical purposes" that is philosophically defendable, I do not know why, with a reduced sub-system that is no longer in a superposition of states, either this sub-system continues to drive away these superposition of states in the whole of the universe, the whole of the universe is in a big wave function, I am a part of it, and I do not know anything anymore.

Hervé Zwirn. Even if we say that the fact that everything remains in a superposed state has as a consequence that the measurement of the global system is not attainable, we can be interested in things which are forever beyond our reach but for which we can see the point philosophically.

Let us place ourselves in a state of the universe 10^{1000} years from now. Nobody knows what that state will be. Some theories say one thing, others say the exact opposite. However, if we suppose that the system is still in a superposed state, we can make predictions that show that a certain number of correlations can appear 10^{1000} years from now which would have been lost if the wave function had been reduced. Admittedly, 10^{1000} years is a length of time beyond our reach. Nevertheless, the predictions of what will happen in 10^{1000} years' time are not the same if we suppose a wave function that is reduced in reality compared to one where we suppose a wave function that is not reduced in reality. This is what we mean when we say they are not the same.

Jean-Michel Raimond. This is because the Poincaré recurrence time applied to the wave function of the entire universe extends far beyond 10^{1000} years, i.e., probably closer to 10 to the power of 10^{1000} years.

The point would be to arrive at a philosophy that would limit itself to what is "in practice" because that is where we are at. We are debating things that make no sense.

Bernard d'Espagnat. My dear friends, it is getting late. We must put a provisional stop to this discussion. Otherwise, we will be kicked out. Let me remind you that the next session is on the 14th of March. Jean-Michel Raimond will talk about decoherence.

Jean-Michel Raimond. From a "practical" point of view!

Bernard d'Espagnat. From an experimental point of view, which amounts to the same thing. We will then try as a group to draw out the significance of the experiment in question. This will be interesting and extremely important.

References

1. See session I, "The Inescapable Strangeness of the Quantum World".
2. *Idem.*
3. Hervé Zwirn, *Les limites de la connaissance*, Paris, Odile Jacob, 2000; in Jean Bricmont and Hervé Zwirn, *Philosophie de la mécanique quantique*, Paris, Vuiber, 2009.
4. Michel Bitbol, *De l'intérieur du monde. Pour une philosophie et une science des relations, Flammarion*, 2010.
5. Immanuel Kant, *Critique of Pure Reason,* 1781.

6. Collection of philosophical texts that form the theoretical basis of Hinduism.
7. Session I, "The Inescapable Strangeness of the Quantum World".
8. See *Matière à penser*, Odile Jacob, 1989.
9. See sessions V "The Pilot-Wave Theory of Louis de Broglie and David Bohm" and VI "Pilot-Wave Theory: Problems and Difficulties".
10. *Quantum Theory, Concepts and Methods*, Kluwer, 1995.
11. On this expression, see the discussion at the start of session VI, "Pilot-Wave Theory: Problems and Difficulties".

Chapter 3
Experimental Investigation of Decoherence

J.M. Raimond

Bernard d'Espagnat. We will now have the pleasure of hearing a presentation on decoherence from Jean-Michel Raimond himself who—as we all know—is one of the principal authors of the pivotal experiment on this matter: the one that showed that, after his unfortunate predicament, the mesoscopic Schrödinger cat is indeed in a state of quantum superposition, "dead plus alive", and that it is only after a very brief but measurable lperiod of time that it becomes, to our eyes, either dead or alive.

The theme of decoherence and its interpretation are sufficiently important for the discussion it will generate to not be limited to today's session. There will be a follow-up session in May (session IV) where, if we have time, we will also come back to the philosophical question of the definitions of realism, which was left hanging last time.

Dear friend, the floor is yours.

3.1 Presentation by Jean-Michel Raimond

Thank you very much for your invitation. I will speak, as planned, about decoherence and in particular about the experiment we carried out with Serge Haroche, Michel Brune and many others at the École Normale Supérieure (ENS) in 1996 [1].

However, I would like to associate decoherence, superposition, entanglement and complementarity, and show you how in my opinion, at least in the very simple way I understand this, superposition, complementarity and decoherence are three

---

J.M. Raimond (✉)
Université Pierre et Marie Curie, Paris, France

B. d'Espagnat and H. Zwirn (eds.), *The Quantum World*,
The Frontiers Collection, DOI 10.1007/978-3-319-55420-4_3

sides of a the same "die" (insofar as we can speak of three-sided dice). I mention on the first slide all the funding bodies that have contributed to the development of our experiment. It has to be said that experiments in fundamental physics are sometimes very costly.

This experiment is far from being the only one on this topic. There are many other experiments on decoherence that I do not have the time to describe, but which have many points in common, at least conceptually.

These experiments are a part of the renewed interest in fundamental quantum mechanics in the area of experimental physics. For a long time, the questions of decoherence, entanglement, and non-locality were considered pure intellectual curiosities for philosophers. The subject has become very fashionable again over the last twenty years with, first of all, the advent of experimental techniques that have allowed us to carry out a number of the "thought-experiments" of the founding fathers by manipulating a single atom, a single photon, a photon box, Schrödinger cats…as we will see. And underlying all this—this is particularly important—is everything that pertains to quantum information. How does quantum mechanics combine with information theory in order to eventually do things that cannot be done by classical computers? All this sheds a new light on fundamental quantum phenomena and in particular on the phenomenon of decoherence.

Just to go back to what Édouard (Brézin) explained in his introductory seminar [2], we know that quantum mechanics allows the superposition of states. If we imagine state $|x\rangle$ "particle in x" and state $|y\rangle$ "particle in y", state $|x + y\rangle$ (with a normalizing factor $1/\sqrt{2}$) is equally admissible as a quantum state and describes a particle located both in x and y.

This superposition manifests itself of course essentially in interference phenomena and in particular in the simplest one: Young's double-slit experiment. There is maybe no need for me to describe it in detail here. If there is a modulation of the probability of presence on the screen, it is precisely because during an intermediate step the particle is on two paths at once, suspended in a quantum superposition of one position and another. The interferences are direct manifestations of the principle of quantum superposition. As first year quantum physics students discover, to their great amazement, to carry out a simple interference experiment in an optics laboratory is a demonstration of quantum superposition.

I cannot resist showing you a neat photon by photon interference experiment carried out by my colleague Jean-François Roch at ENS Cachan [3]. It is the Fresnel biprism experiment, where a beam is split in two and the two parts are made to overlap: we therefore expect interference. However, Roch does this with a single-photon source and registers their impact with a camera capable of seeing them arrive one by one. After a 10 s exposure, we see an apparently random distribution of photon impact points. After 100 s, we begin to see patterns. After 500 s, we see a modulation of the distribution of the impact points. After 2000 s, we clearly see fringes.

There is another example of interference of single particles that I like to describe to my first-year students. It is an experiment on the interference of electrons carried out in the Hitachi laboratories by Tonomura [4] in 1989 with an electronic microscope. Conceptually, it is equivalent to Roch's experiment. I found a video of this on the internet [5]. We can see electrons arriving one by one with randomly distributed impact points, then by widening our field of view, and by gradually accelerating and accumulating events, we start to make out a certain regularity, and then we can see the regularity, the fringes. We have interference with electrons which are particles of matter. We can say to ourselves that maybe Louis de Broglie was right.

These experiments were repeated using objects more complex than electrons. We are able to make a C_{60} molecule interfere, and even extremely complex molecules. Richard Feynman used to say that this experiment in itself revealed all the mysteries of quantum mechanics. All the mysteries? We will see that that is not completely true.

Let us now speak of superposition and complementarity. I will recall the discussion between Einstein and Bohr at the 1927 Solvay conference, but perhaps historians will correct me on this. We can see the wave-like nature of a particle in interferences. Can we also assign a corpuscular nature to it? Can we ask the particle which path it took? At least in principle we can. I cannot remember if it was Einstein or Bohr who suggested the clever design where one of the slits is suspended from a spring. When a particle travels through the bottom slit, which is fixed, nothing happens. When it travels through the top slit, it is deviated, and by conservation of momentum, it gives an impulse to the slit, and therefore to the whole mobile device which is set in motion. Thus theoretically, we should be able to know whether the particle passed through the bottom slit, in which case the mobile device is at rest, or the top slit, in which case it is in motion, by looking at the interferometer at the end of the experiment. Bohr shows by following Heisenberg's uncertainty principle that if the slit is light enough to be detectably set in motion (and therefore to enable us to say which path the particle took), then the position of the slit is so uncertain that we do not have interference fringes on the screen. Either we know which path the particle took, by having a very light slit and by looking at whether it moved or not, but we then have no indication that the particle can behave like a wave, or we give up knowing which path it took, which is what we do with standard interferometers, by having macroscopic slits that weigh many kilos, in which case we effectively have an interference signal that shows the wave-like nature of the particle, but we have no indication of any corpuscular characteristic.

There have been many variations of this experiment. We can imagine coding the information of the path taken, called "*which-path information*", either on the interferometer—the slit that moves—, either with an external observer who looks at the passing particle, for example by shining a laser beam on it and registering the emitted light with a detector, or either by coding this information on the particle itself: for example, if we place two crossed polarizers behind the slits, we will no longer have interference because the final polarization of the photon will tell us which path it took. This is a very standard practical experiment, there is no Michelson interference pattern when crossed polarizers are placed in the two arms, but it is also, strictly speaking, a complementarity experiment.

To discuss this quantitatively, I will consider not the Young interferometer, as the detailed analysis is relatively complicated, but the Mach-Zehnder interferometer. You probably know the general set-up: we have two beam splitters and two mirrors delimiting two paths (forming a rectangle). Following the principles of Bohr and Einstein, we imagine that the first beam splitter is on a mobile device. If the particle takes the top path, nothing happens. If the particle takes the bottom path, it drives the mobile device backwards, setting it in motion. Can we analyse this set-up in detail?

In the manner of Einstein-Bohr, we can do this in terms of Heisenberg uncertainty relations. What is the condition for knowing which path the particle took? The impulse given by the particle to the mobile device needs to be greater than the quantum fluctuations of the mobile device when it is at rest. Obviously, I suppose that everything is ideal: everything is at 0 K and the mobile device suspended from a spring is likened to a harmonic oscillator. Therefore the impulse P given by the particle (which is, at $\pm$a numerical factor of little importance, its total momentum) must be greater than the fluctuations Δp of the momentum of the mobile device. However, this means that the uncertainty Δx on the position of this beam splitter, which is because of Heisenberg relations greater thatn $h/\Delta p$, is necessarily greater than h/P. Therefore we can see that we have which-path information provided the uncertainty of the position Δx of this beam splitter is greater than h/P, *i.e.*, greater than the wavelength λ of the related particle. If the uncertainty of the position of the beam splitter is greater than the wavelength λ, there cannot be any fringes.

Of course, we can make this argument entirely quantitative. That said, I find that it is a somewhat simplistic description of the experiment, which warrants a much deeper interpretation in my opinion in terms of entanglement. We shall see that complementarity and entanglement are necessarily linked.

You know what entanglement is: two quantum systems that interact are, after entanglement, in a global non-separable state. It is not possible to assign a specific quantum state, a wave function, to one or the other of the two sub-systems. We can at best define for each a reduced density matrix. This is the essence of the Einstein-Podolsky-Rosen situation. Édouard Brézin has given us a very clear presentation of this [6], I will not go over it again, nor the fact the Alain Aspect has shown in his rightly famous experiments that quantum mechanics is right and common sense is wrong: it is unfortunate but that is how it is.

Where is entanglement in our experiments? Let us describe it in a somewhat technical way. The initial state of the beam splitter, ideally likened to a harmonic oscillator, is the resting state $|0\rangle$, meaning a vacuum. The final state if the particle takes path a is also a vacuum. If the particle takes path b, the beam splitter has a small oscillatory movement, described by what we call a "coherent state of the harmonic oscillator". This state is completely described by classical amplitude; we will come back to it later. I have named this state $|\alpha\rangle$. This means that somewhere, in the middle of the interferometer, I cannot speak of the state of the particle on its own, I need to speak of state $|\Psi\rangle$ of the particle and the beam splitter. This global state is a quantum superposition of the state "the particle is following path a and the beam splitter is at rest" and the state "the particle is following path b and the beam splitter is in motion". It has all the characteristics of an entangled state. The particle and the beam splitter have interacted and are in an entangled state, the sum of two different states, or of two different wave functions.

From that point, it is relatively easy to calculate the fringe signal. Ultimately, this fringe signal is (more or less) the sum of the squared modules of two wave functions describing the states mentioned above, plus one term that is the scalar product of these two wave functions and which contains the entire interference signal. If states $|0\rangle$ and $|\alpha\rangle$ are very different from each other, we can distinguish the two paths taken by the particle. However, their overlap (or scalar product) is close to zero and we no longer have any fringes. By contrast, if the two states cannot be easily distinguished, we have a very small displacement in front of the quantum fluctuations in state $|0\rangle$. The scalar product $\langle 0|\alpha\rangle$ is practically equal to 1 and we have a normal fringe signal. Therefore, fringes disappear when the particle is strongly entangled with the motion state of the beam splitter [7].

In fact, this is a very general phenomenon. Interferences are destroyed as soon as a particle becomes entangled with something, *i.e.*, if the two states of this something are different depending on whether the particle took one path or the other. This can be the beam splitter of an interferometer, as described above, or an external path detector–we have a detector somewhere that switches on when the particle takes path *b* but that does not switch on when it takes path *a*, or it can be another degree of freedom of the particle itself. If we take a photon, we have one degree of freedom which is its "position" (I insist on the quotes!) and another which is its polarization. It can also be a thermodynamic environment. If the interfering particle modifies the thermodynamic environment, with which it is strongly coupled, in such a way that there are two perpendicular states of the environment associated with the two paths, then there will be no fringes. In other words, if the particle leaves information on its position somewhere, anywhere, even if it is not read–there is no need for an observer, or to invoke consciousness—, either within the particle itself or in the interferometer, in the explicit detector or in the thermodynamic environment, then there will be no fringes, there will be no interference phenomenon.

Photons are more or less insensitive to what they pass through and thus they are not easily entangled with the environment, which explains why we can build interferometers with kilometre-long path difference like the ones at LIGO or VIRGO [8]. By contrast, material particles interact strongly with the environment. The slightest collision of an atom with a residual gas molecule in an atomic interferometer changes at least its state of vibration or rotation, and this is sufficient for the environment to know which path was taken by the atom. It is therefore extremely difficult to create interferences with material particles. It is all the more difficult the larger they are. While it is still possible with a C_{60} molecule, it would be very difficult to make billiard balls interfere, since it is extremely difficult to make sure that the ball does not collide, over its entire path, with at least one other particle of residual gas. Even in the emptiest interstellar space imaginable, we would not succeed. Even if we did succeed, there would remain subtle things like gravitational wave background that would interact sufficiently with the ball to know which path it took.

This tells us that the more complex the object is, the more it will become entangled with the environment and the less there will be quantum superposition, which leads us to the question of why there is no quantum superposition at our scale, and gives us a qualitative idea of what the answer could be.

This leads us to the measurement problem as presented in Schrödinger's well-known 1935 article illustrated by his famous cat [9]. We must of course define the measurement of a sub-system of the entire universe (the cat in Schrödinger's article). As you may have guessed from one of my interventions during the previous session [10], I am a philosopher of the "*for all practical purposes*" and I prefer to not ask myself too many conceptual questions and rather be able to say that "for all practical purposes, I can define a sub-system".

If this sub-system is complicated, the Hilbert space is colossal. For one mole of spins 1/2, the size of the Hilbert space is 2 to the power of 6.10^{23}! The Hilbert space is gigantic. However, ultimately, I only ever observe a few of these states in a macroscopic object. Why? This is an important problem, closely linked to the strangeness of the quantum world as illustrated in Wojciech Zurek's 1991 [11] article in *Physics Today* by Michael Ramus' beautiful drawing that you are probably familiar with, where the clear classical world is separated by a barbed wire from the fuzzy quantum world, with the cat half dead and half alive. I believe decoherence can help us understand all of this.

Decoherence is therefore important for understanding why we only see a very small portion of the possible states, why I am here and not at the same time in my office, why things are black or white, not black and white at the same time. It is also important for measurement.

Let us take a simple measuring apparatus, which measures the component along Oz of a spin 1/2 (this system is the paragon of quantum systems). There are only two possible results: +1/2 or −1/2. Let us say the needle points upwards if the system is in state $|+\rangle$ (in the specific state of the observable "component of the spin along Oz" corresponding to the specific value +1/2) and downwards if the system is in state $|-\rangle$ (corresponding to the specific value −1/2).

Jean-Pierre Gazeau. I have a question regarding the classical-quantum boundary. We usually hear that a physical system is considered quantum when a quantity such as the action that characterizes it is of the same order as the Planck constant. Where do the characteristic quantities of this system lie here that enable us to call it quantum?

Jean-Michel Raimond. It is very difficult to say. We realize this, and there have been recent debates for example regarding superconducting circuits, which are macroscopic quantum systems. It is not easy to determine what the macroscopic quantum parameter, which describes the macroscopic nature of the system, is. I think one of the advantages of decoherence is precisely to give us an indication of the macroscopic parameters. In decoherence, as we shall see, there are two very separate time constants, and it is the separation of these time constants that gives us the size of the system. We shall come back to it. The problem is very subtle. It is not because we have many particles in a system that it is necessarily macroscopic. A superconducting circuit through which a flux quantum passes is a quantum system. We can excite some coherence between its two states, and this corresponds to thousands or tens of thousands of pairs that circulate in the circuit in one direction and in the other at the same time. However, it is not because there are tens of thousands of pairs that it is macroscopic; there is only a flux quantum. It is not because something is big that it is macroscopic.

Jean-Michel Raimond. Let us return to measurement. If, now, we measure a spin which is in a superposition of state $|+\rangle$ and state $|-\rangle$ and if evolution, that given by the Schrödinger equation, is unitary, then theory tells us that at the end we will have an entangled state of the measured system and the measuring apparatus, a state that will be a quantum superposition of two macroscopic states. This is very surprising, and not at all what the experiment shows, where it is one or the other with probabilities. Therefore there is something more than the unitary evolution predicted by theory. And the unfortunate Schrödinger cat is ultimately only a paradigm for a measuring apparatus. It measures macroscopically whether the atom has disintegrated or not. A paradigm, or a metaphor if you prefer.

Decoherence simply consists in recognizing that a quantum system is necessarily coupled with a complex environment, all the more so that it is itself complex.

We cannot isolate a quantum system. For all practical purposes, we can isolate it up to a point, but there will always remain residual coupling, a thermodynamic environment (which we will avoid considering as the entire universe). There is always a thermodynamic environment, be it gas, residual radiation, gravitational wave background, etc. It is the coupling, therefore the entanglement, with this environment that will destroy the quantum superposition. Zurek and others have popularized this approach.

However, if we are dealing with microscopic systems, like an atom or a photon, and if we work hard for decades, we manage, still for all practical purposes, to make the coupling with the environment negligible during the length of time of the experiment. Therefore we can make interferences, and carry out all the neat quantum information experiments we know of. If the system is mesoscopic or worse, macroscopic, it is extremely difficult to carry out this type of "adiabaticity" (in quotations, adiabaticity is not really the right word) of the quantum world, and we will have to take into account this coupling with the environment. Obviously, this is the case with any measuring apparatus.

Essentially, decoherence amounts to taking into account coupling with the environment… I do not believe there is at this time a general theory of decoherence, which says that in all cases, we can specifically describe coupling of a system without formulating a hypothesis either on the system or on the environment. There are nonetheless extremely general and generic models, where we can more or less explicitly consider evolution. We can, in general, but at the cost of approximations which are extremely well controlled. Such is the case for instance of a massive particle undergoing Brownian motion. There are equations that describe this process very well. It is also the case, of interest here, of the harmonic oscillator, a simple quantum system coupled with any sort of bath.

If we describe the state of the sub-system—the harmonic oscillator on its own—in terms of a density operator, we can write the evolution equations for this operator, *i.e.*, Lindblad equations [12] based on very reasonable approximations regarding the environment. The environment is complex—there are resonant frequencies in the entire or near-entire spectrum—and the environment is large, meaning it is not, or nearly not, modified by the interactions with the system. We can therefore write these equations, which have experimentally verifiable consequences. At a push, when during practicals we build RLC circuits (resistance, inductance and capacitance in series) and we look at the damping effect of this circuit on the oscillatory regime, we verify the Lindblad equation for a damped oscillator, and we do so very well.

All these models have a certain number of general characteristics. The first is the existence of "pointer states". Without going into details, pointer states are stable or quasi-stable states that do not evolve even if the system is coupled with the environment. In a way, these are (quote!) "specific states" of the operator which link the system with the environment. For instance, in the case of Brownian motion, position states are quasi-stable states since more or less all that a particle in a given position does is have a small movement—namely Brownian motion—that diffuses slowly in $\sqrt{t}$.

If we take a harmonic oscillator at 0 K, the vacuum is a pointer state, *i.e.*, strictly a stable state. If the oscillator is in the empty state, it remains in this state. The vacuum, as I have said, is the fundamental state of a harmonic oscillator. The coherent states of the harmonic oscillator, which are classical states, describe as well as possible in quantum physics, considering the Heisenberg relations, the oscillations in a potential well. They are quasi-pointer, quasi-stable states, meaning they remain coherent, they remain of the same nature; simply their amplitude is damped slowly. All these states are practically insensitive to decoherence. They evolve very slowly, or not at all, in the presence of coupling with the environment, and technically they form a base of the Hilbert space of the system.

By contrast, if we have superposition of pointer states, we show in all these models that it is rapidly transformed into a statistical mixture. If I prepare state "pointer *a* + pointer *b*" and plug in the reservoir, I find myself very quickly in a state of statistical mixture "50% of pointer *a* and 50% of pointer *b*". A statistical mixture simply describes something where everything happens as if the system was prepared half the time in pointer state *a* and half the time in pointer state *b*. This evolution from superposition towards a mixture, of course, takes a certain, very brief, time rather logically called decoherence time.

This decoherence time has two main characteristics: it becomes shorter the more different the pointer states that we superpose are. The more distant the pointer states are, within a metric to be defined, the shorter the decoherence time; and this decoherence time is, for different macroscopic states, infinitely short compared to the length of time of ordinary dissipation of the system. If we take a damped harmonic oscillator, for example, the decoherence time of a macroscopic superposition is infinitely short compared to the damping time. The main characteristic of the relaxation of a mesoscopic system is to have two timescales. We have the damping time of the energy, which I will call "a slow timescale", and we have the decoherence time, which is the lifespan of quantum superpositions of pointer states, which is in a rapid timescale.

Of course, this does not constitute an addition to quantum mechanics. All I am doing is applying the standard relaxation theory of quantum mechanics known since the 1940s and 1950s [13]. I am in fact applying the standard quantum relaxation theory to a system coupled with a bath, on which I formulated a certain number of hypotheses.

A simple example, proposed by Anthony Leggett [14], is once again Brownian motion. I take a particle of mass m that is in a gas at equilibrium at temperature T. It is a good model for the position of the needle of an apparatus: the simplest needle is a point particle of mass m.

What is a long timescale? It is the damping time of the speed through friction. It obviously depends on pressure and gas temperature—in short it is $1/\gamma$ and it is calculable.

What is a rapid timescale? I consider a superposition of two wave bundles spatially separate from quantity a, a typical set up within an interferometer or a measuring apparatus. The decoherence time is therefore the ordinary relaxation time multiplied by the square of the ratio of the thermal de Broglie wave length of the particle to twice the separation a. The bigger a is, the shorter the time is, and the parameter $\lambda_T/2a$ will have a marked tendency to be very small, as λ_T is very small. If you take a rubidium atom as a point particle (it is not a very heavy point mass, a mass of 85 atomic units) at 300 K, the thermal de Broglie wave length is 4.10^{-11} m. This means that, for a single rubidium atom, the decoherence time becomes shorter than the relaxation time when the separation of the two wave bundles is greater that 4.10^{-11} m! This is short for small separations, extraordinarily short for macroscopic separations. We must be very aware of the orders of magnitude. If I take a kilo at one meter, I will find decoherence times of 10^{-50} s, which simply have no sense because there is no sense in speaking of Brownian motion over such a short timescale. We go beyond the hypotheses in which we can use the equations I previously mentioned.

What do these pointer states look like? There is a nice term that was introduced by Zurek: "quantum Darwinism". Let us imagine a simple system: a qubit, a spin or an atom with two levels or two states $|0\rangle$ and $1\rangle$. The environment is a large set of identical systems all initially in state $|0\rangle$. I imagine that, during the length of time I am interested in, the interaction is very simple: if my system is in state $|0\rangle$ then all systems of the environment remain in state $|0\rangle$; if my system is in state $|1\rangle$ then all systems of the environment switch to state $|1\rangle$ (it is a "*toy model*"). What are the pointer states then? They are obviously $|0\rangle$ and $1\rangle$. They remain stable, at least with respect to the environment with which they have had this interaction. What is their characteristic? It is that $|0\rangle$ and $1\rangle$ spread their copy in the environment. They are in a sense objectified by these multiple copies in the environment; all observers who can measure the systems of the environment, whatever they do, will agree on the state of the system, either $|0\rangle$ or $1\rangle$. The first observer obtains a certain result and all the others will find the same thing, an objective state $|0\rangle$ or $1\rangle$.

That said, let us ask ourselves first of all what we can say, in general, about the link between decoherence and measurement. We cannot make it say too much or too little. I think decoherence tells us essentially two things about measurement.

The first thing is that ultimately the uncertainty of a measurement outcome is a simple classical probabilistic alternative. If you measure a system in a superposition of equal weight of $|0\rangle$ and $1\rangle$, the outcome is $|0\rangle$ in 50% of cases and $1\rangle$ in 50% of cases. So God plays dice, it is true, but these are not classical dice. This uncertainty, I would say—but I may be stepping outside my area of competence—seems to me of exactly the same nature as that of statistical physics. After all it also describes systems that are in probabilistic superpositions: we do not know what the microscopic state of a macroscopic system is but it exists and we know the probabilities.

Something more subtle, in my opinion, is that it is decoherence that defines the measured quantity. Indeed, the measuring apparatus, at the end of its decoherence time, is in a statistical mixture of pointer states. The states I am measuring (the specific states for my measurement) are therefore states of the system that are entangled with the pointer states.

Let us imagine that we can separate the Hamiltonian evolution phase of the system and of the measuring apparatus (this phase puts this apparatus in a superposition of states) and the decoherence phase. And let us imagine a measurement as previously mentioned of one spin. Suppose the two possibilities are "*spin* up, the needle pointing up" and "*spin* down, the needle pointing down"—or in short "*spin* up" and "*spin* down". If I consider the entangled state "spin up" + "spin down", I can also write it as 1/2 [("*spin* up + *spin* down") ("up + down") + ("*spin* up−*spin* down") ("up−down")]. Am I only measuring whether the spin is up or down or am I measuring the spin in a base {"*spin* up + *spin* down", *spin* up−*spin* down}? What tells me this is that the states ("needle up + needle down") and ("needle up−needle down") are not pointer states, thus are not the states associated with those I am measuring. What I am measuring is "needle up" or "needle down", correlated with "spin up" and "spin down" respectively. Therefore, ultimately, in a measuring apparatus, what defines what I am measuring in a system is the interaction Hamiltonian between the system and the measuring apparatus itself, obviously, but also the base of the pointer states of the measuring apparatus.

Bernard d'Espagnat. I completely agree with you regarding your second point but I am a bit perplexed regarding your first point. When you say that probability here is of the same nature as that of classical physics, that God plays dice but with classical dice, are the dice really classical? No. They are classical to our eyes, they appear classical to us, but it seems to me there is a difference that needs to be introduced between "appears to us" and "really are". Unless you reduce the notion—and this may be what you have been saying—of what really is simply to what appears to us all, in all circumstances. Is there not something like this?

Jean-Michel Raimond. I have, as you probably have guessed, a rather pragmatic philosophy which is *for all practical purposes*.

Bernard d'Espagnat. Yes, that is right.

Jean-Michel Raimond. … and not asking myself philosophical questions, because I am totally incapable of competently posing them, I have a tendency to equate the two points of view, meaning that from my point of view, that of the experiment I am carrying within the current state of knowledge, for all practical purposes, I have a tendency to equate what I use to describe all that I can do in a practical experiment with reality, with what is. It is a pragmatic point of view, knowing of course that my vision of what is will most likely change the day quantum mechanics changes.

Bernard d'Espagnat. It seems to me there is still a difference, because in classical physics you could make this association without any problem. We accept in this case that the things we see are in themselves how we see them. We know it is more complicated in quantum physics.

Jean-Michel Raimond. I am not sure I can see the thermodynamic state of a complex state. For a base particle, I can see the speed, and even then… For a more complex thermodynamic system, I am not sure I can see…

Bernard d'Espagnat. In ordinary classical thought, let us say of materialism at its peak, the association in question went as far as metaphysics, whereas in quantum physics we cannot take it so far. We must take the pragmatic point of view you mentioned and admittedly you are right to do so and dismiss philosophical problems. It is nevertheless interesting to see that taking this pragmatic point of view is now compulsory, whereas at the time of classical physics we could remain in an implicit "chosist" metaphysics.

Jean-Michel Raimond. I am practically sure it would be impossible, within the strict framework of quantum physics, to answer any philosophical question, and therefore not knowing how to answer, I prefer to not aski myself the question; we have discussed this at length during the previous session. My fear is that I will get stuck in solipsisms like "the world does not exist" or the wave function of the universe, which both seem to me as having no way out.

Bernard d'Espagnat. We should not interrupt your presentation on a fundamental experiment with our digressions into admittedly more uncertain territories. However, Hervé Zwirn would like to add something.

Hervé Zwirn. Along those lines, I have a proposition to make in order to give a heuristic for clarifying the debate on what we can or cannot say in quantum mechanics. You have said that "what happens with decoherence is that we go from "and" to "or", and that is crux of the matter: how do we go from "and" to "or"? "And" is clearly superposition, entanglement. "Or" however assumes we have arrived at a state that is a statistical mixture, in the classical sense of the term like in classical statistical physics. In order to be able to say this, the density operator of the system associated with "or" must have been diagonalized. However, this does not occur in a rigorously complete manner. Extremely small off-diagonal terms subsist. This happens—as you say very well—for all practical purposes, which means it is practically impossible to demonstrate that diagonalization is not complete. However, let us place ourselves in a world where we do not have any order of magnitude in mind, meaning we do not know the time it could take for minute off-diagonal terms to become non-negligible. What happens, and we know this through calculations, is that the off-diagonal terms still present in the density operator are in fact so small that they are negligible for all practical purposes, even if they could in theory become non-negligible again after a Poincaré recurrence time which, anyway, does not have any physical meaning. Therefore this numerical estimate allows us to say that everything happens as if the state was diagonalized. But let us imagine that we do not know the numerical values assigned to the decreasing exponentials of the off-diagonal terms and we are simply examining the formalism. We could then say, in the absence of knowledge of the time required for the off-diagonal terms to become important again, that "we are not in an "or" state because off-diagonal terms persist". I believe that is what Bernard d'Espagnat meant. If we place ourselves in a world where we forget the numerical values and we simply look at what things are, well then, we are not in a state associated with "or", we are still in a state associated with "and".

Jean-Michel Raimond. Except that, in decoherence, there is, especially when it comes from a macroscopic object, so many orders of magnitude that we should be able to remain in the "*for all practical purposes*". Recurrence times are absolutely enormous, far longer than the lifespan of the Universe. The Universe will have ceased to exist, it will only be radiation, before any recurrence time can occur, therefore this does not make any sense.

What worries me more, and what is probably more difficult, is defining what a system is. Where do I place the boundary? How can I sufficiently isolate the system from the environment so that I can effectively discriminate the variables of the system from the variables of the environment? This is a delicate problem, and it is precisely for this reason that it is difficult to carry out experiments in quantum mechanics. It is difficult to isolate it sufficiently to be able to speak of a system as such. This is a more serious conceptual problem than the Poincaré recurrence or the existence of off-diagonal terms in the density matrix, which when written with zeros would not fit into the memory of a computer.

Hervé Zwirn. This clearly shows the difference between the point of view of a physicist and that of a philosopher. It is obvious that physicists and philosophers can agree on the fact that, *for all practical purposes*, there will be no measurable effect, but that is said a posteriori, once we have done the calculations and realized what the order of magnitudes we are dealing with are. Now, if we forget the orders of magnitude resulting from the calculations and look at whether the density matrix is really diagonal, we realize that is not the case. The philosopher, at that point, tells himself that the state of the system is still superposed. On a strictly philosophical level, there is a difference between something extremely small but not zero and something that is rigorously zero.

Jean-Michel Raimond. I am afraid that, from the pragmatic point of view of a stubborn layman, this legitimate philosophical preoccupation comes second to other preoccupations, namely: can we speak of sub-systems? Can we speak of quantum mechanics at all? If we do not know how to speak of sub-systems, we do not know how to speak of anything, and I believe these effects of splitting the system into sub-systems are of more immediate concern than the problem of Poincaré recurrences or of the non-zero value sensu stricto of the off-diagonal elements of the density matrix.

Bernard d'Espagnat. Yes. Can we even speak of a system other than as an interpretation?

Jean-Michel Raimond. Can we speak of a system that is other than the whole of the Universe? Here we fall into a solipsism that is not tenable either by an investigator or even by a philosopher, I would think.

Jean Petitot. If I may add to Hervé's (Zwirn) point, I would like to modify slightly what you have just said… The difference between a near zero value and a real zero value is more a problem of mathematical ideality than a philosophical problem. Mathematical objects are by definition idealities and thus at some point we overstep the mark of what is physically interpretable. It is obvious but I would not say it is a philosophical or ontological problem.

Hervé Zwirn. I did not say that that particular problem was philosophical.

Jean Petitot. It is the link between mathematics and experience that is at stake, and what we speak of means that mathematics loses, at these extremes, all effectiveness and no longer has any experimental meaning.

Jean-Michel Raimond. I would not want to sow any confusion, but there have been some remarkable things done recently on the relaxation of entangled quantum systems [15]. We find that in many cases, entanglement does not tend exponentially towards zero but explicitly and suddenly disappears at a certain point. Mathematically entanglement disappears, therefore it is not impossible that effects of this kind, if we knew how to formulate decoherence theory precisely enough, could not provide a solution to the problem of infinitesimal off-diagonal elements. This cannot be done yet, and for practical purposes, it has no meaning whatsoever because these remarks concern the extreme end of the exponential, in an area where there is practically nothing left to see.

I believe nonetheless in the fact that decoherence defines the measured quantity. It is important to know what the dynamics of decoherence are to know what an apparatus measures. This is also true in classical physics: if you do not know the damping of the apparatus, you do not know what it is measuring.

By contrast, of course, it does not tell us anything regarding why and how a unique outcome emerges when we make a measurement, i.e., instantiation. Moreover, it is clear that this is not an addition to the standard postulates. It does not take the place of standard postulates but is equivalent to them. We include explicitly in decoherence theory, and in all of this approach in terms of pilot equations, the notion of partial trace, which is equivalent to the postulate of reduction. We cannot demonstrate the measurement postulates using decoherence without running the risk of circular reasoning.

Bernard d'Espagnat. Decoherence is not a new theory.

Jean-Michel Raimond. No, it is the quantum relaxation used since the very beginnings of quantum mechanics, applied to a mesoscopic system that is sufficiently complex to have two clearly distinct timescales.

Jean-Michel Raimond. How can we investigate decoherence experimentally? We could say: it is easy, we do it in everyday life; however, this is not really satisfactory. If we want to see it functioning, we cannot content ourselves with looking at the final state, the everyday statistical mixtures, we must try to resolve the dynamics and for that, we need to start with a system whose time constant is sufficiently long for us to measure decoherence time which is obviously a lot shorter. We also need a carefully chosen set-up to examine the state of the system, since it is not easy to known whether a system is in a state of quantum superposition or not. A few systems are amenable to experimental investigation, in particular those of the first two experiments carried out in the 1990s on ion traps and cavity electrodynamics. In both cases, we play experimentally with the simplest non-trivial quantum model [16], which is a two-level system where a spin ½ is coupled with a harmonic oscillator.

What mesoscopic object will we manipulate in our experiments? We manipulate a field of a few photons stored in a photon box; it is a modern version of Einstein's famous photon box. Of course, it is not a "photon box" but a cavity, two mirrors facing each other, a Fabry-Perot cavity built with superconducting mirrors of extraordinary geometric quality, which means that at the level of the millimetre, near 50 GHz, 6 mm wavelength, we manage to build the best mirror in the world, a mirror on which a photon can bounce a billion times without being lost. We thus have a cavity, around three centimetres long and five centimetres in diameter that can store a photon for 0.13 s. Of course photons are eternal in open space, but they hate being in captivity and die off quickly. Maintaining a trapped photon for 0.13 s is not easy.

In this cavity, we place a classical mesoscopic field, one of the coherent states of the harmonic oscillator. Of course you know that a classical oscillating field may be described in the Fresnel plane by an amplitude and a phase, or by a complex number. We can describe it in a quantum manner with a type of wave function in the phase plan, which is the Wigner function. Basically, quantum physics tells us that quantum fluctuations are superposed on the classical amplitude and we can consider the coherent state as a density of the probability of presence of the electric field whose amplitude is detectable only in a small circle around the classical amplitude. Let us draw a circle to represent this simply. I have a classical amplitude and a circle with a constant radius that describes the quantum fluctuations which are basically those imposed by the Heisenberg limit.

If I consider a field in the order of the photon, the amplitude of quantum fluctuations is in the order of the field amplitude. It is ultimately a very quantum object whose entire dynamics will be governed by quantum fluctuations. However, if the amplitude is equal to a few tens—in the right units, the square of the amplitude is the number of photons in the field—the classical amplitude dominates the amplitude of quantum fluctuations and we find ourselves with an object that can essentially be understood, once again for all practical purposes, as a classical object. We can prepare these coherent states within the cavity simply with a classical source of radiation, a type of laser, and by turning a button, we go from a very quantum object, containing on average one photon, to a very classical object, containing tens of photons. We can thus adjust the size of the object we will manipulate between a decidedly microscopic scale where there will be no decoherence, and a very macroscopic scale where there will be decoherence.

Bernard d'Espagnat. You go from a state with, on average, one photon to a state with many photons just by turning a button?

Jean-Michel Raimond. Yes. I can prepare these coherent states, which are quasi-classical states, within my cavity by purely electronic means. I know how to couple my cavity with a classical source and I know how to modulate, how to adjust, the amplitude of this classical source.

Jean-Pierre Gazeau. These are the coherent states of the electromagnetic field.

Jean-Michel Raimond. Yes.

Jean-Pierre Gazeau. When you say "*n* photons...", is the number of photons in a coherent state undetermined?

Jean-Michel Raimond. No, no, the *average* number of photons is determined, it is the square of the amplitude, but it is not a whole number. A coherent state of the electromagnetic field has a statistics of photons that is Poissonian, with an average number of photons that is the square of the amplitude with a relative dispersion of 1 over the square root of the average number of photons.

Jean-Michel Raimond. So, how will we prepare a cat? We will place a coherent mesoscopic field in the cavity and couple it with a two-level system similar to a spin 1/2, an atom with two levels. This atom is effectively prepared in what we call circular Rydberg states. These states are both very excited and very close to ionization and are the best quantum realization of the Bohr orbit. They are the maximal quantum orbital and magnetic numbers, which means that the orbital is somewhat like a bicycle tyre centred on the Bohr orbit. They are not easily prepared. They are completely classical states; their entire properties can be calculated classically, all quantum numbers being large. They have a very long lifespan, are easily detectable, can be measured one by one, and we can detect them and manipulate them in very subtle ways, as we shall see.

Jean-Pierre Gazeau. How many levels are placed together?

Jean-Michel Raimond. We will isolate two levels, meaning we will try to make sure that there are two levels that play a role.

Jean-Pierre Gazeau. In order to create a Rydberg state?

Jean-Michel Raimond. No, the Rydberg state is a given state. Here we consider two levels *g* and *e* that are circular states with a principal quantum number of 50 and 51. There is an infinite number of Rydberg states. Within the multitude with a principal quantum number of 50, there are 5000 levels—$2n^2$—but we manage to place electric fields, etc. that lift any degeneracy so that we only see one state, and we only prepare that one state in a selective manner, with a 98 or 99% selectivity. The cavity is tuned to the vicinity of this transition, and only this transition.

Jean-Michel Raimond. What will we do with these? We will make the atom and the field interact in a non-resonant manner. The point is that a non-resonant atom is not capable of emitting or absorbing photons as this would not conserve energy. All it can do is to behave like a small fragment of transparent dielectric. It does not emit or absorb photons but it has a refractive index. The atom that passes through the cavity changes the resonance frequency of the cavity. And it so happens that these atoms are extraordinarily strongly coupled with radiation simply because they are very large. This atom has a size 2500 times bigger than the hydrogen atom, and is around 0.2 micron in diameter. Therefore, being large, they are strongly coupled with radiation. A single atom passing through the cavity can change the phase of the classical field amplitude by a few tens of degrees, a large effect, whatever the field amplitude, and it can change it differently depending on whether it is in level *g* or level *e*.

This looks exactly like a measuring apparatus: we have an amplitude that points in the Fresnel plane. If the atom is in level *e*, the amplitude will travel with a phase in one direction. If the level is *g*, the amplitude will travel with a phase in the other direction. Therefore the amplitude that points in the Fresnel plane is the needle of my measuring device. Coupling with the atom is precisely the dynamics where "if the atom is in one level, I travel in one direction; if the atom is in the other level, I travel in the other direction". I thus have a prototype of a measuring apparatus.

Jean-Pierre Gazeau. Will this not change the number of photons?

Jean-Michel Raimond. This will not change the number of photons because I cannot absorb or emit them as this would not conserve energy.

Jean-Michel Raimond. First of all, I will carry out a complementarity experiment. Let me briefly introduce an atomic interferometer that is found in all atomic clocks: it is the Ramsey interferometer. Initially in level *e*, the atom will interact in succession with two perfectly classical microwave impulses. In this preliminary experiment, we do not yet have a cavity and we make sure that each of the two impulses transfers half of level *e* to level *g*, meaning that from *e*, it prepares the atom in a superposition of both *e* and *g*.

If the atom passes from *e* to *g*, it can do so in two ways: either it makes its transition when it receives the second impulse, or it makes its transition when it receives the first. There are therefore two quantum paths that lead from the same initial state to the same final state. If there are two quantum paths, there must be interference. This looks exactly like the Mach-Zehnder model. The probability of observing the effect "the atom passes from e to g" depends on the relative phase of these two successive impulses or on what happens to the atom between the two impulses. Therefore we have an atomic interferometer, an interferometer of internal state.

We are therefore conducting a complementarity experiment. The cavity, with its coherent field, is as I have said a measuring apparatus where the needle, the field amplitude, is capable of telling whether the atom is in e or g. You can see that e coupled with the state of the field will make the "needle" turn one way, and g will make the needle turn the other way. Therefore ultimately the cavity is a measuring apparatus where the needle is likely to give me information regarding the path taken by the atom in the interferometer. If the information is sufficient to tell me in which state the atom was in the middle, I should not have fringes. If the information is insufficient, I should have fringes.

When is the information sufficient? It is when the phase shift of the field is large on the scale of quantum fluctuations. If the phase shift is very small on the scale of quantum fluctuations, I will know nothing of the path taken and I should observe fringes. It is as before: the momentum of the particle was small on the scale of quantum fluctuations of the beam splitter and I had fringes. By contrast, if the dephasing is large, I can determine at least in principle in which state the atom was and the fringes disappear.

This is effectively what we observe [17]. I am capable of modifying as I wish the phase shift of the field resulting from the interaction with the atom and I notice that, when the two phase states corresponding to the two atomic states are well-separated, I do not have fringes. Once again, this is a complementarity experiment which, instead of being carried out with a macroscopic measuring apparatus, is carried out here with a measuring apparatus that contains three photons, hence a very small measuring apparatus.

Jean-Pierre Gazeau. When you say "phase shift", we expect some uncertainty regarding the average number of photons.

Jean-Michel Raimond. In all this, there is obviously an uncertainty regarding the average number of photons, which is constant throughout the experiment.

Jean-Pierre Gazeau. That is nonetheless related to the uncertainty on the phase?

Jean-Michel Raimond. No, because the interaction with the atom does not at all change the distribution of the number of photons. We only change the field phase, the distribution of the number of photons remains constant; we have verified this. In order to verify this, we send hundreds of atoms through the cavity and we observe that the field intensity does not change. Furthermore, we can verify that the decrease of fringe visibility follows exactly what is predicted by theory.

Jean-Michel Raimond. This is the first stage. However, in fact, what is interesting to look at is not the state of the atom, as once I have detected the atom in state g, what is left in the cavity? Calculations show—this is an exercise from a quantum mechanics Masters course—that it is the quantum superposition of the two states. In short, the atom travels through the field, is entangled with the field, but then receives the second microwave impulse. Through its action, I muddle the information on the state of the atom, I mix the states of the atom just after it has interacted with the field. What remains in the field at the end is either this superposition—if I detect the atom in a certain state—or a superposition with a minus sign if I detect it in the other.

The process I have described produces this superposition of quantum states of the field and I can now look at what happens to this mesoscopic superposition of quantum states. Does it remain in quantum superposition or does it go in a statistical mixture? We shall reveal the dynamics of decoherence.

How do we uncover the dynamics of decoherence? We will send a second atom, an atom probe that is capable of telling us whether, at a given time, the field is in a superposed state or in a statistical mixture. Basically, we perform quantum interference on the field itself.

I will show you later a more recent signal, but here is the signal we obtained in 1996. It is a function of time: we prepared this quantum superposition, left it for a certain length of time in the cavity, measured it and obtained a signal that was proportional to the degree of "quanticity" of the superposition, to the off-diagonal elements of the density matrix. Here is the evolution of the signal over time, in units of relaxation time of the cavity, and for two separate phases of the field state. We can see that effectively the quantum nature of the superposition collapses more rapidly than the relaxation time of energy, and is faster the more the pointer states are separated.

This is what we were doing in 1996. I would say that this signal is somewhat indirect and difficult to interpret. We have much advanced since and in 2008, we were able, with Samuel Deléglise, instead of contenting ourselves with indirect signals, to make a complete reconstruction of the density matrix of the cavity field [18]. We were able to prepare a cat as I explained previously: we sent the atom, created a superposition of phase states and then erased the state of the atom, and there remained in the cavity a Schrödinger cat, meaning this superposition of two coherent states of different phases. We were capable of reconstructing the state of the cat over time. In fact, rather than a density matrix, which is a complicated matrix and not very visual, we looked at the Wigner function, a function describing the field state (it contains the same information as the density matrix). A Wigner function is a distribution of quasi-probabilities in the phase plane. It therefore allowed us to calculate all the average quantum values. It is not a real distribution of probabilities as at certain points it can take on negative values.

So that is the reconstructed Wigner function of a Schrödinger cat. We can see all that we expect from a cat. We can see two "ears". These are the two coherent components we expect, which are in superposition. It has indeed the shape of a cat's ear, a curved shape, as in fact the preparation of our cat is not perfect. We distort the coherent states a little by separating their phases. If we had a statistical mixture, we would only have the two ears. However, since it is a quantum superposition, we have, at the initial time, between the two ears, a "smile" or "whiskers", interference fringes that reveal the fact that the cat we have prepared—we looked at it just after having prepared it—is in a quantum superposition.

In addition, with this method we are capable of following the evolution of the Wigner function over time with a good temporal resolution.

This cat—I forgot to say—contains on average twelve photons. More precisely, the separation between the two ears is of twelve photons. This is what matters for decoherence time. We expect a lifespan for the cat (decoherence time) in the order of 19 ms. Therefore we can resolve it temporally. I will now show you the life and death of a cat, 50 ms in the life of a cat. We slow things down on a factor of 1000. What we observe is that the whiskers disappear relatively rapidly; after 20 ms, not much is left. The ears are still there, the energy is still there, there is still a field, but it is no longer a quantum superposition, it is only a statistical mixture. If we decompose what happens systematically, we are capable of measuring decoherence time and we find that it is congruent, give or take measurement errors, with theory. We are really capable of preparing a cat at the initial time and look at how it decoheres in a very quantitative manner.

A participant. This is a rather complicated evolution in fact.

Jean-Michel Raimond. It is a complicated evolution because of the noise on state reconstruction. The theoretical evolution is a rapid exponential damping of fringes, whereas we have a slow exponential damping of their separation. It is therefore a lot calmer than the experimental events.

Jean-Michel Raimond. I am nearly finished. Just one more thing. There is some confusion in the literature between decoherence and dephasing. What is dephasing? It is something that is extremely important experimentally, namely the effect of classical noise on a quantum system which muddles quantum coherence. For example, this can be the phase noise in an interferometer due to a truck going down the road. This may look a lot like decoherence and can have the same dynamics.

However, sensu stricto it is not the same thing. At least in principle, we can undo dephasing. If we were able to measure the origin of this noise we could, through a system of retroaction, filtration and correction, cancel the effect whereas decoherence is really an entanglement between a mesoscopic system and an environment. We cannot, unless we know how to turn back the clock—but unfortunately we do not—undo entanglement. Decoherence is therefore not "undoable" in the sense that dephasing is. I think this is important. We must be careful to apply the term "decoherence" to situations where there is really entanglement.

Nevertheless, we can undo decoherence with quantum error-correcting codes. This is a very neat result of quantum information [19]. We can use entanglement to fight off decoherence, but that is experimentally difficult.

To conclude, I would say that we are now able to demonstrate decoherence, not its end-product which is classical, but the dynamics of decoherence of systems that are sufficiently large to have two distinct timescales and sufficiently simple for us to believe the quantum description we make of them. This shows us that the dynamics are indeed formidably efficient and incredibly fast, which poses a major problem if we want to build a quantum computer.

We can understand decoherence even better. We are able, by measuring experimentally the decoherence of a mesoscopic object, to carry out a tomography of quantum processes, meaning to go back, using experimental signals, to the equations that govern them and compare them with our model equations. We have already done this, very precisely, for the decoherence of the distribution of the number of photons. We notice effectively that, for example, a reservoir absorbs the photons of a system one by one and not two by two.

We can understand decoherence better, but we can also act against it better, which is interesting. Can we now block decoherence from a quantum system? It may not be interesting philosophically but it is very interesting pragmatically if we want to do something with the quantum physics of a mesoscopic system. It is very complicated, we are trying to do this with quantum retroaction where we take the information on a system and where we react on it to try to fight off decoherence.

Briefly, if I wanted to sum up this plethora of slides in one sentence, I would say that decoherence ultimately results from a "*which path*" information acquired by the environment when it becomes entangled with the system. There is nothing else in decoherence [20].

Therefore there is a nice triangular relationship between decoherence, entanglement and complementarity. A decoherence experiment is a complementarity experiment that demonstrates the entanglement of the system and its environment. Of course, in the middle of the triangle—we can draw a masonic triangle with an eye in the middle—there is quantum superposition which is at the heart of everything.

I believe I have finished. I would like to acknowledge my collaborators and current team members, especially Serge Haroche and Michel Brune. Of course, Serge and Michel were there during the 1996 decoherence experiment. The others were too young to have taken part. There is also the book we wrote, which contains the detailed mathematical explanation of all this, including the specific treatment of pointer states in this case [21].

3.2 Discussion

Bernard d'Espagnat. Thank you for this fascinating presentation of an experiment that clearly has had exceptional impact.

Jean-Michel Raimond. I am flattered, Mr. d'Espagnat, by your assessment of the experiment but I must insist: it is not the only one. There have been experiments by David Wineland on ion traps, and recently there have been very elegant experiments by John Martinis at UCLA, and Robert Schoelkopf [22] at Yale, and many others who are doing experiments that are in different ways complementary to ours, where they substitute atoms with superconducting circuits and our 3D cavities with chip-integrated cavities. They are able to prepare superb Schrödinger cats and watch them decohere in real-time by reconstructing the Wigner function in real-time. There are therefore other validations in other fields.

Bernard d'Espagnat. It was nevertheless one of the first.

Jean-Michel Raimond. It was the first.

Bernard d'Espagnat. That is what I thought. Naturally, the topic will raise many questions, particularly of a philosophical nature like those we tried to discuss earlier. However, your experiment brings into play a process—a "trick"—of quantum technique that arouses the interest and curiosity of the theoretician. I have been asking myself this question for a long time: how does your second atom observe the state of the system while at the same time leave it in a state of quantum superposition? At first sight this is surprising, as we can think of all these (simpler!) experiments where the act of knowing the path taken by a particle in a superposition of states destroys this superposition; we tend to generalize this and think that any knowledge of a superposition of states destroys it. Just by the fact that your experiment is possible shows that this generalization is too hasty and reading your publications, I think I have understood what you did to avoid these simplistic set-ups where it would be valid. Could you, without too many technical details, shed more light on this matter?

Jean-Michel Raimond. As much as the more recent experiment—which showed how to reconstruct the Wigner function—is technically difficult but provides the Wigner function which is visual, things are more subtle for the earlier experiment. Everything is in this Ramsey interferometer, where we place our cavity. What does the first impulse of classical microwave do? It transforms e into $e + g$ and g into $g-e$. What does the second impulse do? It does the same thing. Thus when I detect an atom in g for example, I have no way of knowing whether it passed through the cavity in e or g because it can arrive in g coming from either e or g. That is why when I detect the atom in g, I project the field on a superposition of these two phase states. It is difficult to explain with words. It is much easier to write down with kets but I did not wish to bore you with such formulas.

What does the second atom do? The first atom enters the cavity in a superposition of states. It is entangled with the two phases. Then, we break up this entanglement in such a way that the superposition of these two phases remains in the cavity. What does the atom do if everything is coherent? Exactly the same thing as the first. It enters the cavity in a superposition of states. It is therefore in a superposition of refraction indices. It turns in the two possible ways the two phase components left by the first atom. At the end, we have a cat with legs, where two are overlapping.

Either the two atoms passed through the cavity in the same state—when I say "either" this corresponds to terms in a complicated wave function of two atoms and the cavity—and they added their effects. I thus have a component which has turned by an angle of 2φ, if φ is the rotation produced by level e. Or they were in the other state, g, and the two fields turned by -2φ. Furthermore there are two ways to reach the final state, which is in the same phase as the initial state. Either the first atom has turned upwards and the second atom has turned downwards, e then g, or the first atom has turned downwards and the second atom has turned upwards, g then e. We reach the same final state of the field by two different atomic paths but they are indistinguishable since I do not know in which state the atom was in when it passed through the cavity. There must therefore be quantum interference.

The signal we measure is in fact a conditional probability of detecting the second atom in e when the first was at least in e minus the probability of detecting the second atom in g if the first atom was in e. The difference of conditional probabilities shows this interference. We can write this in a few lines of formalism.

Bernard d'Espagnat. We do not doubt that, but I think we will content ourselves here with the qualitative description you have given us, which is already very enlightening. That said, forgive me for monopolizing the discussion especially with a question that is once again somewhat technical, however, it seems to me to be important philosophically. Up to now, you have carried out experiments with photons. However, photons are a particle with zero mass, and we know that the quantum electromagnetic field, because of the fact that its quanta have zero mass, can be represented in many different ways. These are in particular the coherent states that you use, which are easy to produce with particles with zero mass, but not with particles with non-zero mass. However, if there is a clear qualitative leap between "zero mass" and "non-zero mass", there is not a clear one between "mesoscopic" and "macroscopic". Consequently, it would be interesting to redo this type of experiment with systems that are indeed still mesoscopic but that have particles with non-zero mass. And without actually carrying out these experiments, it seems to me to be an interesting question to know whether this is theoretically possible. By this I mean whether the mechanism you have just described for photons that allows the second atom to test the coherence state of your system without destroying it, could in principle—I do not take into consideration any technical difficulty—work with particles with non-zero mass.

Jean-Michel Raimond. I will be even more affirmative: not only is it feasible–it is being attempted. There has been evidence of signals in the area of cold atoms [23]. We can envisage preparing a small set of atoms—small because decoherence times are short for obvious reasons—in a macroscopic superposition from Bose-Einstein condensates and other methods that are not simple. With two potential wells, we can prepare a set of N atoms in a state we call the *NOON* state, in reference to the famous western [*High Noon* [24]]. This state described N atoms in one well and zero in the other, plus zero in one and N in the other.

Olivier Darrigol. These are bosons. It is more complicated for fermions.

Jean-Michel Raimond. Yes, these are bosons, material bosons. It is harder for fermions but not impossible, because we find it much harder to manipulate fermions and cool them down adequately.

Olivier Darrigol. In the concrete case of a real cat, or more generally of something macroscopic, we do not have superpositions of states which have a relatively simple mutual relationship, whereas in your example, your alive-dead cat, there is a very simple relationship for going from one state to another that is superposed to it. Can we envisage more complex superpositions?

Jean-Michel Raimond. We can do it in principle, it would be rather simple: we can transform our phase cat into an amplitude cat, meaning vacuum + coherent state, or any other superposition. We have even recently proposed to make an arbitrary superposition of coherent states that do not overlap [25]. We can thus make more complicated things, we can envisage it, like eight-legged cats.

Olivier Darrigol. This is harder to probe.

Jean-Michel Raimond. It can be probed. A simple way to probe it is to undo the preparation. The probability of success of this time inversion tells us whether decoherence has taken place during the interval. It is not impossible to make cats with more complicated shapes.

There is another type of experiment in quantum optomechanics that will probably be completed soon [26]. The idea is to couple the vibratory motion of a mirror with the incoming light and use the radiation pressure force to excite in a quantum manner the vibratory motion of the mirror. We are at the stage where people know, by exploiting the coupling between light and position, how to cool down in practice a mechanic oscillator, which is an oscillator weighing a few micrograms with a vibration frequency in the order of the megahertz, in its fundamental state of vibration.

Knowing how to do this, it becomes possible to prepare a Schrödinger cat by transposing our techniques as well as explore the decoherence of the vibration state. It will still be a vibration state in the Fresnel plane, involving not the oscillation of an electromagnetic field but rather the oscillation of a small material object which is macroscopic, although it does not mean that the mass of an object is the right criterion of macroscopicity. Here again, it will be the number of photons in the coherent state, or rather phonons. We must be very careful; there has been much confusion in the literature regarding what is the most appropriate criterion of macroscopicity.

Here is an amusing order of magnitude: a photon stored in our cavity corresponds to an electric field at 51 GHz, which is 1.5 mv/m, a rather macroscopic value for the field of a single photon! We can calculate, from this field and the associated magnetic field, the total current that circulates in the mirrors (the surface current generates a field reflected by the mirror while stopping the incident field to enter the superconductor). This current is tens of nanoamperes. This is a very macroscopic value, but that is not a pertinent parameter to describe to what extent the system is macroscopic. Besides, we can make a detailed model, and observe that the charges of the current are not entangled with the cavity field and therefore do not contribute to decoherence. In short, they are not part of the environment.

Olivier Darrigol. Because it is a superconductor.

Jean-Michel Raimond. Because it is a superconductor, and also because the mirror is so macroscopic that this current it is negligible in comparison. It represents a totally negligible charge shift, and the amplitude of the charge oscillation is within the range of Heisenberg uncertainties. We must be very careful, when we wish to speak of a cat, to choose the right criterion of macroscopicity. I think the best way to define it is to effectively observe decoherence, to observe that we have a time-scale that depends on the state we prepare. The relaxation timescale for coherent states is 0.13 s, for the cat it is 0.017 s. The two timescales are well separated. The ratio of both tells us to what extent the system is mesoscopic.

Michel Bitbol. I would like to come back to the philosophical question asked by Mr. d'Espagnat and Hervé Zwirn, and approach it from another angle to try to stimulate the debate.

My "other angle" simply consists of comparing quantum theory with classical probability theory. You know that one of the most important laws in classical probability theory is the law of large numbers. Contrary to what we often think, this law does not say that the frequency of a certain type of event tends uniformly and inevitably towards the probability that was calculated a priori by using arguments of symmetry or of minimization of missing information. It simply says that the greater the number of draws, the greater the *probability* that the frequency is close to the probability calculated a priori. It evaluates probabilities, the probability that measured frequencies converge towards estimated probabilities, probabilities that these second-level probabilities be corroborated by second-level frequencies, and so on to infinity. In other words, probability theory remains (as we should have expected) strictly confined to the area of probabilistic utterances. At no point does it say something regarding what *is* or what could *be*, but simply regarding the *probability* that something is.

Jean-Michel Raimond. Probability theory can be applied sensu stricto to quantum mechanics. All I am able to calculate is the probability of obtaining a frequency of occurrence in an experimental run. Probability theory completely applies here.

Michel Bitbol. That is exactly what I wanted to say. Quantum theory is also a form of probability theory; a theory that contains within itself classical probability theory in a rather particular way (I am generalizing). Thus, most of what we have learned regarding the status and limits of classical probability theory applies immediately to quantum theory. Let us come back to the example of decoherence theory, which is a part of quantum theory. In the same way that classical probability theory does not say that the measured frequency strictly and inevitably converges towards the calculated probability, decoherence theory does not say that the systems will strictly and inevitably tend towards a classical behaviour. It simply says that the probability to find a system that, all the while interacting with the environment, would present a strong deviation from classical behaviour tends towards zero.

Jean-Michel Raimond. I agree with you, except that it is clear that for quantum systems, it is a probability that tends very rapidly, and very exponentially, towards zero.

Michel Bitbol. Yes, but what I simply wanted to highlight through this comparison was that, in the same way that classical probability theory does not go beyond probabilistic utterances and says nothing about what *is* but simply about the *probability* that something is, quantum theory also does not go beyond probabilistic utterances, and therefore says nothing regarding the properties that we suppose *are* those of objects. Quantum theory only states the *probabilities* of something being or not being measured. Each proposition of quantum theory has no other status than a probabilistic one, and none has an ontological status.

Hervé Zwirn. I completely agree with what you have just said. The question is knowing what the status of extremely low probabilities is. Does a probability of 10^{-50} have any physical meaning? Do we have to consider that in reality it is equivalent to a probability of zero or must we consider it to be different? As for the philosophical interpretation of decoherence—once again, we place ourselves on a philosophical level—I think that the position we adopt simply depends on the position we have regarding the question of very low probabilities. In the case where we liken very low probabilities to zero, and *for all practical purposes* that is what we should do, we can consider that decoherence tells us that the world becomes classical. By contrast, in the case where we consider that a very low probability is not reducible to zero, then decoherence is not sufficient to make the world classical.

Jean-Michel Raimond. Are we hampered in statistical physics by the probabilities of events like "we are all going to die within the next three minutes because all the air in this room will condense itself to 10 cm^3 in that corner over there"? The probability is in the order of the probabilities that we find in the quantum world. This does not hamper us, philosophically, from doing statistical physics despite these probabilities.

A participant. In this type of situation, which consists iof reducing to zero all that is very small, the difficulty for me is where to place the cut-off point, because the fact is that you can carry out experiments or experience these situations, which are *borderline*, where we can still see the decoherence process, but then comes a time… Where do you set the boundary? That is…

Jean-Michel Raimond. Yakir Aharonov wrote many papers [27] where one of the central ideas was that we can have surprising events in quantum mechanics on the condition that their probability is low. What I mean is that "philosophically"—with quotation marks around "philosophically"—I do not see in what way, here, the status of quantum physics is any different from that of ordinary statistical physics, which discards improbable events.

Hervé Zwirn. It does not really discard them. It does so only on a practical level. We all know that there is a non-zero probability that we will die in the next three minutes if the air of the room condenses over there, but this does not worry us much.

Jean-Michel Raimond. Yes, there are other things that worry me much more.

Hervé Zwirn. Nonetheless, even in classical statistical physics, we do not say that it is impossible.

Jean-Michel Raimond. Yes but what I want to say is that the philosophical debate we may have seems ultimately to reduce itself to the questions of "What do we do with very low probabilities?" and "What do we do with infinitely small values?" and both statistical physics and quantum physics seem to me to be concerned with these questions, and in more or less the same terms.

Bernard d'Espagnat. It seems to me there is another aspect to the question, which is the following. Let us consider a system and an environment. There are common observables in the system and the environment, observables that refer to the correlation between the properties of the system and the properties of the environment. If I have everything initially in a pure state (*i.e.*, I describe the system by a vector state and the environment by a vector state, then initially the compound system is also in a pure state), it will remain like this, meaning there will be common observations between the system and the environment. The measurement outcomes, if they were taken, would therefore be extremely different, not close but extremely different, from what they would be if we were dealing not with a pure state but with a mixture. These observables are of course, in the case of a macroscopic system, completely inaccessible to us but philosophically we can think of a Laplace's demon that, like all good Laplace's demons, knows how to calculate infinitely rapidly and to infinitely subtle degrees, who could thus observe these observables, and who would say "no, we are not dealing with a system that *has* these properties. The truth is that the system *appears* to you as having these properties". There remains, of course, that reasoning in this way inevitably leads to a rather subtle discussion on the theme of "What capacities can we really attribute to Laplace's demons?".

Jean-Michel Raimond. There are arguments in one of the books written by Roland Omnès [28] that take into account this very question. Through semi-qualitative arguments, it was suggested that these Laplace's demons should have much more energy that then entire energy of the Universe and should have a measuring time far greater than billions of times the age of the Universe to be able to resolve the value of these observables. Therefore, I think we can do without philosophically.

Bernard d'Espagnat. Yes, but you see, our Universe is like that but things could be different, it is a contingency in the sense that it is not "*law-like*" but "*fact-like*". If we want to define what is real, what is reality itself, we should not proceed from the contingent aspects of this reality at the risk of falling into a vicious circle.

Jean-Michel Raimond. This time we can mention *Gödel, Escher, Bach* by Hofstadter [29]. He imagines that we measure the equations of motion of all the particles in the atmosphere and by inverting them we can hear Bach play the organ. It is possible in principle, despite the frenetically positive Lyapounov exponents underlying this. It is possible in principle; however, we can live very well without thinking about doing it (unfortunately, in a way!). It is as inaccessible as disentangling the variables of the environment, there is an extraordinary dilution of information in the environment.

Bernard d'Espagnat. Absolutely, but my reasoning seems to show us that we have no right to draw arguments from the fact that certain existing quantities are de facto non observable, to justify the idea that such and such property is really intrinsic to the system when quantum physics teaches us that this idea is untenable if the quantities in question were observable.

Jean-Michel Raimond. I believe we could nonetheless set a pragmatic limit to this type of reasoning, here again pragmatic and *for all practical purposes*. If extracting this information requires more resources than are available in the known Universe, then perhaps it is not worth talking about. Be it the extraction of Bach playing the organ or the extraction of extremely subtle observables, by any intelligence, by any technology based on quantum mechanics or statistical physics as we know them, this requires surely more resources than are available in the Universe. Even if we mobilized all the energy and the entire length of existence of the Universe to do the calculation, we would still not hear Bach play the organ. Therefore it makes no sense to talk about it. If it was feasible, it would be feasible by something that transcends the Universe, and there I believe we would be violently entering into the domain of metaphysics. Yes, God could do it, but that is another question.

Bernard d'Espagnat. In materialist, let us say classical, philosophy we did not trouble ourselves with such subtle considerations, centred on what we can *do*. We recognized that, of course, we are not capable of measuring very small, very very delicate, quantities, but we considered that this was a detail of no conceptual consequence, that simply reflected human incapacity, and had obviously nothing to do with the composition of the Universe, where the quantities are what they are, independent of us. And we added that these quantities known by us really had the values that we attributed to them. I consider that this philosophy is not compatible with standard quantum mechanics, even taking decoherence into account. It is clear, however, that it is a logically unprovable conception, in other words metaphysics. I would agree with you by saying that when doing his job, a scientist must not rely on any metaphysics.

Jean-Michel Raimond. When we create a cryptographic code, the goal is for it to be undecodable, not just obviously by enemies with the same technology as us, but by enemies with infinitely more advanced technology based on the same physics as us. That said, these resources have a limit, which is the size of the Universe. In order to reconstitute Bach or measure an observable, it is not a linear increase but an exponential increase of the resolution of the measurement. My excellent colleague Peter Knight expressed this very well: "*The Hilbert space is huge, really huge*". Reconstructing these observables requires making measurements that resolve everything, all the states of the Hilbert space of the environment, and that is exponentially difficult. I really believe that we can easily do without considering these types of questions.

Jean Petitot. I would like to make a more mathematical point and come back to what you were saying earlier about mathematical idealities. We have the impression, in the discussion, that there are two very different problems that converge: firstly, the problem of low probabilities, you cited Borel, the problem of infinite precision in hyperbolic systems, etc., that refer back to the structure of the continuum—what do we allow in the structure of the numerical continuum?—and secondly, the problem raised by Mr. d'Espagnat regarding metaphysical realism. But the problem of metaphysical realism, it seems to me, is very different from that of the ultimate structure of the continuum.

We have used repeatedly the term infinitesimal. We should maybe look at what mathematicians have said regarding the fine structure of the continuum, and indeed regarding the notion of infinitesimal. All this has been used within the framework of what we call non-standard analysis and I would like to comment on this subject. There is first of all non-standard analysis, which we would call classical, the one that Abraham Robinson developed from Leibniz where he showed how infinitesimal Leibnizian calculus could be a perfectly coherent calculus. To this end, he had to use the tools of logic.

Other, and I think much more interesting, points of view on non-standard analysis were then developed. First, the intuitionist point of view stressed the fact that classical mathematics is ideal in a very strong way, precisely because it is classical and not intuitionist, and it introduces therefore countless operations that would be concretely impossible to do and, if we want effective mathematics, we need to a use an intuitionist logic which is not at all a classical logic.

However the continuum, seen from an intuitionist logic, becomes very different from the classical continuum sensu Weierstrass, Cantor, Dedekind, etc. we learned at school. The intuitionist continuum is subjected to constraints of effectiveness. This was developed, for example, in Strasbourg by Georges Reed and Jacques Harthong. Their points of view tend toward our discussion, toward finitist points of view regarding the continuum, not only intuitionist points of view but radically finitist points of view, meaning that very large numbers function, by law, not by fact —this is not a pragmatic problem Mr. d'Espagnat—like infinities and very small numbers function, by law, like infinitesimals.

Some distinguished mathematicians, the most famous being probably Pierre Cartier, an important figure of the Bourbaki group, have developed in recent years radically finitist points of view on the continuum in relation to what you were saying regarding physics, namely that numbers that are too small or too large, which completely transcend physical limits, have absolutely no significance, even mathematically, meaning we cannot make mathematical ideality go beyond that. There is a sort of horizon phenomenon, and if we try to go beyond this horizon, we fall into a form of mathematics that is so ideal that it cannot have any real link with physics.

Jean-Michel Raimond. Physics has two lengths of this type which are the Planck length—we know that physics fails at that scale because it is not consistent between general relativity and quantum mechanics—and the size of the Universe—because we do not know what there is beyond that.

Jean Petitot. Yes, but this is a physicist's point of view, and I could retort, as Mr. d'Espagnat has done, that it is in fact somewhat contingent. What is interesting is that when we look logically at the structure of the continuum, we find ourselves dealing with questions of law; thus to call upon a finitist approach of the continuum that is even more radical that the intuitionist approach could be interesting for our discussion.

Hervé Zwirn. It may be interesting, however I believe that it is nothing more than a position of philosophical principle at the start. Intuitionists proceed from a position of principle which is different from those who do not accept intuitionist logic.

Jean Petitot. Yes but this is an effectiveness problem.

Hervé Zwirn. We cannot accept effectiveness. Anyway, most mathematicians have not accepted it. Therefore I think that you are right to stress this because it provides an analogy for our debate, but it will not resolve…

Jean Petitot. … it is not a debate to say that in physics we must use effective theories.

Hervé Zwirn. No, I say "debate" meaning we can agree with this position of principle or not. There are schools that agree with this position of principle and that will refute the notion of continuum in the full sense of the term. It is true that in physics, we can say that we do not know what there is below the Planck length. Perhaps space-time is discrete. A length of 10^{-100} meters may simply not have any meaning. At present nobody knows. It is the same in mathematics, we can be intuitionist while refuting even the infinity of N, it is not even a continuous infinity; even what is countable is suspicious to intuitionists. All that is actual and not potential infinity is suspicious. It is enlightening because it provides a type of analogy for the debate we are having on the "*for all practical purposes*" approach, however I do not think it will resolve the question in the sense that it will always be there, even when we frame it in this way. It is actually an interesting way to formulate it. There will always be people who are on one side and those who are on the other, because it is a question that is beyond discussion, it is a matter of feeling. Those who refuse to admit that the positions demanding effectivity are restrictive because they lead to the limitation of obtainable mathematical theorems and to the abolition of a certain set of things that matter—including the entire modern set theory—well, they have strong feelings, coming from within, but they cannot prove it. Those who, conversely, absolutely want to claim that this position is ridiculous because we need to go much further than what is allowed by effectiveness, and who in particular defend the entire modern set theory, have a position that also cannot be proven. We can simply argue in a non-decisive manner.

I think that in our physics-related discussion, the situation is the same. What I mean is: it is true that, *for all practical purposes*, it is stupid or completely unrealistic to consider energies far greater than the known energy of the universe or timescales that would be billions of times greater than the age of the Universe. This makes no sense, *for all practical purposes*, and we can tell ourselves that in that case, the rest of the discussion has no sense. Conversely, we could tell ourselves that there are worlds (sensu the possible worlds theory of David Lewis [30]) where there are other possibilities. This is neither science fiction nor fantasy; the semantics of possible modes is something serious used by logicians to provide an interpretation of modal logic. Within this framework, we can imagine possible worlds that are completely different from ours. In one of Lewis' possible worlds, it is perfectly conceivable that the measurement of the observables we were speaking about earlier, which in our world is completely impossible, is possible. If we place ourselves on a philosophical level and accept this framework, then it becomes legitimate to say that: "Claiming that decoherence leads to making the world classical would mean in modal logic: "decoherence necessarily leads to a classical state."" However, this is wrong because there are possible worlds that are not ours and where decoherence does not lead to a classical state because we could take measurements that would allow us to distinguish between the diagonalized state from the non-diagonalized state. Of course, this is no longer physics but pure philosophy.

Jean-Michel Raimond. It is pure philosophy. It is clear that this possible world would have laws of physics that would probably be very different from ours. In this sense, its quantum mechanics, or rather its basic physics would be very different from ours and perhaps the problem of decoherence would not even pose itself.

Hervé Zwirn. It is a possibility, however I speak of another possible world than would be like ours in every way except in the order of magnitudes we would be allowed to manipulate. I know that if we wish to give a rigorous meaning to this, it is difficult since when we seek to manipulate the values of physics constants, we rapidly end up with consequences that render the world in question very different from ours. There, obviously, I am not capable of giving the precise description of a world where, everything being equal, decoherence would lead to something measurable or would not lead to a classical state, including *for all practical purposes*. However, in philosophy, we can ask ourselves this, but once again no one would be able to settle the question. The question is therefore whether we allow ourselves to ask these questions or not.

Jean-Michel Raimond. In that case we can image things so large that the mind falters. We are in a similar, although more modest, situation as the superstring theoreticians who must sift through an unimaginable number of "vacuums," of different worlds. They perhaps have too much freedom in their theories. If we start imagining worlds that do not obey our laws, then I give up... Or I become a science-fiction author, because we can do some really neat things in science-fiction on these types of parallel universes.

Hervé Zwirn. Of course, but there, you cannot allow just anything. It is the very precise difference in modal logic between what is necessary and what is possible. In the semantic interpretation of modal logic, "necessary" means what is true in all possible worlds, and "possible" means what is true in one possible world. Obviously, the goal is not to allow just anything, otherwise this would be of no interest.

Jean-Michel Raimond. I fear that on reflection we realize that this alternate world has such aberrant properties that we were talking nonsense all along. That it is not like ours in any way, not even slightly.

Bernard d'Espagnat. It is getting late, and the subject has far from been covered. We will come back to it during our next session.

References

1. Michel Brune et al., "Observing the progressive decoherence of the "meter" in a quantum measurement", *Phys. Rev. Lett.*, 1996, p. 4887–4890.
2. Session I, "The inescapable strangeness of the quantum world".
3. Vicent Jacques et al., "Single-photon wavefront-splitting interference. An illustration of the light quantum in action", *Eur. Phys. J. D 35*, 2005, p. 561–565.
4. Akira Tonomura et al., "Demonstration of single-electron buildup of an interference pattern". *Am. J. Phys.*, 57, 1989, p. 117–120.
5. See the "Single electrons build up interference pattern" video on www.youtube.com/watch?v=ZUI3lhRje_0.
6. See session I, "The inescapable strangeness of the quantum world".
7. We carried out a few years ago a version of this experiment with atoms and a cavity: Patrice Bertet et al., "A complementarity experiment with an interferometer at the quantum – classical boundary", *Nature*, 411, 2001, p. 166–170.
8. LIGO: Lasser Interferometer Gravitational-Wave Observatory, American interferometer designed to detect gravitational waves. VIRGO: Franco-Italian interferometer designed for the same purpose.
9. Erwin Schrödinger, "Die gegenwärtige Situation in der Quantenmechanik", *Naturwissenschaften*, 23(48), 1935, p. 807–812.
10. See session II, "Quantum physics, appearance and reality - exchange of views".
11. W.H. Zurek, "Decoherence and the transition from quantum to classical". *Physics Today*, 44 (10), 1991, p. 36–44.
12. G. Lindblad, "On the generators of quantum dynamical semigroups", *Comm. Math. Phys.*, 48, 1976, p. 119–130.
13. William H. Louisell, *Quantum Statistical Properties of Radiation*, John Wiley, 1973.
14. Amir O. Caldeira & Anthony J. Leggett, "Influence of Dissipation on Quantum Tunneling in Macroscopic Systems", *Phys. Rev. Lett.*, 46, 1981, p. 211–214.
15. Marcelo P. Ammeida et al., "Environment-induced sudden death of entanglement", *Science*, 316, 2007, p. 579–582.
16. Serge Haroche, "Nobel Lecture: Controlling photons in a box and exploring the quantum to classical boundary", *Rev. Mod. Phys.*, 85, 2013, p. 1083–1102; Serge Haroche & Jean-Michel Raimond, *Exploring the Quantum: Atoms, Cavities, and Photons*, Oxford, Oxford University Press, 2006.
17. Michel Brune et al., "Observing the progressive decoherence of the "meter" in a quantum measurement", *Phys. Rev. Lett.*, 1996, p. 4887–4890.
18. Samuel Deléglise et al., "Reconstruction of non-classical cavity field states with snapshots of their decoherence", *Nature*, 455, 2008, p. 510–514.

19. Andrew Steane, "Error correcting codes in quantum theory", *Phys. Rev. Lett.*, 77, 1996, p. 793–797.
20. The experiment has been conducted since: C. Sayrin et al. "Real-time quantum feedback prepares and stabilizes photon number states", *Nature*, 477, 2011, p. 73–77.
21. Haroche & Raimond, *op. cit.*, 2006.
22. Respectively: Christopher Monroe et al., "A "Schrödinger cat" superposition state of an atom", *Science*, 272, 1996, p. 1131–1136.; Max Hofheinz et al., "Synthesizing arbitrary quantum states in a superconducting resonator", *Nature*, 459, 2009, p. 546–549; Gerhard Kirchmair et al., "Observation of quantum state collapse and revival due to the single-photon Kerr effect", *Nature*, 495, 2013, p. 205–209.
23. Markus Greiner et al., "Collapse and revival of the matter wave field of a Bose-Einstein condensate", *Nature*, 419, 2002, p. 51–54.
24. High Noon, directed by Fred Zinnemann (1952).
25. Jean-Michel Raimond et al., "Quantum Zeno dynamics of a field in a cavity", *Phys. Rev. A*, 86, 2012, 032120.
26. Tauno Palomaki et al., "Entangling mechanical motion with microwave fields", *Science*, 342, 2013, p. 710–713; Jasper Chan et al., "Laser cooling of a nanomechanical oscillator into its quantum ground state", *Nature*, 478, 2011, p. 89–92; Aaron D. O'Connell et al., "Quantum ground state and single-phonon control of a mechanical resonator", *Nature*, 464, 2010, p. 697–703; Olivier Arcizet et al., "Radiation-pressure cooling and optomechanical instability of a micromirror", *Nature*, 444, 2006, p. 71–74.
27. For example Yakir Aharonov et al., "How the result of a measurement of a component of the spin of a spin-1/2 particle can turn out to be 100", *Phys. Rev. Lett.*, 60, 1988, p. 1351–1354.
28. Roland Omnès, *The Interpretation of Quantum Mechanics*, Princeton, Princeton University Press, 1994.
29. Douglas Hofstadter, *Gödel, Escher, Bach: an Eternal Golden Braid* [1979], Basic Books, Inc., New York.
30. See David Lewis, *On the Plurality of Worlds*, Oxford, Blackwell, 1986.

Author Biography

Jean-Michel Raimond has been a professor at the *Université Pierre et Marie Curie* since 1988. He was a junior member at the *Institut Universtaire de France* from 1994 to 1999, and a senior member from 2001 to 2011. His scientific research has focused on the exploration of fundamental quantum properties. He has contributed to the development of experimental techniques, based on Rydberg atoms and micro-wave cavities, allowing the direct observation of the interaction between a single atom and a single photon. These experimental results can be interpreted with the fundamental postulates of quantum physics.

Chapter 4
Theoretical Aspects of Decoherence

Round Table

Bernard d'Espagnat. We have the pleasure of welcoming among us Alexei Grinbaum. During our last session, thanks to Jean-Michel Raimond's brilliant presentation [1], we saw experimentally that physical systems which have a typically quantum behaviour at time 0+ have a classical behaviour later on. This constitutes a very strong indication, not to say an experimental proof, that there are not two types of systems, one obeying quantum physics and the other obeying classical physics, but on the contrary that there is only one physics, even if depending on the circumstances, physical systems can appear to us either under a quantum or a classical aspect. The interaction between macroscopic systems and the environment, and the resulting decoherence seem indeed to be the cause of this progressive passage from the quantum to the classical world, classical mechanics being derived under certain conditions from quantum mechanics. Today, it is logically the theoretical and particularly the conceptual aspects of decoherence that we propose to explore together. A number of us here have thought about the subject. However, perhaps it is right to ask those newly present among us to speak first.

In order to make this read more informative, this report has been divided into four sections:

1. Overview of a non-standard conception of decoherence and discussion;
2. Discussion of the standard conception of decoherence;
3. Arguments for and against realism;
4. Mathematical aspects of the conflict between quantum mechanics and spatio-temporal localization.

B. d'Espagnat and H. Zwirn (eds.), *The Quantum World*,
The Frontiers Collection, DOI 10.1007/978-3-319-55420-4_4

4.1 Overview of a Non-stardard Conception of Decoherence and Discussion

Alexei Grinbaum. I have a rather unorthodox theoretical view of decoherence, which I associate with the problem of the observer. Some of you here have already heard me present this idea. To summarize, I think that decoherence seems, from a theoretical perspective, to be a phenomenon that is linked to the analysis of the complexity relation that exists between the observer and the observed system.

In my opinion, the main characteristic of the quantum observer (I summarize, of course, with words what can be described with mathematical formalism) is that it must necessarily be much more complex than the observed system—and when I say complex, we can think either of the degrees of freedom used by the observer to store information, like a sort of memory, or, on a more formal level, of the Kolmogorov complexity of the observer, since to my mind the observer is essentially a system identification algorithm, and therefore has an invariant formal characteristic in its physical implementation, which is its Kolmogorov complexity [2]. I leave to one side the mathematical details of this description.

In my opinion, decoherence appears when the number of degrees of freedom of a system observed by the observer come close to a certain threshold characterizing the observer. By that I mean that each observer is characterized by a complexity threshold below which it can always observe the quantum system, i.e., it describes the system completely without doing any coarse graining. In other words, decoherence is linked to the complexity of the observer. The threshold for a quantum description of a system is not the same for two observers where one is much more complex than the other [3].

Ultimately, all this results from the analysis, which I think is necessary but has not been done in the history of quantum mechanics, of the notion of system which, in a way, is pre-acquired and predefined: quantum mechanics "begins" when the observer and the system are already in place. I think it is possible, using informational language, to go one step back and analyse the system in informational terms and draw from this analysis a theoretical explanation of decoherence.

Jean-Michel Raimond. If I may ask a question, does decoherence as you define it do without the notion of environment?

Alexei Grinbaum. Not necessarily. Indeed, if the observer observes a system whose number of degrees of freedom is fixed and is much less than that of the observer, then we speak, in everyday language, of a closed system. Then quantum mechanics usually works without any problems. That said, what does the statement mean that the number of degrees of freedom can increase and come close to a certain threshold? It means that the observer is beginning to take into account, or attempts to take into account, other degrees of freedom than those initially identified. We can say that before, these other degrees of freedom were classed as those of the environment.

Jean-Michel Raimond. If there are two observers, can there be an objective description of the system that the two agree on? In other words, if two observers are in the same room, what do they do?

Alexei Grinbaum. Indeed, the main result of my article is a theorem where observers with not very different complexities—and I give a mathematical criterion of what that means—would agree on the characterization of the systems. This result, on the possibility of agreement between two observers, seems to me to be very important. Among other things, it means that two observers that are different will not characterize quantum systems in the same way.

Jean-Michel Raimond. There is therefore a system that is more or less big and totally decoupled from the Universe, and in this system there are one or two observers who are decoupled from the Universe and who interact together?

Alexei Grinbaum. I do not use the term Universe because I use an informational language. I do not take a realistic, or antirealistic, position, in the sense of the realism of physical systems. The observer observes what he identifies as a quantum system. This observer is a system identification algorithm. It is implemented physically, and can be implemented in different ways, but the informational description on its own provides interesting theoretical results.

Jean-Michel Raimond. Is this quantum system a perfectly isolated system, apart from its interaction with the observer? I was thinking of the problem of the environment in that sense, in the most standard view of decoherence.

Alexei Grinbaum. My vision does not begin with the theatre of nature as physics frequently does. I do not say that there is first of all the theatre of nature with its objects and that we use physics to account for it. My point of view regarding quantum mechanics stems from the epistemic observation that the observer interacts with systems and observes them, thereby obtaining information. "What really exists around us" has no place in this vision; consequently the notions of universe or of a system isolated from the environment are not defined. That is why I specify that this point is not orthodox, even if in reality, it is close to Bohr's position.

Bernard d'Espagnat. Does the observer, in your vision of things, differ by some qualitative trait from what is being observed?

Alexei Grinbaum. The observer differs from the observed system by the label observer, meaning a feature related to complexity, not written down as such in nature, but implemented in the form for example of memory size. The difference between observer and observed system resides in the characterization of informational exchanges.

Bernard d'Espagnat. Your thesis that the observer is different from what is observed in a feature "not inscribed in nature as such", reminds me of Niels Bohr: according to him, an instrument is an instrument not because of its physical composition, but because *we* use it as an instrument.

Alexei Grinbaum. Exactly.

Bernard d'Espagnat. Is there a strong analogy between Borh's approach and yours?

Alexei Grinbaum. Absolutely. In my opinion, any informational approach, at least the one I defend, falls into a neobohrian framework. It is a "neo-Bohrism".

Michel Bitbol. I like your neobohrian connection, which I also relate to if perhaps in a slightly different way. The crucial characteristic of this group of positions is that the observation process belongs to a profoundly different category compared to what is involved in the description of physical systems. For Bohr, this categorial leap is represented by the quantum/classical boundary (the process that is observed being explained by quantum mechanics, the process of observation being explained by classical mechanics). For you, this categorial leap amounts to going from a quantum language to an informational language. In neobohrian positions, the reason underlying these categorial leaps is be found not in physics itself but in the necessities of the act of knowing: it amounts to bringing to light the conditions for the possibility of knowledge, in a typically Kantian manner. I would like to specify some of the consequences of your catergorial leap.

You have insisted on the importance (particularly for decoherence processes) of the degree of complexity of the observer. You have highlighted that the degree of complexity you had in mind regarding the observer was not directly linked to its first order constitution as a physical system, but in its capacity of a higher order to memorize, analyse and conceptualize. Does this dual characterization of the observer as a physical system and as an information processing device not lead you to a *functionalist* conception of the mind, where there is on the lowest level a certain *material* device, and on a higher organizational level the equivalent of *software*?

Alexei Grinbaum. I believe I can indeed avoid falling into the trap associated with the word "mind". When I speak of the complexity of the observer as an observer, I speak of a rather precise notion which is the Kolmogorov complexity, the algorithmic complexity, namely an invariant mathematical characteristic—the only one in fact of the system identification algorithm. First of all the observer determines what the quantum system is. He determines the degrees of freedom of the quantum system, to take them into account. This process can be described as an algorithm in a very abstract sense.

Whatever the physical medium of the observer, this algorithm has an invariant characteristic which is the Kolmogorov complexity. The physical content does not play a role in this description. Indeed, we know (these theorems have been proven) that, up to a constant, it does not depend on the physical content of the given system—in the same way that a computer can run the same programme on different physical media.

In my opinion, the question of medium does not necessarily lead to dualism, and is simply not relevant to this description. This level of description has nothing to say about the concrete physical system determined by the observer—who can be a human being, a butterfly, the entire planet, etc. Once the observer identifies a quantum system, there are invariant characteristics in the form of an algorithm that characterize the systems. That is sufficient.

Bernard d'Espagnat. An "old school" realist—a school of thought I personally do not adhere to, but I know and have known many of its followers, in particular John Bell—would probably ask you if your analysis as a whole is compatible with his views and can be accepted by him. "Old school" realists consider that things, like atoms, exist in themselves. They exist with their properties, completely independently of the question of knowing whether there are conscious beings that could know them. Can we transpose your general conception to a language that would call upon only the *be-ables* [4] of John Bell? That is the question.

Alexei Grinbaum. Indeed. In the same way that a word processor can be completely described in the language of atoms that make up the computer, likewise transistors and their physical states—which is useless for understanding how they work—an observer can be described, to quote Einstein, by a "constructive theory"—however, this does not help us understand quantum mechanics.

Bernard d'Espagnat. It does not help, but do you think it is compatible? Therein lies the question.

Alexei Grinbaum. I think each observer, while being a physical observer, can be described as a physical system. This does not provide us with information regarding its capacities as a quantum observer. In the same way that the description of a computer as a physical system does not tell us anything about the word processor it operates.

Jean Petitot. That is exactly the definition of functionalism in cognitive theories.

Alexei Grinbaum. There is the question of the medium. What carries the function?

Jean Petitot. The medium on which we implement the algorithm is not relevant to what the algorithm does as an information processing algorithm.

Alexei Grinbaum. I agree on this point.

Jean Petitot. That is really the definition of functionalism. This brings us back to question Michel Bitbol asked earlier.

Alexei Grinbaum. Not entirely, because there is no mention of the mind.

Jean Petitot. The mind is simply the mental process acting in a functionalist way compared to the neuronal process.

Michel Bitbol. It appears that there are here different conceptions of the mind. Alexei, you probably fear (justifiably perhaps) that to invoke the mind necessarily leads to associate with it consciousness. However, in the functionalist paradigm, the mind is only defined in the first instance as a set of information processing and decision-making functions that are implementable on all types of physical media. The term "mind" is used specifically as a marker of a functional and informational level of organization which is opposed to the level of its basic substrate.

Alexei Grinbaum. The element that allows us to distinguish these two levels is, I believe, the effectiveness of a given description for constructing theories. The only reason that could make us take this step is to explain things that could not be explained otherwise.

Michel Bitbol. What you have said is very important. It means that a process as important for quantum physics as decoherence cannot be explained if we describe the observer only in terms of a physical system: we must assign functional properties to it, properties that pertain to the software rather than to the hardware. We must perhaps even assign to it projects, *goals*, for example that of extracting a fraction of what appears to be and treat it like a physical system.

Alexei Grinbaum. I would not go as far as saying "goals".

Michel Bitbol. Nevertheless, is there not a form of circularity in your approach? On the one hand, to derive the process of decoherence, we must from the start call upon a non-physical level of description of the observer (of an informational and algorithmic order). On the other hand, we ask that decoherence accounts for a level of organisation above what is described by quantum physics: i.e., precisely that which allows an informational and algorithmic description.

Alexei Grinbaum. This circle is part of the explanatory circles known elsewhere, which appear each time we are dealing with a principial theory, meaning a theory based on principles or postulates. I'm thinking about the distinction between principial theory and constructive theory; it has been the subject of discussion for over a century, and you have no doubt much to say about it.

Constructive theory begins with a fundamental physical level that we believe is part of reality, and constructs a theory from this reality, whereas principial theory always works in circles, since where do the principles on which we base our theory come from? They stem from our own analysis of the systems around us. We then "raise" them, as Einstein would say, to the rank of principle. In the same way, I raise certain things to the rank of principle—things which come from the physical experiences we all have.

Bernard d'Espagnat. From the physical experiences we all have, in other words, without any preconception regarding the nature of what creates the experience in us. Is that what you mean?

Alexei Grinbaum. In a pragmatic sense and not necessarily in an empirical sense.

Bernard d'Espagnat. OK.

Alexei Grinbaum. This is not to defend a type of empiricism, but a pragmatic reason for favouring one principle over another.

Hervé Zwirn. The concept of "the Kolmogorov complexity of the algorithm that defines the observer in the system" seems unclear to me. It is something that seems to me to be extremely difficult to define rigorously. We know how to define accurately the Kolmogorov complexity of a string of characters or bits.

However, it seems to me that we will encounter great difficulties in defining what characterizes an observer as a system to which we would attribute a Kolmogorov complexity. I have my doubts concerning even the meaning of the expression "the Kolmogorov complexity of the observer".

Alexei Grinbaum. What Kolmogorov complexity are we talking about? That of the observer as an algorithm defining the observed system. What does "defining the system" mean? The image I use of this process is the following. Imagine a long strip of tape, like in a Turing machine, with all possible degrees of freedom. This algorithm consists of putting a cross next to the relevant degrees of freedom. This vision is obviously very abstract, as abstract as a Turing machine.

For example, a human observer who indicates observing an electron means that this electron has a certain number of degrees of freedom. However, does a fullerene define a quantum system, like a photon, in the same way a human would? We are currently debating on this subject with colleagues in Vienna. Can we observe differences at the thermodynamic level? We have already expressed certain ideas on this matter.

The memory of the observer can constitute a certain number of degrees of freedom achieved in certain ways. What the observer does, as an algorithm that defines systems, is to indicate which degrees of freedom he will observe. That is the algorithm whose Kolmogorov complexity is invariant, at an abstract level. A man does not define the degrees of freedom in the same way as a fullerene would, but a fullerene can also do this.

Hervé Zwirn. Indeed, but I believe there is a significant difference between an algorithm that would consist iof putting crosses on a list of degrees of freedom (which can effectively be done with a Turing machine with the right programme) and what we mean by "observer". It is not the same thing. I've somewhat lost the thread of the discussion regarding the question we were trying answer.

Let us come back to the end of Jean-Michel Raimond's presentation from the previous session. We set the problem surrounding decoherence, not experimentally, but philosophically, as being a necessity, firstly, to explain the appearance of the classical world to human observers, and secondly, to know whether this classical appearance was simply an appearance or whether the world had really—for the realists—become classical. The debate, at least as it was presented during the previous session, consisted of identifying two questions: why does the world appear classical and is it really classical since it appears as such?

I am struggling to see which question is being addressed with what you are proposing and whether there are links with the questions we were asking, which are the usual questions when decoherence is mentioned. These are the questions that Wojciech Zurek himself mentioned at the start of his articles [5]. Zurek's first position was extreme in the sense that he concluded his first articles by stating that the world does become classical and the problem is, in fact, solved. Mr. d'Espagnat discussed this with him—and he was not the only one to do so. Zurek's second position was consequently a moderate conception of decoherence. Let me repeat, I struggle to find a link between this and what you propose.

Alexei Grinbaum. I can see the irony here. Indeed, what I propose corresponds rather well, even if I propose some changes, to the work Zurek did much later, around 1994.

I will attempt to answer your first question, leaving aside the second question which requires a different type of reasoning. In the first question, you mention the human observer. I was not at the previous session, but I think the question of the human observer needs to be broadened. We must, first of all, ask ourselves whether the observer is necessarily human or not. Then there is the question of knowing what its minimal characteristics are. How can we understand this notion of observer? Scientifically, we need to give the minimal informational characteristics of what an observer is. My answer would be to say that the observer is an algorithm of systems. Period. That is my definition.

From there, I try to conceptualize the notion of decoherence. That is my line of reasoning.

4.2 Discussion of the Standard Conception of Decoherence

Bernard d'Espagnat. The research on algorithmic complexity applied to the observer is undoubtedly promising and we will have the opportunity to come back to it. However, I believe, like Hervé Zwirn, that it must not make us forget the questions we were asking concerning the standard conception of decoherence, founded on the notion of environment. You have mentioned Zurek and his recent papers. This may be the occasion for me to tell you what I made of his publications, in particular of his important 2003 paper [6].

Jean-Michel Raimond. The one in *Review of Modern Physics*?

Bernard d'Espagnat. Yes, that one. I have three points to make regarding this article.

First of all, Zurek writes on page 51 of the ArXiv version: "*many conceptual and technical issues* (*such as what constitutes* "*a system*") *are still open*", backing up the comment he makes as early as page 4 that we must *accept* the existence of the environment, in other words, the distinction between system and environment. This shows that this article does not claim to solve the issue which, during our previous sessions, appeared to be still open to more or less all of us, i.e., the problem of the existence of systems.

Jean-Michel Raimond. Except that if we abandon this existence, we have a slight problem…

Bernard d'Espagnat. Yes indeed, as decoherence is based on this distinction, and therefore ultimately, so is the theoretical resolution of the cat paradox. On this first point, I would like to say here that, in my opinion, within the framework of standard quantum mechanics, the existence of systems is an appearance to us, they must not be considered as existing per se, and I find in Zurek's article an echo to this idea since he writes on page 4 that: "*Einselection delineates how much of the Universe will* appear *classical to observers who monitor it from within using their limited capacity to acquire, store, and process information*".

This was my first point, which amounts to highlighting that apparently we are not capable of knowing reality per se, and that we can only know the appearances that are valid for everyone. The two go in the same direction.

The second point deals with the way that, implicitly, Zurek brings a type of solution to the "conceptual" problem raised by some of us last time, namely the passage from an "and" to an "or", and more precisely the passage to an "or" understood as the *choice* of *one* eventuality among others. How and where does he do this? He does this in his paper at the same place where, with the aim of proving Born's rule, he focuses on probabilities. He shows first of all (from quantum principles not involving probabilities) that the measurement outcome of an observable can only be one of the eigenvalues of this observable. Then (p. 37) he examines the case where multiple outcomes are possible, and in particular the case where the coefficients of the different components of the wave function of the system, developed according to the eigenvectors of the measured observable, have the same absolute value. He still speaks, of course, of the singularity of the measurement outcome, since he accepts implicitly (but which is for me the essential point!) the evidence that a measurement has only one outcome. If two outcomes are possible, then it needs to be either one *or* the other. He shows, explicitly this time, that in that case it will be with a 50/50 probability. As you can see, for this passage from the "and" to the "or" (and I stress once again that "or" is used in the strong sense of the term, namely implying a truly random choice between multiple eventualities), the notion of measurement as such, implying the idea that an outcome is necessarily unique, plays an essential role in his article. From my point of view, I cannot see what could replace, to this end, the notion of measurement carried out by an agent.

My third point deals with the notions of event and history, for which Zurek introduced that of "*relatively objective past*". He writes: "*When many observers can independently gather compatible evidence concerning an event, we call it relatively objective. Relatively objective history is then a time-ordered sequence of relatively objective events.*" It seems to me that we have here the well-known antirealistic interpretation: to say that an event took place at a given time means that many of us have documents in support of this, but nothing more.

As you can see, all this leads to what is sometimes called "weak objectivity". Zurek calls it "relative objectivity" but this is not a question of words. As Pascal wrote in the *Provinciales*: "I never quarrel about a name, provided I am apprised of the sense in which it is understood". Therefore Zurek, despite appearances to the contrary, and reading his work quickly may give the impression he wants to return to traditional realism, does not come back to this type of realism, nor does he want to.

Such is my conception. It seems to me that in certain ways, that of Alexei Grinbaum is not far off.

Alexei Grinbaum. I try to go a bit further, all the while accepting Zurek's reasoning. When he says we do not know what the notion of system means and the question remains open, he uses an algorithmic argument, the Kolmogorov complexity, to study the change of state of the system. He asks the question of knowing what the complexity of this algorithm, which passes from a given state to another, is. This is where he uses algorithmic ideas.

I think that prior to that, before speaking of states of a quantum system, along the same lines as Zurek, we must apply an algorithmic reasoning to the question of the notion of system. As for everything else, I think that we can perfectly follow Zurek, and that we are not talking about realism in the sense of Bell. The keyword, with Zurek, is "*relatively*": everything is relative to the observer.

Bernard d'Espagnat. That is right, everything is relative to the observer.

Jean-Michel Raimond. The important contribution of this paper is to explain that there is a relative objectivity with multiple independent observers, which are all part of the Universe and share parts of the environment. All these parts of the environment provide the same information on the system if the latter is in a pointer state. There is therefore a common objective reality for these observers.

Alexei Grinbaum. Absolutely.

Jean-Michel Raimond. It is an interesting point.

Alexei Grinbaum. From there, as I have tried to show in my paper, we can speak of objectivity in relation to a class of observers. The question then arises of the boundaries of this class for shared objectivity to have any meaning.

Jean-Michel Raimond. In your opinion, a single spin of the environment is not an acceptable observer because it is not complex enough to have unambiguous and classical information on the state of the system?

Alexei Grinbaum. Indeed. For instance, how far can we go while maintaining the same notion of objectivity?

Bernard d'Espagnat. It seems to me there has been much technical, and even conceptual, progress in what has been achieved by many people—Zurek, you, and many others. However, ultimately, from a philosophical point of view, all this is closer to the line of reasoning of Bohr than of traditional realists.

Alexei Grinbaum. Yes.

Jean-Michel Raimond. If all the observers of the Universe, whatever they are provided they are sufficiently complex, agree on a reality, then this reality takes on an objectivity that appears strong rather than weak.

Bernard d'Espagnat. I called them "weak" in my writings because in themselves they are not capable of conferring any meaning to seemingly obvious claims such as: "the Sun would exist even if no observer had ever existed"; and as a consequence I needed an adjective to distinguish it from the—probably illusory!—objectivity of conventional realism, which considers the claim in question to be sensible and even self-evident. Basically, my aim was to show that there are two possible conceptions of objectivity, not one as commonly thought, and it is wrong to think they are the same. That is why I gave the name of strong objectivity to conventional realism. Zurek calls the objectivity I call weak "relative", implying relative to the observer. That is all very well. For the sake of clarity, it might even be better since, as you pointed out, this objectivity is nonetheless extremely strong.

Olivier Rey. I have a problem: if we speak on the one hand of objectivity arising from an agreement between observers, and on the other hand, if we extend the class of observers to include a huge number of things including molecules, how do you observe alongside or agree with a molecule?

Alexei Grinbaum. I think this agreement is neither given nor something obvious to be proven. I think we need to first ask whether we can categorize observers into classes for which we can have an agreement. Frankly, I think that as an observer myself, a fullerene does not provide the same idea of objectivity as a human observer.

I then asked myself whether we could imagine a physical experiment that would corroborate the idea that a fullerene is an observer. I have tried to do this—but this is not the topic. Anyway, the agreement between different observers is not an agreement between all observers but between certain classes of observers.

Bertrand Saint-Sernin. I would like to take part in this discussion to ask you for information or advice. I must speak tomorrow morning in front of the French Society for Plant Biology about the difference between nature and artifice, and about the particular problem of genetically modified organisms (GMOs). You know there is a position that is specific to France, which is the refusal of GMOs.

Now, one of the arguments used touches upon the notion of realism. In essence, the problem is the following: can something obtained artificially or through modifications in the laboratory have the same properties and be considered identical to something that has evolved under so-called natural conditions?

The problem is therefore not that of the observer, but that of the user. We notice something quite peculiar: anti-GMO advocates who have cancer, heart disease or diabetes have no problem using insulin or blood-thinning drugs, etc. Yet they are anxious when it comes to food.

The first question I would like to ask is this: what is the point of view of the chemist regarding synthetically-derived compounds, of which there have been, from what I have heard, 22 million types since 1928? In my opinion, the problem of realism is especially relevant to something that has been conducted on massive scale in terms of quantity within the world's population. The question is to know how to provide theoretical justifications to say either you are right to be realist, or you are wrong. Are they right from the perspective of chemistry? Are they right from the perspective of biology? I do not know.

Oliver Rey. We must distinguish between situations. In the case of insulin for diabetics, it is not a GMO that is absorbed but only what it produces. Meaning we use genetically modified yeast, grown and maintained in the laboratory, for producing insulin molecules that are exactly identical to those synthesized directly by the human body. In the case of GM food, corn for example, we eat the genetically modified organism itself, which has a different molecular composition from that of traditional corn since its DNA is different. The impact of this difference on the consumer or the environment is another matter.

Bertrand Saint-Sernin. Yes indeed, but the question I ask is this: is it right to claim that synthetically-derived chemical molecules, proven to be identical to the naturally-occurring molecules, have the same effect as those? Is this line of reasoning valid, or does quantum mechanics change this perspective?

Jean-Michel Raimond. A molecule with the same chemical composition and the same conformation is the same molecule. Quantum mechanics does not say anything different. Unless you are an animist, a molecule produced by whatever means (a GMO, explicit organic synthesis, or natural synthesis) has exactly the same functions and properties.

Bertrand Saint-Sernin. This is accepted in America, but not in France.

Bernard d'Espagnat. Jean-Michel Raimond is of course completely right. However, I would like to specify the relation between Bertrand Saint-Sernin's question and what we are debating today. I would say that, following the terminology we arrived at earlier, if we believe that quantum mechanics is a universal theory, the realism of chemists as well as of pro- or anti-GM camps necessarily refers back to the objectivity I called "weak" and which Zurek called "relative".

All phenomena, like this table or all the elements present in this room, are appearances that are the same for all human beings and probably for all conscious beings. They belong to the reality that is relative to the notion of the observer. What I suggest here is only that, in the light of the constitutive principles of standard quantum mechanics, they cannot be considered as objective in what I called the "strong" sense, meaning in the sense of the term "objective" given by common realism.

Consequently, I do not believe that quantum mechanics has anything particular to say regarding the problem you have raised. That is my first conclusion; my second conclusion being, I repeat, that on this point I completely agree with Jean-Michel Raimond. There is no difference.

Hervé Zwirn. If we come back to the questions we were asking previously, I would like to present a very simple point of view to see if we share it or not. It is about the description of the problem as we defined it at the end of the last session.

We asked ourselves whether the problem of measurement (consisting, in orthodox quantum mechanics outside of decoherence, of not being able to come out of this chain of successive entanglements for each new interaction between the initial system and a measuring device, the observer, etc., and which seems to suggest that the observer is in a superposed state), is resolved by decoherence by breaking the chain. We asked ourselves whether we needed an observer that was conscious or not, and whether the outcome can be interpreted as being real in the sense of strong realism, or whether we must consider that beyond appearances, the world remains profoundly quantum. These are the questions we were asking ourselves.

Jean-Michel Raimond had strongly insisted, and rightly so, that the phrase "*for all practical purposes*" meant that we can consider all that happens as if it was classical. The debate we had at that point consisted of saying that on a practical level we obviously agreed, while asking ourselves whether it was possible to claim, philosophically speaking, that the conclusion is that the world is classical or not.

I have a proposition to make. It is very simple. I would simply like to ask whether we agree on the following process. When a system is coupled with a measuring apparatus and to the environment, we have a big system S + A + E comprising of the system itself, the measuring apparatus, and the environment. This big system is in a superposed state and its density operator contains off-diagonal elements. The usual rules of quantum mechanics state that if an observer takes a measurement without measuring the degrees of freedom of the environment (such a measurement deals with observables that are unattainable to the observer), the system S + A will be described by making a partial trace on the environment of the density operator of the big system, and calculations show that, in general, the off-diagonal elements of this trace become very small very rapidly (this is "decoherence time") and remain that way for an extremely long time. The description of the system S + A is practically equivalent, for an observer that would not be taking measurements that are unfeasible, to that given by a diagonal density operator. Last time, we debated on the off-diagonal elements, which can eventually become non-negligible again after a time that is possibly greater than the age of the universe—thus we left this topic to one side.

Since the density operator that describes the system as it is accessible to the observer is practically equivalent in all its observable consequences to a diagonal density operator, do we agree to say that in reality, in the strong sense of the term, the world remains quantum and can be superposed, but that this is not important because we cannot see it that way. This is somewhat equivalent, as a situation, to what happens with relativity: the world is not a classical world, it is a relativistic world. However, at low speeds, the relativistic effects being completely invisible, the world seems classical to us. We would have an identical situation here: the world is in fact quantum, but the quantum effects being, at our scale, for this type of measurement, unverifiable, it appears classical to us.

We could therefore reconcile the points of view from our previous discussion by considering that the world is quantum and never becomes classical in the old sense of the term, but that this quantum effect has no visible effect—the world thus appears classical to us. One of the problems, it seems to me, in this philosophical discussion, is that very often we think that if something is considered quantum, then this must have a surprising visible effect, different from the classical world. In fact, what decoherence shows is that a system can be totally quantum while having a behaviour that appears classical to us, without this posing the slightest problem.

Do we agree or not on this point?

Bernard d'Espagnat. Let us vote! I for one am for it.

Jean-Michel Raimond. I must say I would not like a theory where the observer is necessarily defined by consciousness. That the observer must be a complex system, that he is a part of the environment seems to me to be a reasonable approach. In which case, this part of the environment can be the 60 × 3 degrees of freedom of a fullerene or a normally constituted Doctoral student. Anything between the two would seem a reasonable observer.

Effectively, I have the impression that you propose a "decoherence plus Everett" approach. We know that everything is superposed. We know that we are in the wave function of the Universe. But I the observer, you the observers, observing the same phenomenon and all other reasonable observers observing copies of the same phenomenon scattered in the environment agree on the fact that this phenomenon has an objective reality and that the measured observable has an objective value.

Bernard d'Espagnat. A relative value.

Jean-Michel Raimond. In our Universe, we completely agree on everything, such that if everything in the Universe agrees to say that the electron spin is positive, why not say that it is the real reality?

Bernard d'Espagnat. It's a matter of convention. Indeed, why not say that when everything in the Universe that possesses the quality of observer agrees with a certain observation, what is observed is a "real" reality. That's what many, including I, call "empirical reality", in which all living things are immersed, and what many philosophers just call "reality". That being, it is worth differentiating between this notion and that of a reality "per se" that would exist even if no observer existed. It is a priori conceivable that these two notions have the same field of reference, but this is a postulate and not a truism. And a postulate that is difficult to reconcile with standard quantum mechanics, even complemented with decoherence.

Jean-Michel Raimond. If I may say so, the very "Zukerian" notion of the existence of a complex environment split into multiple parts that all agree on the state of the system (this is what Zurek says, quite rightly I think, in his papers from the 2000s) means that it is not defined until there is an observer, but that the Universe in its entirety agrees to consider that this sub-part of the Universe is in that state.

Bernard d'Espagnat. The Zukerian notion you speak of appears as a consequence of his demonstration, effectively very enlightening, of the fact that the non-isolation of macroscopic systems leads to the publicly known existence of robust observables, in the sense that once measured by someone, all other observers know they will be able to measure it again, possibly in rotation, without changing the values. And that is true, even if we take only indirect measurements on intermediate objects, hence this universal agreement you mentioned. It is undeniable that this agreement provides observers with a very strong *feeling* for the reality of what they observe. Nevertheless the problem of the passage from "and" to "or" is still not resolved as it is a more fundamental question, which arises prior to this. Zurek resolves this question implicitly by calling upon the notion of branches of the Universe, in other words Everett's theory ("Distinct memory states label and "inhabit" different branches of the Everett's "Many Worlds" Universe" [7]). It is effectively a possible solution, but I personally do not adhere to the theory of multiple worlds in which I see an attempt at a metaphysical explanation comparable in its detail to many previous metaphysical attempts, and like them, lacks credibility precisely for that reason.

Hervé Zwirn. On this point, there is an alternative reasoning.

We can either say that decoherence takes place "in the manner of Everett", meaning there is only one wave function with coexistence of all the possibilities, or we can say that there is at a given time a choice in the "or "of all possible states, without this coexistence—even though this brings about many problems. This is the first alternative. We can make either one choice or the other. Both pose problems of a different nature.

In addition, even if we accept Everett's definition where everything coexists and where each branch of the Universe corresponding to a choice is in agreement with itself, it does not imply that reality as it is described becomes objective—this takes us back to our previous discussion. Simply because the off-diagonal elements, unattainable to us but not strictly null, come back to the fore. We can then think that there is a difference between the fact that the spin is rigorously up and the fact that is practically up with off-diagonal elements that will (even in a very, very long time) become important again. There is a nuance here. The debate we had last time on small probabilities regains its significance: either we consider that small probabilities have no meaning and there is a cut-off point, below which we set them at zero; or we consider that no matter how small, they still have meaning.

Jean-Michel Raimond. From the observers' point of view, the two possibilities (either a global wave function and we are on one of these branches, or there has effectively been a choice) are indistinguishable. The question of course is to know whether we can devise experiments to discriminate between these points of view. There have been many proposals based on stochastic quantum mechanics and others that perform a reduction of the wave packet which are not experimentally detectable in the current state but which can become so. As of yet, we have not managed.

Alexei Grinbaum. Decoherence is part of physics.

Hervé Zwirn. We agree.

Alexei Grinbaum. Then any interpretation of quantum mechanics among all the interpretations we have known for decades will do: each accommodates perfectly well the existence of decoherence phenomena. I think Zurek confuses matters—and not only him, but Murray Gell-Mann and James Hartle for example. Indeed, in my opinion, their description of physical phenomena and their philosophical interpretation of quantum mechanics are too close to each other. What Zurek describes is his interpretation of quantum physics. It is not something required by the existence of the decoherence phenomenon. Consequently, I am not sure we need to seek an agreement among us: each can have his own favoured interpretation.

Hervé Zwirn. Allow me to repeat myself, I just wanted to know if we agreed on the fact that the decoherence mechanism, as I described it in a simple manner earlier, provides an explanation for the *appearance* of the world to a human observer. This is the first step. Many other questions ensue. Simply, regarding the question of why the world appears to us as it does, which was problematic without decoherence (many hypotheses, including the reduction of the wave packet by consciousness, have now been mostly abandoned by physicists), do we agree that this problem is practically resolved by the decoherence mechanism? Of course, this does not provide a definitive solution to the problem of realism.

Alexei Grinbaum. Provided that we agree on considering this problem of appearance or occurrence of the world as we see it as different from the measurement problem. The measurement problem, for me anyway, is not resolved by decoherence. The problem of understanding what happens *for all practical* *purposes* is resolved by decoherence. However, it is not the same thing as the measurement problem, which is not resolved. The choice of the preferred base, the passage from "and" to "or", etc., all these reformulations of the measurement problem, are not…

Hervé Zwirn. …the choice of the preferred base is settled.

Alexei Grinbaum. Yes, the choice of the preferred base depends on the observer…

Jean-Michel Raimond. …who depends on the environment.

Alexei Grinbaum. Who depends on the environment, but…

Jean-Michel Raimond. … I believe the rather clever idea of Zurek's article was to say that the observer never directly interacts with the system. The environment is between him and the system, the observer being in fact only a part of the environment, which interacts indirectly with the system. The dynamics of the system/environment interactions are what determine the preferred base. It is both experimental and, I think, it provides *for all practical purposes* a solution to the question of the preferred base.

Hervé Zwirn. The problem of the "or" at the final stage remains unresolved. Unless we remain in Everett's model where we say that everything coexists but we do not take that into account.

Jean-Michel Raimond. It is a way of "sweeping things under the carpet"…

Bernard d'Espagnat. As I was saying earlier, I think the use of the notion of conscious observer allows us to do better.

4.3 Arguments for and Against Realism

Michel Bitbol. I have a question I would like to ask to you as a group, and in particular to decoherence specialists. We have mentioned two different aspects of the measurement problem: (1) we have spoken of the disappearance for all practical purposes of off-diagonal terms in the density matrix, and (2) we have also spoken of the ability, or rather the inability, of decoherence theory to resolve the so-called "and/or" question, meaning the passage from a superposition of states of an observable to a disjunction of singular values of this observable. The question I would like to ask you is the following: is there a link between the two?

To formulate my question more precisely, I would like to ask you this: if decoherence was really capable of making off-diagonal terms of the density matrix disappear, if it could really impose a value strictly equal to zero, would you consider the "and/or" problem resolved?

Jean-Michel Raimond. I am very pragmatic and very "*for all practical purposes*" (because something is wrong if you are not "*for all practical purposes*" in the lab). I would say that for all practical purposes, decoherence resolves the "and/or" problem. For all practical purposes and for any experiment conceivable by man or any normally constituted extra-terrestrial. It claims that the density matrix is, for all practical purposes, diagonal.

Michel Bitbol. Bear in mind, I was pushing the problem to its limits, by saying: "let us accept that the density matrix is *really* diagonal, not just for practical purposes, give or take negligible values, but *strictly* diagonal". Would your answer to the question "is the and/or problem resolved" be affirmative? Allow me to ask you again.

Jean-Michel Raimond. This clearly does not resolve the problem of choice.

Hervé Zwirn. We agree.

Jean-Michel Raimond. We can perfectly reason "*à la* Everett" where all the branches are resolved. Either in the entire Universe, if there is not an interpretation "*à la* Everett", or in each of the branches, everyone agrees on what took place.

Michel Bitbol. We clearly agree on this point. But please note that in my opinion, the "or" problem is not really different from what you call the problem of choice. Indeed, if one "or" another possible measurement outcome of an observable is obtained, this means there is only one, chosen among all possible outcomes, which is achieved although we do not yet know which one it is.

Bernard d'Espagnat. I have wondered, as I mentioned earlier, how Zurek would resolve this problem. Ultimately, in his previously cited 2003 article, he resolves it implicitly through this obvious observation—provided that we introduce the notion of observer and that of measurement—that a measurement has only one outcome. When there are N possibilities, as is the case once the density matrix has been diagonalized, Zurek says explicitly that: as a measurement outcome can only have one of the values that is specific to the observable, it is necessarily identical to one *or* the other if $N = 2$ (or to *one* of these in the general case) with equal probabilities for each when the coefficients of the wave function are equal. It is the first step in his attempt to demonstrate Bohr's rule.

I found that presented in that way, his reasoning was correct. However, I can see that it is also fundamentally derived from the notion of measurement per se. Unless we resort (somewhat problematically) to Everett, no "purely physical" interaction would give you this, it seems to me. We need measurement.

Michel Bitbol. I would like to add a point to make you understand what my motivation was when I asked about the resolution (or lack of resolution) of the problem of choice through decoherence. This question seems almost naive when taken on its own, but it takes on another dimension when you relate it to a certain probabilistic idea of the status of quantum formalism. Let us suppose therefore that quantum formalism, through decoherence theory, produces a rigorously diagonal density matrix (with strictly null off-diagonal terms). In that case, the quantum formalism is exactly like classical probability theory with its Kolmogorovian axiomatic. Yet no one has ever asked classical probability theory to designate the choice that is effectively observed in the laboratory: we do not even ask it to justify that a particular choice is made among all those possible, as this is taken for granted from the moment we accept that one *or* the other is achieved; we only ask of it to determine a priori the probability of each possible choice. It is therefore surprising that, even at the boundary where quantum theory meets classical probability theory, we still ask of the former to justify on its own why a particular choice is made. This is why I am perplexed.

Jean-Michel Raimond. I think that this vision of things gives quantum mechanics exactly the same status as ordinary statistical physics. We know there are an infinite number of microscopic realities for the same macroscopic state. We do not which one is achieved, but we know only one is achieved. I believe we could manage by simply suggesting that a first postulate is added to all the postulates of quantum mechanics that says that if all the observers of the Universe agree on one physical reality, then this reality is unique. Period.

Bernard d'Espagnat. I like your "first postulate" all the more that it is the one made by all the antirealist philosophers starting with… Schrödinger. He added that under these conditions, there was no need, for science to be done rigorously, to implicitly add (just by using the word "reality") the metaphysical, unverifiable and now problematic (non-separability) postulate where this physical reality exists "per se", meaning it would exist as we apprehend it even if there was not, and never has been, an observer.

However, I am less convinced by the equivalence of status that you are suggesting following Michel Bitbol's comment and within the context of his hypothesis. Even within this framework, I do not see it being achieved, and that because of the "ontological" realism implicitly postulated in many presentations of classical statistical physics (with the exception, I think, of Gibbs-like presentations).

Admittedly, nothing stops us from conceiving a mechanics that would be both realistic (i.e., ontologically interpretable) and fundamentally non-deterministic. In such a theory, certain events would have an intrinsic probability. For example, during a measurement and for certain states of the measured system the pointer would have, independently of any consideration of the environment and of decoherence, a certain probability to move "as one block" to the right and the complementary probability to move "as one block" to the left (this is similar to the idea of "propensity"). In such a theory, there would be no need to explain the "or" using the notions of observer, measurement, etc. There would be no need because in my theory, the "or" is introduced in a way "by hand" as an integral part of the axiomatic. However, this theory does not comply with either quantum mechanics or experience (your experiments show that at time 0+ it would be impossible to assign a position or even a determinate form to the pointer). With quantum mechanics, we are dealing with a completely different theory, where it is in principal always possible to move back the "or" (i.e., non-determinism) and at the same time assign a credible form to the system, up to the point where an observer takes a measurement. It is therefore not surprising that the problem of the "or", which is non-existent in a theory that sets the "or" at the start, is in a theory with such characteristics a real problem that is furthermore linked to measurement.

Jean-Michel Raimond. We need to have choices without determinism.

Bertrand Saint-Sernin. Could I make just one point? Historically, the 19th century philosopher [8] who was the first to explicitly investigate the notion of realism using experiments of synthetic chemistry was a probability theoretician. He thought that contingency was part of nature's make-up. He was not at all a determinist. He criticized Laplace's demon using very strong terms.

Alexei Grinbaum. From this point of view, Jean-Michel (Raimond), it is absolutely true that there would be no difference between statistical physics and quantum mechanics when off-diagonal elements are equal to zero. That being, what is different, is that when we say "statistical physics" we think of "a system with multiple components". We do not think of a gas with a single molecule, which is an extreme example—which besides can be studied, and is studied, and is even rather interesting. Whereas when we say "quantum mechanics", we also want to study not the statistical aspect of things but a single photon (which is an example of a single system). There is not physical (since the mechanics works) but conceptual friction: to say that quantum mechanics is a statistical theory is, in my opinion, akin to what was said in the 1920s or 1930s. Nowadays, when we use quantum mechanics to describe single systems, we want to say that there are things which are probabilistic and not statistical.

Jean-Michel Raimond. I agree. I spent my life manipulating single systems! Decoherence provides probabilities that are not of the nature of probabilities of statistical physics, because they do not concern a large number of sets of systems, but the description of a single system or a single object, yet whose conceptual status —or philosophical status if you prefer, supposing I understand it—does not seem to me so different from that of a probability of statistical physics, which is a probability of ignorance. This can concern a spin, a photon, or a molecule.

Alexei Grinbaum. Yes, but what is interesting is that when you consider single systems in quantum mechanics, this leads to a number of paradoxes. In quantum mechanics, paradoxes (which are not logical paradoxes, but very strange counter-intuitive phenomena) are linked to post-selection. The nature of these paradoxes is different from the conceptual problems raised for example when considering a gas with a single molecule.

Jean-Michel Raimond. Most of these paradoxes are linked to a non-trivial description of non-trivial experiments where we suppose entirely quantum behaviour and where, at the end, we analyse the measurements. We take a measurement first and only acknowledge it at the end. This is simply determined by complicated quantum behaviour. I am not saying that quantum behaviour is not complicated. I simply say that if we introduce decoherence, at a given point probabilities appear that do not have a conceptually different role from those of statistical physics. I am not saying that quantum mechanics is statistical physics.

Jean Petitot. Yes, it is definitely not statistical physics.

Olivier Rey. In fact there are two ways of considering statistical physics in a classical framework. One approach consists of trying to construct a physical theory from what we can know empirically of reality. The other consists of thinking that all is determined within reality, yet the number of elements forces us to treat them statistically. There are thus two ways of considering statistical physics, both leading to the same outcomes, but which are philosophically very different.

The hypothesis of underlying determinism is not necessary for statistical physics. Recently, Jean Bricmont argued for a so-called "Bayesian" formulation of statistical physics. The main question is then: what is most rational way of thinking about reality taking all the available information into account?

Alexei Grinbaum. Which we can do, as well, with quantum mechanics.

Olivier Rey. Yes, precisely. The "Bayesian" approach is much more in line with quantum physics than any statistical approach that supposes an integral determinism, and deploys the statistical arsenal from this hypothesis.

Michel Bitbol. If I may comment, I find it amusing and paradoxical that it would Jean Bricmont who would put forth this view of statistical physics. It apparently supposes that we devise a global stochastic description by excluding any preoccupation concerning hypothetical underlying microscopic processes. And yet Jean Bricmont has made himself the advocate of Bohm's "ontological" interpretation, the same one that claims to use mechanisms supposedly underlying quantum probabilities…

Olivier Rey. He is part of a long tradition.

Jean-Michel Raimond. I will be once again outrageously *"for all practical purposes"*! It seems to me that we can agree on the fact that independently from everything, all the observers of the Universe can agree on the fact that there are objective realities in the physical world that are one, unique and indivisible. Perhaps this should be the common denominator of all the physical theories we attempt to formulate, be it in statistical physics or in quantum physics. In other words, there is an objective reality. We should not ask of physical theories to extract their own formalism—at least not the physical theories we have now.

Bernard d'Espagnat. With nonetheless some reservation regarding the term "there is", which is too close to "per se" to my mind. What you have defined is an objectivity that is relative to all possible observers or all possible conscious beings.

Jean-Michel Raimond. And even to all possible fractions of the environment.

Bernard d'Espagnat. I am not sure about this last point. It depends on what we mean by "fractions of the environment". We have a tendency, when we speak of the environment, to keep in the back our minds a physicalist notion of reality made up from I do not know what, perhaps atoms linked by forces, or something more complex yet similar (made up for example of objects and fields existing per se and scattered here and there in space). We must do away with this, to say the least, questionable image (non-separability). In fact if we try to go beyond the "*for all practical purposes*"—which is sufficient for science, we all agree on this point—we have no valid mental representation at our disposal of what could be "an environment per se" made up of "fragments". As we are readily quoting Zurek, I would say that the quotation, which we have already mentioned, where he says that the Universe "*will appear classical to observers who monitor it from within, using their limited capacity to acquire, store and process information*", shows that that is also what he thinks.

Hervé Zwirn. Moreover, it seems to me that the agreement reached by observers pertains to phenomena. The problem of realism is to agree on the existence of something per se so that we agree on phenomena. We are not contesting that we agree on phenomena. We can use whatever vocabulary we wish. I for one call this not empirical but phenomenal reality: we all agree on phenomena as they appear to us and this constitutes for me phenomenal reality.

The problem of realism arises later. Given this reality, which I call phenomenal (but which is sometimes called empirical reality), is there an underlying reality per se which "causes" it? That is the problem of realism. As for the fact that there are phenomena on which everyone agrees, I think no one denies that.

Jean-Michel Raimond. The link between phenomenology and reality per se seems to me to be outside the grasp of physics.

Hervé Zwirn. It is philosophy.

Jean-Michel Raimond. I am not sure that this link is any different for any type of science, in particular for classical physics compared to quantum physics.

Hervé Zwirn. It seems, and that is the debate we are having here, that for reasons linked to what we were talking about, the relatively simple link (without mentioning other problems) that can exist between the two in classical physics is more complex to establish with quantum formalism. The reasons previously mentioned include non-locality, contextuality, etc. This is what the debate hinges on. Does quantum formalism, or quantum mechanics in the largest sense, enable this transition or not?

Jean-Michel Raimond. What we have been saying, nonetheless, is that formalism and decoherence, despite their notorious inadequacies, give quantum physics the status of a physics of classical probability—thereby making the link easier.

Hervé Zwirn. It appears that decoherence has been a major step forward from the beginnings of quantum physics (with the debates of the founding fathers) up to the discovery of decoherence, when some of the world's greatest physicists formulated some rather far-fetched hypotheses to resolve this problem. Is this a definitive paradigm shift? In that case, do we consider that the philosophical problems that allow us to bridge the two are resolved?

Jean-Michel Raimond. They are no more or less resolved than in the other branches of physics.

Hervé Zwirn. There. It is getting closer. Of course.

Bertrand Saint-Sernin. Historically, the problem of realism is linked to a very classical theological question: the problem of divine guarantee. In other words, the first definition of realism, the one we find in Antiquity, is: "have we access to divine reason when it created the world and as it maintains it?". The nature of the problem of realism changes the moment we say we need to construct a science without divine guarantee. The founders of modern science, be it Descartes, Newton or Leibniz, etc., think we can achieve, with more or less difficulty, a sort of vision of God. However in the 18th century, the nature of the problem changed completely. What does it mean to create a science by strictly human means and without referring to the idea of an infinite spirit with which we could communicate? From that point onwards, this profoundly changed the nature of the notion of realism.

Jean-Michel Raimond. I would really like us to do physics without consciousness and without God.

Bertrand Saint-Sernin. Of course. But historically, this is what happened. That is all.

Jean-Michel Raimond. I do not know what consciousness is, but the fact that we must bring in thinking objects for the physical description of the world is not at all to my liking.

Hervé Zwirn. This is a central point for our discussion. We all seek to avoid resorting to consciousness, which is akin in a way to resorting to God. God was eliminated and no one supports the idea anymore that the reduction of the wave packet is due to the direct action of consciousness on the system. Nevertheless, it seems to me that it is possible to take consciousness into account in the following sense: consciousness has no physical effects for reducing a system, but what we observe is done, in a Kantian sense, through a set of "filters" so that what we observe may not be totally independent from what we are. It seems to me to be something to consider, and is less troublesome than resorting to a divine idea or a consciousness that has a direct action. New theories that bring in information theory are close to this idea.

Jean-Michel Raimond. I hope that, if we have to bring in consciousness, it is more like the minimal degree of complexity of what is doing the observing.

Hervé Zwirn. Allow to me quote Zurek, from one of his articles from 2003 [9] reformulating his earlier publications: "*Hence, the ontological features of a state vector—objective existence of the einselected states—is acquired through the epistemological information transfer.*" This means that he links the ontological aspect of the vector state with some form of epistemological information. However, epistemological information…

Alexei Grinbaum. … the keyword in "*objective existence*" is "*objective*" not "*existence*". For Zurek, existence is a philosophical term. What concerns him is to put objectivity in the description.

To reiterate what Jean-Michel (Raimond) was saying, I think we can ask exactly the same question regarding the observation of a quantum system by a fullerene. We are not saying that a fullerene has a consciousness, which would be a bit strange; however we can ask ourselves how a rather complex molecule, like C_{60}, observes photons. With the support of evidence, I can show you that a fullerene can observe up to ten photons and keep this information in its memory. There is no reason, in my opinion, to think that there is a fundamental difference between a fullerene and a human being as quantum observers.

Bernard d'Espagnat. Unless perhaps when you consider probabilistic events and decide to exclude all hidden determinism (of the Bohm type). I think there is a difference. If you consider your fullerene as a purely physical, yet quantum system, then I believe it will not be able to tell you by itself that there is *one* single answer. In other words: that a given observable that could have taken on a number of different values has in fact taken on one of them and not the others. I believe that for this to be possible you need to consider your fullerene as classical, in the same way Bohr considered his instruments as classical because they were *used* as instruments.

Alexei Grinbaum. By stating that "it will not be able to tell you", you are assuming there is a communication or interaction problem between observers. It is not the same framework.

Bernard d'Espagnat. But the two problems are linked because you, as a conscious observer, know there is not (or rather, in my opinion, "are compelled by human mental processes to say that there is not") a single answer. In the sole light of quantum principles, we cannot see how a fullerene, a simple quantum system, could be brought to make such a "choice".

Hervé Zwirn. What does this mean? What meaning do you give to what a fullerene feels when it observes spin up or spin down?

Jean-Michel Raimond. It does not feel anything. However, when my computer saves the results of an experiment, it stores is in a RAM. In modern RAMs, to store one bit of data, you need twelve electrons. These twelve electrons carry the data I am saving from my experiments. It is neither very big nor very small compared to a fullerene. It is in the same order of magnitude. That is what ultimately carries objective reality. Countless conscious minds can look at these electrons, which, until I inadvertently press on the "start" button, will hold this information. If we consider the way a modern computer works, what holds information in an objective and verifiable manner, is a small set of quantum particles.

Michel Bitbol. That is true, but as long as you have not observed this system of electrons which holds the information you speak of, you have to describe it by a *non*-diagonal density operator.

Jean-Michel Raimond. Or a completely diagonal one, because these twelve electrons are very strongly coupled with a very complicated environment that stops them completely from being in superposition.

Michel Bitbol. Mr d'Espagnat would say "Diagonal indeed, but only *for all practical purposes*". We always come back to this!

Alexei Grinbaum. I think that is not quite right, because you cannot know what outcome was read by the fullerene. However, I think you can devise an experiment that will show you it acted as an observer. The measurement outcome is not accessible, because you are outside the observer-observed system pair. However, you can observe the thermodynamic consequences of the fact that the fullerene has acted for a time as an observer.

Michel Bitbol. The ease with which you consider a molecule to be an observer troubles me. What exactly is being an observer?

Alexei Grinbaum. It is to keep something in memory.

Michel Bitbol. To keep something in memory… What exactly is *memory*? What does keeping something in memory mean?

Jean-Michel Raimond. It means to be classically correlated with the state of the measured system. It is the classic correlation between the state of the measured system and the state of the meter, meaning a purely classical correlation, on a classical probabilistic superposition of classically entangled states.

Alexei Grinbaum. There is a problem with photons, which are absorbed—thus are no longer there.

Jean-Michel Raimond. Indeed, I agree.

Michel Bitbol. Why do you speak of "classically" entangled states? It is the adverb "classically" I do not understand here.

Jean-Michel Raimond. I mean described by a probabilistic alternative: either this state where the needle is in this position, or this state where the needle is in another position.

Michel Bitbol. The problem is that the alternative is, in theory, of a quantum nature, meaning that the off-diagonal terms can be extremely small but are not strictly null.

Jean-Michel Raimond. Yes, I agree. But *for all practical purposes*, no one will ever see them, not the entire Universe itself.

Michel Bitbol. I concede you this, of course. However, what I wanted to say was that the problem remains. The alternative has not tipped, as Alexei (Grinbaum) said, on the side of strict determination. A *true* observer would see a single, strictly determined outcome whereas a fullerene molecule remains in theory in a superposed state, (more or less intensely) entangled with its correlated system.

Jean-Michel Raimond. The problem is the same whether we are dealing with a fullerene, an electron in a RAM or a postdoc.

Olivier Rey. You would concede that this is a physicist's definition of the act of observing. It is not the usual definition of the word, which tends to suppose that there is a subject carrying out the observation.

Jean-Michel Raimond. Yes.

Olivier Rey. Words have a certain meaning.

Jean-Michel Raimond. I would not like the results of my experiments to depend on my state of consciousness, on knowing which consciousness is looking and whether that consciousness has had too much whisky or not!

Olivier Rey. What I wanted to highlight was that we need to be careful... When doing physics, we tend, for practical reasons, to use words from everyday vocabulary rather than create new ones. Therefore, we can easily be led to believe that physics always aims in the same direction as everyday language, when, words having taken on a new meaning, it speaks of something else. The definition of observation you have just given us is not in the dictionary for example.

Jean-Michel Raimond. I agree. In my opinion, observation is in a way a classical recording. That is what observation is for me.

Bernard d'Espagnat. We all have, I think, the impression that we are verging on an agreement without having completely achieved it, and consequently the debate is still open. That is all the truer since we have not yet tackled the specific spatio-temporal aspects of decoherence, although these raise particular problems that Jean Petitot, I believe, would like to address now.

4.4 Mathematical Aspects of the Conflict Between Quantum Mechanics and Spatio-temporal Localization

Jean Petitot. I prepared a comment on the link between problems of decoherence and my spatio-temporal localization—It's my geometrician side!—and more generally on the conflict between quantum mechanics and the localization of measurements. For us, human observers, the macroscopic world is characterized by its spatio-temporal localization. In *Decoherence and the Appearance of a Classical World in Quantum Theory* [10], Eric Joos broaches this question in the following way (p. 63): one of the fundamental characteristics of macroscopic objects is that are spatio-temporally localized. In particular, Joos cites the debate between Born and Einstein, where Einstein warns that spatio-temporal localization (or the fact that there is a very well-localized wave packet that does not disperse itself "*with respect to the macro-coordinates*", namely space-time macro-coordinates, positions, momenta, etc.) is in contradiction with the axioms of quantum mechanics. I would like to come back to this point, reprising the mathematical reflections on the work of the Gelfand, Naimark and Segal school, followed by Mackey, who tried to compare the mathematical formalisms of classical mechanics and quantum mechanics, to really "pinpoint" their fundamental difference. I am using as a template Jerrold Marsden's presentation in *Applications of Global Analysis in Mathematical Physics* [11].

In quantum mechanics, state space is a phase space (coordinates p and q of Hamiltonian mechanics). There is a differentiable variety of states (call it P for phase space). The observables are defined functions on this phase space P that take their value from a set of values. In general, these are functions with real values, meaning functions with complex values equal to their conjugates (in quantum mechanics they become self-adjoint operators). The measurement of an observable f on a state represented by point x in phase space P is simply the value of function f (x). It is the evaluation. $F(x)$ is equal to the value on f of the Dirac delta distribution in x. This means we have a duality between space and function: we have points (states) in a space of representation; the observables are functions on it; but we can equally start with the functions and recover the points as Dirac measurements, i.e., as certain linear operators on the commutative algebra of observables.

Mathematically, it is extremely important to note—this is the heart of Gelfand's theory—that there is perfect equality between the space where we can localize phenomena and observables, and algebraic properties, in particular the fact that the points are in bijective correspondence with the maximal ideals of the commutative algebra of functions, i.e., the ideals of functions that annul themselves at a certain point. This is a fundamental property of commutative algebras which disappears completely in the non-commutative algebras of observables we find in quantum mechanics.

In quantum mechanics, we know the situation is completely different: we have as space state a Hilbert space H; we have a non-commutative algebra of operators for the observables; and we have the measurement of observables: if A is an observable, i.e., an operator, and if ψ is a state, a scalar product in H $\langle A\psi,\psi\rangle$ is defined.

The problem is comparing classical statistical physics to this scheme of quantum mechanics.

Very early on, I believe as far back as the 1930s, Bernard Koopman tried to find a Hilbertian and operatorial formalism for Hamiltonian mechanics, in order to compare quantum mechanics and Hamiltonian mechanics. He proposed to formulate Hamiltonian mechanics in the closest possible way to what we find in quantum mechanics. It is rather easy: we take statistical states (thus we have a distribution on phase space P, a state ψ now being a distribution on phase space), from which we derive a measurement (in the mathematical sense) for the values of measurement (in the physical sense) of the observables. The formalism is then exactly the same. The fundamental difference comes from the fact that the measurement, essentially the square of the module of ψ, $|\psi|^2$, multiplied by the Liouville measure in phase space, has a huge group that leaves it invariant: you can multiply ψ by $exp(i\alpha(x))$ where $\alpha(x)$ is any function on phase space P. Therefore you have a huge group, and the quotient of Hilbert space H (which is the space L^2 of the functions on the phase space) gives back phase space P. As it is phase space that guarantees localization, localization is linked to the fact that a huge group operates on the Hilbert space of states. The incoherence of classical mechanics, meaning its decoherence, is fundamentally linked to this type of "localizability".

By contrast, in quantum mechanics the invariance group is miniscule: it is the group $U(1)$. The quotient of the Hilbert space H by this group is simply the projector of H, what we call *rays*. This is what, in this perspective, formulates coherence, the possibility of interferences, etc. In short, it is the magnitude of the invariance group in a Hilbertian formalism of classical mechanics that explains the characteristics of classical mechanics. With such an approach, it becomes easy to show (this dates back to von Neumann) the fact that it is impossible to have theories of local hidden variables whose idea is to try to reuse the formalism of quantum mechanics while adding a space where we could localize things in a somewhat analogous manner to what happens in Hamiltonian mechanics.

This theorem, revised by Mackey, dates back to von Neumann: any formalism of this type, which tries to enrich quantum mechanics by saying that there is an underlying variety, whatever it may be, that allows us to localize the measurements, is necessarily commutative. This was the first fundamental result of impossibility that highlighted the obstruction we face when we try to make quantum mechanics complete in this way.

I consider it to be interesting to see that even at the level of the foundations of mathematics, the problem of measurement "localizability" is a fundamental obstruction. As pointed out by many philosophers (including Husserl), if spatio-temporal localization can no longer be a principle of individuation, then we destroy the classical world. One of the characteristics of the classical world is that spatio-temporal localization is individuating. However, it is really in contradiction with the non-commutability of observables in quantum mechanics. To my knowledge, nowadays only non-commutative geometry attempts to resolve this problem.

That is the somewhat mathematical comment I wanted to make. If we try to "coin" the irreducible difference between classical and quantum, it is essentially linked to that. Can we, "underneath" Hilbertian formalism, introduce geometrical substrates that allow the localization of phenomena and their measurements? I think it would be interesting to discuss this from a philosophical perspective.

Obviously, this does not stop wave functions from being defined on space-time. What I have spoken about has nothing to do with that.

Alexei Grinbaum. My first comment is that von Neumann's theorem is wrong, as we know. Von Neumann's hidden variables theorem needs to be modified.

Michel Bitbol. It is not wrong. Simply, it is very partial, very incomplete. It does not prove, contrarily to what von Neumann claimed, that *no* hidden variables theory is compatible with quantum mechanics. It excludes only a very specific (but at the time the most likely) type of hidden variables theory. All subsequent theorems (Bell, Kochen and Specker, Leggett, etc.) are of this type: they lead to the exclusion of an increasingly large number of hidden variables theories, without excluding *all* theories of this kind.

Jean Petitot. I was speaking of Mackey's demonstration, which is correct [12].

Alexei Grinbaum. The question has been asked many times, especially since the resurgence of approaches that use C-star algebras…

Jean Petitot. Absolutely. All I have talked about, from Gelfand's theory to that of Connes, is expressed in terms of C-star algebras.

Alexei Grinbaum. The problem becomes more complicated when we consider field theory. The general problem is the following. You have a C-star algebra: how do you know it is classical, quantum, or something else? In the literature, we have proposed systems of axioms for understanding how an extremely general algebra becomes quantum, by specifying the constraints that must be added. It is a very active area of research.
In 1927, during the Solvay Conference (I would like to recommend, on this topic, a book that has just been published that contains all the abstracts and notes from all the participants of this conference [13]), Schrödinger asked the same question at the start of his presentation, without using mathematical language. He stated that we do mechanics in three dimensions, then in four dimensions (by adding time), then in $3\,N$ dimensions—we did not know exactly what that meant (that was before Hilbert space, before von Neumann). What does this slide mean? Why go from three to $3\,N$ dimensions, i.e., from a Euclidian space to a space in the abstract sense of the

term? That is the reason why I think the term "localization" has changed meaning. Regarding the discussion we were having earlier, in my opinion the problem of localization consists of knowing how decoherence takes place with the variable position. That is really on a basic level. However, when we say "localization" in this way, we raise the problem of system composition: in algebraic language, how can we conceive of the fact that there is a geometric link between two systems? It is a tremendous problem for algebraic approaches.

Jean Petitot. Yes. Individuation and separation of systems are fundamentally linked to localization in the sense I was speaking of.

Alexei Grinbaum. I would like to make a distinction. There is the fascinating algebraic problem, which is that of separation. It is not the same as the more down-to-earth problem of Euclidian space. We use Euclidian space, which in fact has nothing to do with this story, to understand an algebraic aspect of the structure of quantum mechanics. Euclidian space, which has nothing to do with all this, has been used, in a way, as a starting point.

Jean Petitot. That is what I said. The impossibility of localization does not stop wave functions from being functions of space-time.

Bernard d'Espagnat. Problems of "and/or", "impossibility of localization"… Is the objectivity of the classical world only "weak" (or "relative")? On this philosophical question, we have not, as expected, reached an agreement; however, our respective views have been refined. Each of us now sees the outlines better. I have no doubt we will have the opportunity to come back to it.

This meeting is the last of the academic year. We will meet again in the autumn, probably at the end of September. Our next session will be dedicated, as you know, to an old theory, not one of ours, but that of Louis de Broglie and David Bohm. It is not the double solution theory, but the one that Louis de Broglie presented during the Solvay Conference of 1927 and which was rediscovered by David Bohm.

This theory has the particularity of being ontologically interpretable, like classical physics. It has been snubbed by physicists since its inception, yet there have been people that, even today, say: "Why is this theory so criticized? Is it because it is not relativistic, etc.?". I think, honestly, that we should examine this question.

Franck Laloë has very kindly agreed, although he is not supporter of the theory, to present it to us. We await him for our next meeting at the start of October [14].

References

1. See session III "Experimental investigation of decoherence".
2. The author comes back to this topic during session VIII "Exchange of views on the relational interpretation".
3. I describe in detail the mathematical aspects of this approach in my article: "Quantum observer and Kolmogorov complexity: a model that can be tested", arXiv: 1007.2756.

4. In English.
5. *Cf.* for example Wojciech H. Zurek, "Environment induced superselection rules", *Phys. Rev. D*, 26, 1982, p. 1862–1880.
6. Wojciech H. Zurek, "Decoherence, Einselection and the quantum origin of the classical", *Rev. Mod. Phys.*, 76, 2003, p. 715–775.
7. Zurek, *op. cit.*, 2003, p. 5.
8. Antoine Augustin Cournot.
9. Wojciech H. Zurek, "Decoherence and the transition from quantum to classical", *Los Alamos Science*, 27, 2002.
10. Erich Joos *et al.* (eds), Springer-Verlag, 2nd ed., 2003.
11. Carleton University, 1973.
12. Those who are interested will find a very good summary of Mackey's work in: Veeravalli S Varadarajan, "George Mackey and His Work on Representation Theory and Foundations of Physics", 2000. Section 7 is dedicated to hidden variables and the Mackey-Gleason theorem.
13. Guido Bacciagaluppi & Antony Valentini (eds), *Quantum Theory at the Crossroads: Reconsidering the 1927 Solvay Conference*, Cambridge University Press, 2010.
14. See session V, "The pilot-wave theory of Louis de Broglie and David Bohm" (and session VI, "The pilot-wave theory: problems and difficulties").

Chapter 5
The Pilot Wave Theory of Louis de Broglie and David Bohm

Franck Laloë

Bernard d'Espagnat. Our think tank on the contributions of contemporary physics to the theory of knowledge begins its second year of existence and I have the pleasure of welcoming among us Alain Aspect and Michel Le Bellac who have come to listen to Franck Laloë in order to learn more about the theory of Louis de Broglie and David Bohm. It goes without saying that we are grateful for their presence and we would be delighted to welcome them again.

Franck Laloë has kindly accepted to enlighten us on this old theory, which has the rare quality of being ontologically interpretable and which, despite that, does not appeal to many among us, it must be said—without us always really knowing why. Today's presentation and the discussion that will follow may enable us to progress in resolving this enigma. Dear Franck, the floor is yours.

Franck Laloë. Thank you very much. I am delighted to speak here and be an advocate of the de Broglie-Bohm (dBB) theory for the duration of the presentation.

After a general introduction, I would like to recall the general principle of this theory in Sect. 1. You are all probably familiar with it, but it does not hurt to start with a little reminder. I will then speak in Sect. 2 about Bohmian trajectories, which have certain peculiar and interesting characteristics. Section 3 will cover measurement in Bohm theory, which seems to me to be one of the particularly interesting strengths of dBB theory. In Sect. 4, I will speak briefly about field theory. Up to that point, I will act as advocate of dBB theory, by trying to convince you that it is successful on many fronts and presents numerous advantages. From Sect. 5 onwards, I will nuance this point of view and ultimately explain why, to my mind, it does not contribute as much as we could have hoped.

I will provide a few references, but I have to say that I was somewhat at a loss with those of Louis de Broglie; he published many short notes where he explains

F. Laloë (✉)
ENS Kastler Brossel Laboratory, Paris, France

B. d'Espagnat and H. Zwirn (eds.), *The Quantum World*,
The Frontiers Collection, DOI 10.1007/978-3-319-55420-4_5

his ideas on such and such a phenomenon. There are few complete texts where he gives a general overview of his theory. Let me suggest nonetheless "La mécanique ondulatoire et la structure atomique de la matière et du rayonnement" [1]; *Une tentative d'interprétation causale et non linéaire de la mécanique ondulatoire: la théorie de la double solution* [2]. I will then cite the two articles by David Bohm that many of you are familiar with: "A suggested interpretation of quantum mechanics in terms of "hidden" variables" [3]. I strongly recommend the relatively recent book by Peter R. Holland, *The Quantum Theory of Motion* [4], especially for its illustrations. There are countless articles on dBB theory, particularly in ArXiv, some of which are interesting, others less so, and others which are wrong. Among the better authors is notably Sheldon Goldstein, one of the great advocates of this theory who wrote many articles [5].

During this presentation, I will speak of a relatively well-known paper by Berthold-Georg Englert, Marlan Scully, G. Süssman and Herbert Walter, "Surrealistic Bohm trajectories" [6]. It is an article that is critical of dBB theory, and while in my opinion it does not quite achieve its critical objective, it is still, I think, very interesting.

Quantum mechanics presents difficult problems of interpretation—otherwise our think tank would not exist. We all agree on that point! The approach of dBB theory is not at all to try to avoid these. On the contrary, it aims to "take the bull by the horns". It genuinely speaks of "things that exist" and follows no avoidance or exclusion strategy in the way of Bohr. It is really, from this point of view, an extremely direct and interesting approach.

As Bernard d'Espagnat has said very well, it is fashionable to reject dBB theory, because we consider it old, outdated, some would say unsightly, reactionary… We hear many adjectives. That said, in fact few physicists have bothered to really study it. What strikes me is that we find a considerable number of articles in the literature rejecting dBB theory that mostly illustrate the lack of understanding the authors have of this theory rather than provide a true refutation. Typically, these authors use a combination of their version of dBB theory and standard quantum mechanics and come to the conclusion that "things do not add up". We will come back to it later and see that these criticisms generally do not apply to the true dBB theory but to a hybrid version made up by the author for the occasion.

Many of you know and have already used the quotation by Richard Feynman: "*We choose to examine a phenomenon which is impossible, absolutely impossible, to explain in any classical way, and which has in it the heart of quantum mechanics. In reality, it contains the only mystery. We cannot make the mystery go away by explaining how it works.*" So doing, Feynman describes the interference phenomenon particle by particle and considers that it is really the great mystery of quantum mechanics that no one can explain. It makes Bohmians or neoBohmians smile, as for them this situation is almost trivial and presents no great mystery in particular.

Alain Aspect. May I comment?

Franck Laloë. Please do.

Alain Aspect. Twenty years later, Feynman claimed there was a second mystery in quantum physics, called entanglement.

Franck Laloë. And which may be, by the way, as deep as the first mystery.

Alain Aspect. The quotation you used is from an article written at the end of the 1950's. Twenty years later, in his article from 1982 which is considered the pioneering article on quantum information [7], Feynman writes explicitly that there is a second mystery that he has tried to solve all his life but which remains: entanglement. When I started experiments on Bell's inequalities, everyone was telling me: "It is the wave-particle duality, we know this, we know it is the fundamental problem of quantum mechanics". Thanks to you in particular, Franck, I understood it was of a different nature, meaning that entanglement was the wave-particle duality, if I may say so, for two particles. Let me simply add this. Feynman did indeed write that "it contains the only mystery", but he later accepted to "make some concessions", if I may say so.

Édouard Brézin. If I take the single electron bond theory, it is electronic entanglement that creates the bond between the two atoms. It is still entanglement with an electron, is it not?

Alain Aspect. No. The best proof, as we will see in Franck Laloë's presentation, is that dBB theory allows us to give a local interpretation of single-particle phenomena, whereas when we try to give interpretations of two-particle phenomena, there no longer is any local interpretation. We can talk more about this.

Bertrand d'Espagnat. Bohm theory has never claimed to be local.

Alain Aspect. No, but it seems to me this shows the different nature of these problems.

Franck Laloë. As I have said to Bernard d'Espagnat, I am not very keen on the matter, in particular of the "double solution" theory. Louis de Broglie introduced it just after his dissertation. The general idea is that a wave function does not directly represent a particle, but a field that guides its movement.

Initially, the particle itself was represented by another wave which was a localized wave. It is a second solution to the same equation. De Broglie considered that the two waves must be in phase. Schrödinger's wave propagates itself following the linear equation that we all know and the second wave, the singular wave representing the particle, is a small clock that also oscillates—and the movement of these two waves is such that their phases become synchronized.

Personally, I never understood whether this was a general idea that was being proposed or whether it actually enabled us to write a mathematical formalism and precise equations. I do not really see how Schrödinger's equation could allow localized supplementary solutions (the double solution), hence I imagine the equation—or the type of solution (a distribution?)—must be changed. Nonetheless, de Broglie must also have been uncomfortable with this, at least in 1927 when he was invited to the Solvay Conference to present his theory. Because of the mathematical difficulty of the double solution theory, he presented a "truncated" version (I am only repeating what he wrote—I would not dare say it myself!).

He then considered that it would be preferable to present a simpler theory, whereby the second wave is replaced by a single-particle position. That is when he devised the pilot wave theory, which is the one he presented at the Solvay Conference with moderate success, as you know. It even seems that no one was convinced. With hindsight, what is amusing is to realize that it was often for the wrong reasons. When Wolfgang Pauli strongly attacked de Broglie's theory, he put forward arguments of inelastic collisions, which the pilot wave theory would be incapable of accounting for. If we think about it now, his arguments did not hold up. But at the time, the equivalence between the predictions of the different formalisms and their relationships were not very clear!

Then in 1952, Bohm (who was not familiar with de Broglie's work) proposed a theory that was very similar to the second version of the pilot wave theory. I learned just recently that Bohm was an American physicist. For a long time I thought he was English, even a Londoner. He did his Ph.D. at Berkeley, under the supervision of Robert Oppenheimer. An extremely brilliant physicist, he performed calculations on the collision of protons and deuterons. However at the time they were working on the nuclear bomb. His work was immediately classified. Because of his political views, he did not have access to classified documents—his doctoral work became instantly inaccessible to him. He was even barred from writing up his thesis dissertation. Then, this young physicist faced rather complicated tribulations. He was brave during the McCarthy era, and refused to denounce colleagues, which led him to lose his post at Princeton. He then fled to South America, before establishing himself in England where he finished his career.

Alain Aspect. May I relate a near first-hand account? He was a professor in Rio de Janeiro. Moses Nussenzweig recounts that when he attended Bohm's lectures, the latter had already changed his views on quantum mechanics from what he wrote in his book, which is a standard book on quantum mechanics [8]. Nussenzweig tells that—this is really surprising—when speaking about his book, Bohm would say: "In his book, he says that…", the "he" being himself, that is the Bohm of standard quantum mechanics. He had clearly changed his point of view from the author of the book, who was himself. This story is quite interesting!

Michel Bitbol. He even changed his views a third time, taking exception to his 1952 theory. He proposed in its place a holistic theory where particle trajectories are only appearances.

Alain Aspect. Anyway, it is interesting to have at least one first-hand account of Bohm's lectures.

5.1 General Principles

Franck Laloë. Yes, the work described in his last books was clearly of a philosophical nature. But let us return to a more standard physics and to the general principles of dBB theory.

The duality of standard quantum mechanics is replaced by a type of coexistence: wave and particle always coexist (there is no particle without an associated wave and vice versa). Moreover, these two objects—as they are objects considered to be real—interact.

For a single particle, the evolution of the position R, point-like, is given by a formula that Bohm called the "guidance formula", which you are all familiar with, which describes the variation of the position over time:

$$\frac{d}{dt}\mathbf{R} = \frac{1}{m|\Psi(\mathbf{R},t)|^2}\mathrm{Re}\left[\frac{\hbar}{i}\Psi^*(\mathbf{R},t)\nabla\Psi(\mathbf{R},t)\right] = \frac{\hbar}{m}\nabla S(\mathbf{R},t)$$

Alain Aspect. Is S the action?

Franck Laloë. No, it is the phase of the wave function. Ψ is the wave function, m is the mass of the particle.

Édouard Brézin. This means that if the wave function is real, the particle is immobile.

Franck Laloë. Yes. The wave function evolves following the usual Schrödinger equation.

Franck Laloë. For N particles, the generalization is simple, except that we must reason in a configuration space with 3 N dimensions. We then write a formula that immediately generalizes the one above, which seems simple but as Alain Aspect said, can lead to interesting results in terms of locality.

Allow me to make a few points in passing. First of all, the formula directly gives the velocity as a function of an external field, which is rather surprising for a physicist. Indeed, since Galilei, we are more used to having acceleration as a function of external conditions. And moreover, to have a formula that is more like classical mechanics, Bohm proposed an equivalent version of the theory introducing not a "guidance formula" for velocity but a "quantum potential":

$$V_{\mathrm{quantum(r)}} = -\frac{\hbar^2}{2m}\frac{\Delta|\Psi(r,t)|}{|\Psi(r,t)|}$$

With adequate initial conditions, we can show it is the same thing to write Newton's law for a Bohmian position with this additional potential or to use the guidance condition. If we really wish to have forces, we can do so within this theory, or interpretation.

Another very important point, which I will come back to later, is that to make this theory strictly equivalent to standard mechanics (with guidance or with an additional potential), we must specify the initial conditions regarding the distributions of positions R. We must assume that at the initial time, whatever that may be, the initial position R of the particle is not known, but randomly distributed in space with a probability given by the square of the module of the wave function at the initial time. For N particles, this gives the following distribution of probability D for the Bohmian positions R_1, R_2, ..., R_N:

$$D(R_1, R_2, \ldots, R_N) = |\Psi(R_1, R_2, \ldots, R_N)|^2$$

In 3 N dimensional space, we assume therefore that the positions R_1, R_2, etc. are distributed randomly according to the square of the wave function.

This condition is sometimes called "quantum equilibrium". Once we have postulated that at time $t = 0$, any student in quantum mechanics knows how to demonstrate that the condition is verified at later times. This way of seeing things is reminiscent of an approach devised at the same time as the de Broglie theory, the theory of Erwin Madelung on quantum hydrodynamics. If we assume this, we can show that at any time we find exactly the conditions of standard quantum mechanics.

Each of the N particles is piloted by the wave function of the entire set of particles —this is important—but maintains at all times a position that is perfectly defined as a function of its initial position.

Michel Le Bellac. And if we do not suppose Born's rule, what happens? Do we encounter problems?

Franck Laloë. The mechanics of de Broglie and Bohm does not need Born's rule. It deduces it from the initial distribution of the positions. We postulate quantum equilibrium; it cannot be demonstrated and replaces Born's rule. What is easily demonstrated, however, is that if we assume this at time zero, it remains valid for all time. This is therefore a very strong postulate.

Édouard Brézin. Is there a generalization to field theory?

Franck Laloë. Yes, I will speak of it later.

Alain Aspect. This means that implicitly, we can say that at time zero, whatever happens, we cannot know the state of the particle better than we can do in quantum physics, since we exclude having a probability law that would not be that of the quantum wave function.

Franck Laloë. Exactly. This is a fundamental point to guarantee the compatibility not only with standard quantum mechanics, but also with relativity. If we assume that we are able to prepare initial distributions of particles that are different from the quantum equilibrium, we can show that it would then be possible to transmit signals faster than the speed of light. That would therefore be relatively catastrophic. We are really forced to assume quantum equilibrium if we do not want to call relativity into question.

Franck Laloë. So if we make these hypotheses that, I must admit, are strong hypotheses and are not self-evident, then we arrive at a theory which is practically equivalent in its predictions to standard quantum mechanics—even if Pauli did not realize this and many authors repeatedly forget this. We always find (on ArXiv in particular) many articles where authors claim to have proven that dBB theory is wrong and that only standard quantum mechanics gives the right predictions. This is absurd as the predictions are the same.

We explicitly assume that the measurements are always of positions. It is indeed a restriction, but not a serious one. If you think of all the types of experiments we can do physics, ultimately we always have a needle that will move across a dial…

A few other points in passing. The "guidance formula" shows that velocity depends only on relative variations of the wave function. It does not matter whether the module of this wave function is big or small, it must only not be zero. This may seem to be a problem, especially if the particle positions can reach areas where the wave function becomes null and where velocity is indeterminate. However, thankfully we can show that particles that are initially distributed in the probability cloud can never come out of it. We never have contradictions, particles never leave their wave function and everything "goes to plan".

In addition, as I have already said, wave functions pilot these particle positions. Conversely however, positions have no retroactive effect on wave functions, which is not forbidden, but is a blatant violation of the usual symmetry between action and reaction that is common in physics.

Jean Michel Raimond. How do you place potentials between the particles if the particles interact?

Franck Laloë. Exactly like we do in standard quantum mechanics. The Schrödinger equation remains exactly the same, and the pilot equation does not change either. We do not place direct interactions between the Rs. The interaction between particles is taken into account by the wave function Ψ, which then pilots the particles so that their position at any time gives exactly the same statistical results as quantum mechanics.

Édouard Brézin. Therefore, at this stage, we have introduced new variables, which we call positions. But for now, the formalism of quantum mechanics is in no way modified. We can forget,or not forget about these positions, this does not alter the dynamics of the system.

Franck Laloë. Exactly. And that is probably one of the very simple reasons why most physicists never use this theory. Regardless, we must perform a standard calculation of quantum mechanics, which remains automatically viable in dBB theory, and it is only afterwards that we can add the positions should we want to. Using standard theory requires less work, provided obviously we abandon recovering trajectories.

Édouard Brézin. The image of trajectories can help some people. However, I will come back to my idea. In the fundamental state of the hydrogen atom, for example, the wave function is real, as we know. That means that the position given to neutrons by this formalism is an immobile position. That is not exactly how I imagine an electron turning around a proton. Yet that is what a position is. Those who call this a "position", following de Broglie, prefer this point of view.

Franck Laloë. I intend to speak in detail about the paradoxical trajectories of Bohm and de Broglie. In standard quantum mechanics, in the fundamental state of the hydrogen atom, the probability current is null everywhere and we can say should we want to that the electron is immobile.

Bernard d'Espagnat. According to dBB theory, in the fundamental state of the hydrogen atom, the electron and the proton, as electrified as they are, remain effectively immobile; we could say they are "staring at each other". And of course, this is not how we physicists usually visualize the hydrogen atom. But why not? If these theoretical calculations are congruent with experiments, as is the case, then there is nothing more to say.

Édouard Brézin. Forgive me. If I joined in the discussion, it was because these calculations do not depend on these variables. It has often been said that we can give them whatever name we like, for now the dynamics, or the mechanics, remains unchanged. It is a question of personal taste!

Bernard d'Espagnat. As long as we remain within fields where the measurement problem—with its corollary the "reduction of the wave packet"—does not come into play, I would say that indeed, like you, the choice between standard quantum mechanics and dBB theory is a question of personal taste. In these fields, the wave function does not raise more fundamental conceptual problems than do classical fields. The problem is that quantum mechanics cannot be limited to these fields. The problem of "reduction" arises very rapidly, and cannot be bypassed. And as soon as he tries to deal with it, the theoretician can only preserve his instinctive realist ontology at the cost of implausible means such as Everett's theory [9] … unless he adheres to dBB theory! Unless of course he allows small, more or less ad hoc modifications to Schrödinger's equation, or he concedes explicitly that physics itself is nothing more than a source of excellent recipes for predicting observations. That is the difference, if you want. Bohm insisted on that point. The point of his theory, he would say, is that it is *ontologically interpretable*.

Jean Michel Raimond. Does this not pose a problem with the correspondence principle? If we take a state that is sufficiently excited, we should recover classical mechanics. However, in Bohmian mechanics, the electron is immobile. All the excited states always have a real wave function.

Alain Aspect. This is interesting. It then means that even if we tried to apply the correspondence principle, the R would not follow the classical trajectories we would expect to find.

Jean Michel Raimond. Only real wave functions bring about zero velocity. There is no reason for them to follow the correspondence principle.

Franck Laloë. In the fundamental state of the hydrogen atom, the electron and the proton attract each other through the Coulomb potential. In dBB theory, the force of attraction is exactly compensated by the Bohm potential that I described earlier. The two particles do not move; the sum of all forces is zero. In standard quantum mechanics, the probability current is zero at all points; in this way, the electron does not move either.

Alain Aspect. Exactly.

Franck Laloë. We often say that the "atomic image" representing electrons revolving around the nucleus like satellites around the Earth is not at all in the spirit of standard quantum mechanics, and that is true. Particles have neither well-defined positions nor well-defined velocities, and it is misleading to say that the electron revolves around the proton. Why would the electron revolve better in dBB theory? There is no reason that it should. Furthermore, for a state of quantum number m, we can choose between taking this state and finding a rotation, or (in the absence of a magnetic field) superpose it with—m and discard the rotation. The standard image is also not exempt of paradoxes.

Alain Aspect. You are right. State S does not rotate in orthodox quantum mechanics either.

Franck Laloë. We will come back to trajectories shortly. I would like to insist on the fact that the wave function is a real field, somewhat like an electromagnetic field with two components: an electric and a magnetic component. As John Bell said, "nobody can understand this theory [Bohmian mechanics] until he is willing to think of Ψ as a real objective field rather than a 'probability amplitude'". However, as we have seen, when there are multiple particles, this wave function travels in configuration space. Therefore for a gas with 10^{23} particles, physics takes place in a space of enormous size! The real field of Bohm is thus very different from a classical field.

5.2 Trajectories

Let us start with the Mach-Zhender experiment. I will present first of all what happens from Bohm's point of view on a single splitter blade.

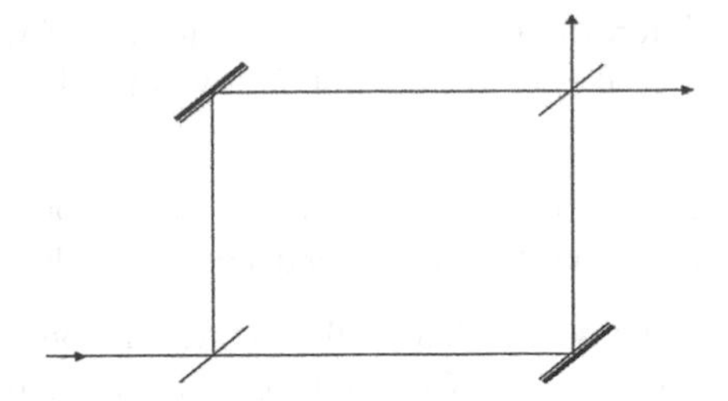

Suppose a wave packet associated with a particle arrives on the semi-reflecting blade. Holland has represented a certain number of trajectories, with positions of the incoming particle over time. We associate with each position the probability current according to the pilot wave equation. We can see that something relatively simple happens: the particles on the left-hand side cross over to the right-hand side and those who were on the right-hand side are reflected. As soon as this is observed, of course, our first reaction is to say is that it is not possible and it certainly does not reproduce the conditions of quantum mechanics. However, that is wrong! In fact, the positions of the particles at all times match exactly the predictions of quantum mechanics.

Alain Aspect. It reminds me of the delayed choice experiment. If we suddenly insert a semi-reflecting blade, we go from an initial situation where we had tubes or lines going up and down to a situation like that.

Franck Laloë. That is right. Precisely at the moment when I insert the splitter beam, that is what happens. The particle then follows these lines.

Michel Le Bellac. The quantum potential you presented earlier changes when you change the experimental set-up and therefore the evolution of the wave function. This potential then guides the position.

Franck Laloë. This set-up shows that it is possible to make a retrodiction about the position of the particle, since the particles' trajectory after their passage on the splitter beam provides information on the possible positions before they are reached. However, it is not possible to make a prediction. We cannot know in advance the position of a particle and know whether it will bounce or not. Some see in this the indication that Bohm theory must be considered only as a purely retrodictive theory.

Alain Aspect. Certain trajectories also come from above in the Mach-Whender experiment. Here you only speak of those coming from below.

Franck Laloë. Indeed, I was talking here about a single semi-reflecting blade. Let me come back to the Mach-Zhender set-up and how beams recombine upon exit. The phenomenon I have just described occurs on the first splitter blade, with trajectories that go in two directions. Particles from above or below arrive on the second blade. On this blade the wave function, i.e. the real object which propagates itself, interferes—as would an acoustic field—and we end up with a more or less large intensity depending on the two possible paths. The way the particle is guided on the second reflecting blade depends on this interference.

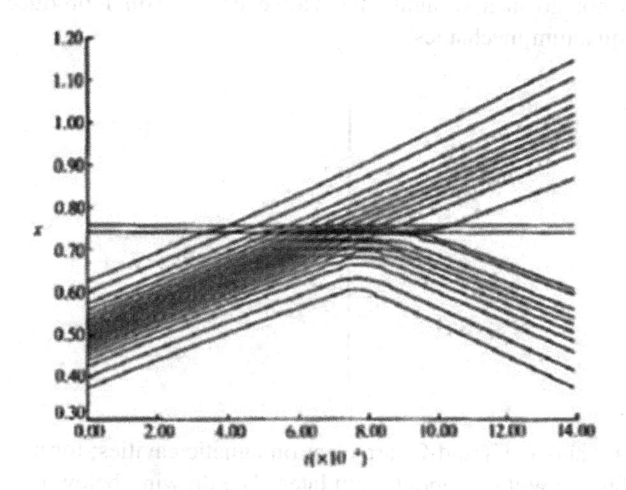

If interference is completely constructive in the upper exit, then the particle is carried by this interference and necessarily takes that exit. If it is destructive and constructive in the lower exit, the particle is guided towards the lower exit. In intermediate cases, you will find that the probability that it reflects upwards or downwards is exactly what is predicted by quantum mechanics.

We therefore have an object particle + wave that reproduces exactly what quantum mechanics says during successive reflections while keeping a trajectory that is perfectly defined at all times. Feynman's "*only mystery*" is no longer a mystery.

Let us go further and study more paradoxical trajectories. I suggest we discuss a classical interference experiment that you are familiar with, shown on the figure. What does Bohm theory predict in this case?

The fact that Bohm particles travel either through the upper slit or through the lower slit no longer surprises us. Indeed, when it leaves its source, the particle follows one of the two wave packets but not both at once. When the particle travels through one of the slits, as its wave function diffracts on the screen, the particle is perturbed indirectly and no longer goes in a straight line. It can thus be deviated as shown in the next figure.

When the wave packets from these two slits cross over, there is an interference phenomenon. This interference changes the probability current (Ψ* gradient Ψ) containing terms from the interference between the two wave packets. Consequently, the trajectory of the particle starts to oscillate in that region. A more detailed study then shows that the particle never crosses the horizontal symmetry plane P.

As you can see, we are in a rather amusing situation because it is unusual. We are all used to the situation where a particle travels in a more or less straight line in space (insofar as a quantum trajectory can be defined). It is not the case here: the trajectories are not at all rectilinear. The fact that there is interference means that the particle does not go in a straight line. However, we still reproduce exactly the positions of quantum mechanics.

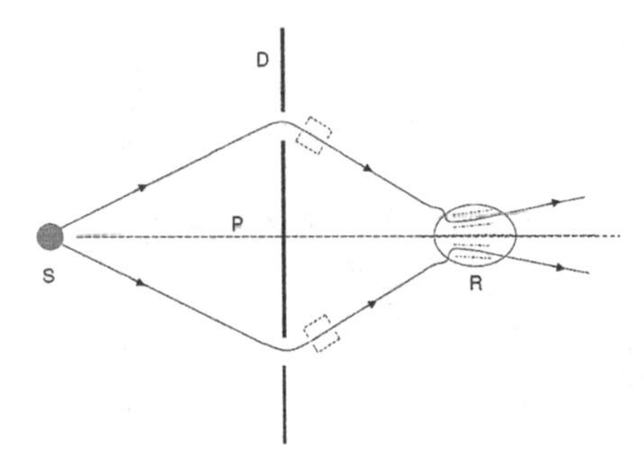

In the figure above, C_1 and C_2 are electromagnetic cavities; for now they do not play a role, but we will talk about them later. The drawing below is more realistic.

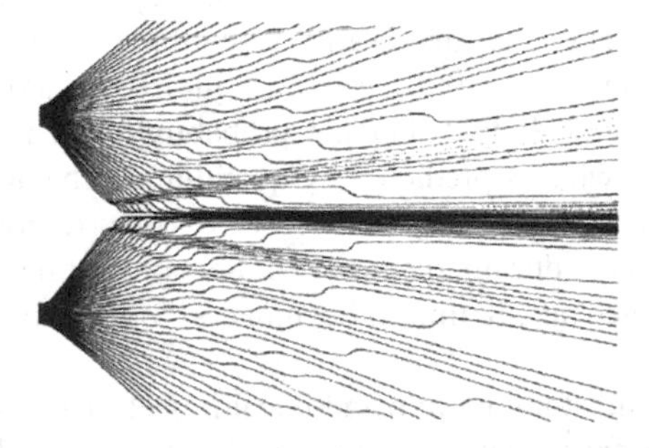

Édouard Brézin. Is that the solution for the variable R from the previous equations? Is that what is transcribed here?

Franck Laloë. Yes, we have resolved the Schrödinger equation, and then we calculated the trajectories of the Bohmian positions guided by the wave function, giving the result represented above. These oscillations are not “inventions” of Bohm theory, they exist in standard theory for the probability current. However, they are taken more seriously in Bohm theory.

Jean Michel Raimond. How does the standard discussion of complementarity turn out? Do we know which way the particle went? Do we know which side it exited from?

Franck Laloë. Once again, I must ask you to allow me to delay answering your question, as I intend to cover this topic later on.

During the next part of my presentation, I had planned to talk about the hydrogen atom. However, as this has already been discussed, there is no need for me to go over it again. For any time-reversal invariant system, we can show that Bohmian velocities are zero in the stationary states of the system. For the hydrogen atom, even when the quantum number m is non-zero, in the absence of a magnetic field we can superpose the wave functions m and—m to obtain real wave functions, and thus Bohmian velocities of zero.

Jean Michel Raimond. If we superpose them, they have no reason to obey the principle of correspondence.

Franck Laloë. Yes. There is no longer a correspondence principle between wave and particle; there is no need for it if trajectories already exist in the theory. It is enough to show that these trajectories have a good classical limit. There is no longer complementarity between wave and particle, only juxtaposition. We must give up a certain number of the notions we are used to having.

Jean Michel Raimond. What the correspondence principle says is that we must recover classical, or quasi-classical, trajectories. Bohmian trajectories should therefore be centred on classical, or quasi-classical, trajectories. Otherwise, quantum mechanics and Bohm mechanics would not agree.

Franck Laloë. Absolutely. In quantum mechanics, we deal with wave packets and show through the Ehrenfest theorem [10] that they move with good approximation following classical trajectories. Bohm theory is conceived from the start so that R follows the wave packets. Therefore, we automatically avoid all conflict.

Édouard Brézin. If I may, I would like to understand the ontology that Bernard d'Espagnat spoke of in this case. If I take an electron that is in a real state and immobile in the sense you described, I would like to understand why the electron does not fall on the proton, which would minimize potential energy. However, we know since Heisenberg that if it does not fall, it is because in reality the closer it would get to the proton, the greater its speed would be—that is the standard image—and under these conditions, we would lose potential energy and gain kinetic energy. This image which explains why there is equilibrium so that the electron does not fall and the hydrogen atom has a finite non-zero volume, this situation, is entirely due to this equilibrium between velocity and potential energy as explained by Heisenberg, and I see nothing of this image here.

Franck Laloë. The same equilibrium is due here to quantum potential. As an aside, the standard position operator of quantum mechanics is different from the Bohmian position (R). The Heisenberg ("uncertainty") relation between this operator and the momentum operator is not applicable to the Bohmian position (if we applied the Heisenberg principle, we would of course have infinite energy). What affects R is solely what is guided by the wave function, or if you prefer, Bohm's quantum potential. We must not apply to the Bohmian position what we usually apply to the position operator in state space.

Édouard Brézin. So, to my mind, the position of a particle is not really the Bohmian position.

Franck Laloë. Exactly: the position operator we are used to is not the Bohmian position.

Édouard Brézin. We must therefore imagine a complementary force.

Bernard d'Espagnat. There is no need to imagine it, it emerges from the wave function.

Alain Aspect. The electron is immobile and does not fall...

Bernard d'Espagnat. As Franck said, it ensues from the axioms of Bohm theory that the existence of the wave function—taken as physically real as Franck said—results from a *quantum potential*, a force generator like any potential. Calculations show that in the case we are considering this force compensates exactly the electric attraction. It is unexpected but simple and devoid of mystery.

Édouard Brézin. The equation for R shows the competition between the Coulomb potential and quantum potential. Is that right?

Bernard d'Espagnat. Yes, exactly.

Franck Laloë. Generally speaking, the danger here is to try to keep on applying the ways of doing things of quantum mechanics. If we try to equate to a one and the same object the Bohmian position and operators in Hilbert space, we will have problems. We must really accept that these are two different types of quantities: a position operator that always exists as in standard theory, and an additional Bohmian position. All that we take as standard in quantum mechanics is applied to operators, vector states, wave functions. I must admit that the notation R I chose for the Bohmian position, although it is traditional, is perhaps not the best one as it seems to generate some confusion with the position operator, which is often also designated as R.

Édouard Brézin. Fine.

Jean Michel Raimond. Anyway, as the predictions of both theories are the same, we will not catch one out like that!

Édouard Brézin. No, but I am simply trying to have a representation, as this is what dBB theory is all about.

Franck Laloë. All that we have said is general of course. If the Hamiltonian is invariant under time reversal, we can find it a base of stationary states whose wave functions are real, and which correspond to zero velocities at all points in space. We can thus construct a base of states where particles have no velocity; all stationary states are states where the Bohmian positions are static. Which is not to say (we will come back to it) that their correlation function is independent of time.

5.2.1 Multiple Particles

Let us take the example of two particles, which is the situation mentioned by Alain Aspect earlier. The variations of the Bohmian positions R_1 and R_2 are given by:

$$\frac{d}{dt}\mathbf{R}_1 = \frac{1}{m_1|\Psi(\mathbf{R}_1,\mathbf{R}_2;t)|^2}\mathrm{Re}\left[\frac{\hbar}{i}\Psi^*(\mathbf{R}_1,\mathbf{R}_2;t)\nabla_{\mathrm{R}_1}\Psi(\mathbf{R}_1,\mathbf{R}_2;t)\right]$$
$$\frac{d}{dt}\mathbf{R}_2 = \frac{1}{m_2|\Psi(\mathbf{R}_1,\mathbf{R}_2;t)|^2}\mathrm{Re}\left[\frac{\hbar}{i}\Psi^*(\mathbf{R}_1,\mathbf{R}_2;t)\nabla_{\mathrm{R}_2}\Psi(\mathbf{R}_1,\mathbf{R}_2;t)\right]$$

Of course, when the wave function Ψ is a product, it is possible to simplify the numerator and denominator, and we can instantly see that each particle evolves independently from the other.

Things become particularly interesting of course if the total wave function is not a product. Remember that at all times we have a single R_1 and a single R_2. What we are interested in, to know the evolution of R_1 and R_2, is the value of the wave function at that point in six-dimensional space, since that will give us the velocity. If the wave function is not a product, then the velocity of each particle will depend on the position of the other, as we must calculate the derivative at a point of six-dimensional space that depends on the two positions. We can start, to familiarize ourselves, with the case of a total wave function with only two constants.

$$\Psi(\mathrm{r}_1,\mathrm{r}_2;t) = \alpha\varphi(\mathrm{r}_1,t)X(\mathrm{r}_2,t) + \beta\varphi'(\mathrm{r}_1,t)X'(\mathrm{r}_2,t)$$

If at time t, one of the two wave functions is cancelled out at a point of six-dimensional space, only one of the variables of the second term will play a role so that the wave function of particle 2 disappears from the expression of the velocity of particle 1 at point R_1. Locality is therefore satisfied. However, if the two wave functions are simultaneously non-zero for the Bohmian positions, then we no longer have the same simplification. Non-local effects generally appear.

The important point is that, in situations where the two wave packets associated with a particle do not overlap, we are always in the first situation. It is then impossible to find Bohmian positions that are not cancelled out for at least one of the two terms of the superposition. We will always have a decoupled evolution between the particles. The wave functions of the wave packets of the two particles must overlap in order to not be in a situation where the evolution is decoupled.

The wave that plays no role when the wave packets do not overlap is what Bohm refers to as an "empty wave". It is still present in the Schrödinger equation but plays no role in the evolution of the system. It provides Bohm with his measurement mechanism. However, before that, I would like say a few words on spin particles.

5.2.2 Spin

Let us take the Pauli spin theory to have a calculation that is as simple as possible. Each particle is described by a spinner with two components. The velocity of the particle (there is only one velocity, not one for each spin component) is defined by the local probability current, thus the sum of the two velocities associated with each spin value.

We obtain an image where not only does the particle have a trajectory, but where we also have a spin direction since for each point R we have a spinner with two components that allow the calculation of the spin direction. The particle moves in free space, and the spin direction can rotate along its trajectory.

Let us take the example below where two wave packets cross and let us suppose that the wave packet coming from above has a spin up, and the one coming from below has a spin down.

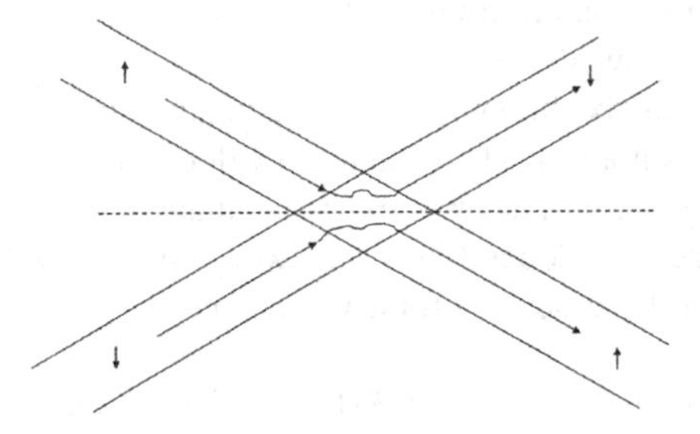

What happens in the interference area? We will have a type of bounce: because of interference, the particle with the spin up will turn its spin and bounce upwards, and vice versa. We are therefore in situations we are not at all familiar with.

Alain Aspect. And yet we obtain the usual outcome for the predictions of physics. If the spins are perpendicular, they do not interfere.

Franck Laloë. Absolutely.

Alain Aspect. At the end, if we look at the exit, they have not seen each other. Or at least, from the outside, it seems that way.

Franck Laloë. Exactly. We obtain the same outcome, including in the area of overlap of the wave packets.

Édouard Brézin. If the two particles are fermions, starting from the Pauli principle…

Franck Laloë. The negative interference of fermions will occur in the same way in dBB theory and in standard theory for the same reasons. The particles will follow the interferences induced by statistics and will reproduce exactly the Pauli principle.

5.2.3 The EPRB Experiment [11]: Emission in Opposite Directions of Two Spins in Singlet States

Franck Laloë. The experiment must be described in configuration space. We consider two particles, each with one R—this is the space of configurations R_1 and R_2 of interest here.

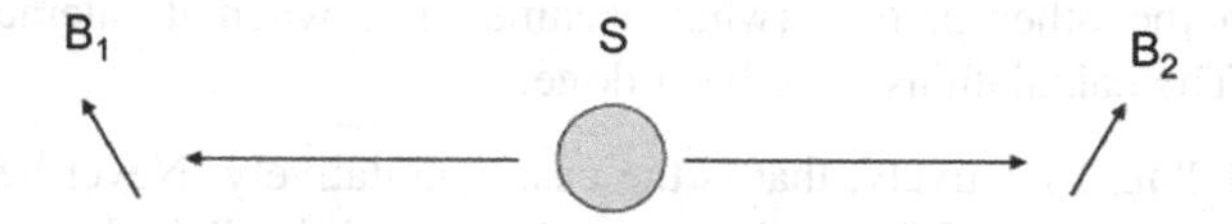

The two spin particles propagate towards the gradients of magnetic fields B_1 and B_2 created by two Stern-Gerlach magnets, which allow the measurement of the spin components. We suppose the wave function is not a product, a singlet for example. In that case, deviation of the first particle by B1 changes the position of the point in six-dimensional configuration space and reacts on the velocity of the second particle in a manner dependent on B1. The direction of B1 impacts the direction that particle 2 will take and conversely. We are therefore not surprised to see that we are able to reproduce quantum mechanics and violate Bell's inequalities.

Jean Michel Raimond. Nonetheless, it poses conceptual problems that seem nearly as critical as those of quantum physics.

Franck Laloë. The dynamic guidance equations of Bohm theory are explicitly non-local. If you introduce locality in your reasoning at this stage, then it becomes impossible to recover the results of quantum mechanics.

Édouard Brézin. Obviously. However, we can see in this image that what happens to particle 2 is very much dependent on what we did to particle 1 (e.g. if we placed a magnetic field on its path or not).

Franck Laloë. Absolutely. In standard quantum mechanics as well, if we apply the reduction of the wave function postulate, the measurement taken by Alice instantly projects the state of the spin for Bob. Thus, to my knowledge, no one is capable of providing a truly local standard quantum description of this experiment, whatever the interpretation. More precisely, if we follow John von Neumann, as I have just said, we will use a non-local reduction postulate. If we follow Bohr, we use the global description of the experiment, which is not an event of space-time and is therefore also non-local. The theory of consistent histories will not provide a more local description. To sum up, no one knows how to really describe this experiment in a completely local manner, from start to finish. Supporters of Bohm claim that one advantage of their theory is to emphasize this non-locality.

Michel Le Bellac. That said, the complete mathematics of this experiment, as for example in Peter Holland's book, is absolutely terrible. I was not able to get to the end.

Alain Aspect. Really? Even you, Michel? I do not believe it!

Michel Le Bellac. The equations are absolutely dreadful.

Bernard d'Espagnat. There is in Bell's book, *Speakable and Unspeakable in Quantum Mechanics*, as you probably know, a rather explicit treatment of this problem of Bohm theory. Bell's calculations show that the second particle, the one that interacts later on with its apparatus, obeys not to its hidden variable but to what happens with the other particle (what became of it when it interacted with its instrument). The calculations have been done.

Michel Le Bellac. Effectively, that is the case qualitatively. Nevertheless, finding the minus cosine factor of the angle is very hard work in dBB theory.

Franck Laloë. I disagree. If we want to recover the results of quantum mechanics in Bohm theory, there is no need for additional calculations. We know this in a general manner, the usual standard calculation gives the average of all initial positions. However, if we want to go further and do like Holland, i.e. show explicit trajectories and therefore go beyond standard quantum mechanics, then additional calculations become necessary, and may be difficult. If we simply want to recover what is already known in standard quantum mechanics, then there is no problem and it requires no additional efforts.

Alain Aspect. What you have just said it is a tautology. If we want to recover quantum mechanics, as we started from quantum mechanics, then there is nothing more to do.

Michel Le Bellac. I am not convinced.

Michel Bitbol. To clarify: you said there was no truly local description of this correlation phenomenon. However, under certain conditions, this difficulty can be overcome. Matteo Smerlak and Carlo Rovelli will give us a presentation at a later date, in which they will claim that quantum mechanics does indeed provide a form of local description. To arrive at this conclusion, they will call upon a truly extreme form of operationalism. We will have the opportunity to discuss this [12].

Édouard Brézin. And thus causality is abandoned from the start.

Michel Bitbol. Exactly.

Bernard d'Espagnat. On the other hand, the theory of Smerlak and Rovelli implies a weakening of reality.

Michel Bitbol. Yes, that is right.

Bernard d'Espagnat. Whereas Bohm theory is completely realist. It is a theory that describes nature as it is, independent of our knowledge.

Franck Laloë. Let us study another simple case:

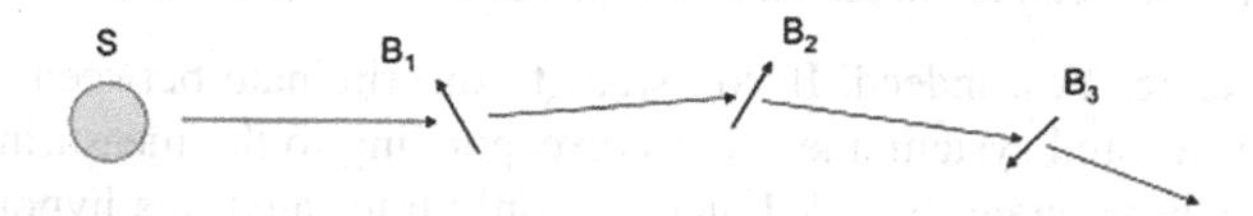

We measure the spin of a particle according to the successive directions B_1, B_2, B_3, etc. As we have seen (retrodiction), each time we take a measurement, from Bohm's point of view we refine the information on the initial position of the particle. Thus we could think that, after many measurements, this information would be so precise that we would be able to predict with certainty the outcome of ulterior measurements. Is that really the case? The answer is of course no: no matter how many measurements are taken, the outcome of ulterior measurements will always be random and congruent with standard quantum predictions. The reason is that in dBB theory the more you increase the number of measurements, the more chaotic the situation becomes. We arrive at a classical chaotic situation where sensitivity to the initial conditions increases with the number of measurements taken. Thus, ultimately, we recover the impossibility to ever predict with certainty the outcome of a subsequent measurement.

5.3 Measurement

Let us apply the von Neumann model of an ideal measurement. Let us call the wave function of the measured system Φ and the wave function of the measuring apparatus X. I have included a large number of positions as I suppose that there a great number. After measurement, the total wave function is a sum on j:

$$\Psi(\mathrm{r}, \mathrm{r}_1, \mathrm{r}_2, \ldots, \mathrm{r}_N; t) = \sum_j \Phi_j(\mathrm{r}, t) X_j(\mathrm{r}, \mathrm{r}_1, \mathrm{r}_2, \ldots, \mathrm{r}_N; t)$$

I describe here the quantum state of the measuring apparatus and the measured system after their mutual interaction—they are therefore entangled. This state is a sum on the possible states of the measuring apparatus corresponding to the different positions of the pointer. We know the positions R_1, R_2, etc. within configuration space must necessarily fall within the area where one of the elements of a superposition is non-zero. However, the different functions X have disjointed supports. It becomes obvious when we recall that for two functions of N variables to not have spatial overlap, it is sufficient that the two supports of *any* of these variables be disjointed; in this case, there is necessarily a position variable, for instance that of the centre of the mass of the needle of the measuring apparatus that, after interaction between the system and the apparatus, finds itself in a given interval of the dial and in no other. At time t only one X is not non-zero. The Bohmian position in configuration space cannot make two X non-zero at the same time.

Michel Le Bellac. Yes, indeed. If you want to discriminate between the different states of the measured system, the states corresponding to the measuring apparatus must be spatially separate. Franck Laloë has only translated this hypothesis.

Alain Aspect. I agree with you on this point.

Franck Laloë. We arrive at a situation where all bar one of the waves at the exit correspond to states, and are “empty waves”. In my opinion, this is one of the strengths of Bohm theory: we have demonstrated, in a way, the reduction of the wave packet postulate. We do not reduce the wave function, however all the other branches play no role and are empty.

From Bohm’s point of view, there is a single position of the system in configuration space, even if we do not know what it is. There is a perfectly precise position, and with it, only one member of the solution can play a part. The others still exit (as in Everett’s theory) but we leave them in limbo in a virtual state; they no longer have a role in physics.

Thus, Bohmian dynamics renders the postulate of the reduction of the vector state an effective rule, a convenience. There is no need to introduce two different postulates for quantum dynamics. The measurement process is a completely ordinary process of physical interaction, which reintroduces a unity in the theory and avoids any distinction between the measured world and the world that measures.

This property of Bohm theory also guarantees contextuality: each ensemble of experimental set-ups acting simultaneously on the measured system is associated with a different dynamics, thus a different “empty wave” formation. From this point of view, what Bohm says is rather Bohrian if you think about it, since what matters is the whole of the measuring apparatus, including all the associated Bohm positions.

To sum up differently this great success of dBB theory: with it, decoherence [13] automatically guarantees macroscopic uniqueness, unlike standard theory where it is necessary to apply a postulate to eliminate the vector branches.

Hervé Zwirn. Then we no longer need decoherence, in a way. This is rather troublesome.

Franck Laloë. Why troublesome, since the predictions are the same? As in standard quantum mechanics, we must take all the variables into account; if we forget this, we arrive at the same difficulties as in quantum mechanics.

Hervé Zwirn. The analysis of the measurement problem conducted here is the same we traditionally conduct in quantum mechanics. We come across a stumbling block since the von Neumann chain [14] never stops. To call upon the environment does not resolve difficulties. However, we can come out of the von Neumann chain without invoking the environment with the simple postulate that the supports being disjointed, we only have one outcome. Thus we no longer need the environment. That is rather strange.

Franck Laloë. I would rather say that, in dBB theory, the environment made up of the measuring apparatus is sufficient to ensure macroscopic uniqueness. There is no need to go further and call upon "*pointer states*" or "invariance" like Wojciech Zurek [15] has done. However, the usual rule remains the same: each time the environment plays a role in quantum mechanics, it will also play a role in Bohmian mechanics. Put simply, we always need the environment, but we no longer need it like a *deus ex machina* that would magically produce a macroscopic uniqueness.

Hervé Zwirn. That is it. Thus, at a push, without an environment, we arrive at decoherence.

Jean Michel Raimond. The measuring apparatus with its configuration space includes the environment, but we no longer strictly need it.

Franck Laloë. If decoherence is the loss of coherence, then we do not need to use Bohm's quantum mechanics to understand it. Any interpretation accounts for it easily. It is the uniqueness of the result, the macroscopic uniqueness, which emerges from Bohm's mechanism. We find it very hard to come out of the uniqueness of decoherence, and even, in my opinion, we are not able to it.

Michel Bitbol. That is very interesting. In standard decoherence, we can lose the effects of interference without the appearance of uniqueness. We have not yet demonstrated (perhaps we never will) that decoherence can make uniqueness emerge from the experimental outcome. In Bohm theory, conversely, we can have uniqueness without loss of coherence. That is remarkable.

Franck Laloë. Absolutely!

5.3.1 "Surrealistic" Trajectories in DBB Theory

Franck Laloë. Let me introduce the surrealistic trajectories of dBB theory by presenting another interesting experiment, similar to the previous one, and described in a very interesting article by Englert et al. [16] which Zurek mentioned to me—which he considered, wrongly in my opinion, to be a refutation of Bohm theory.

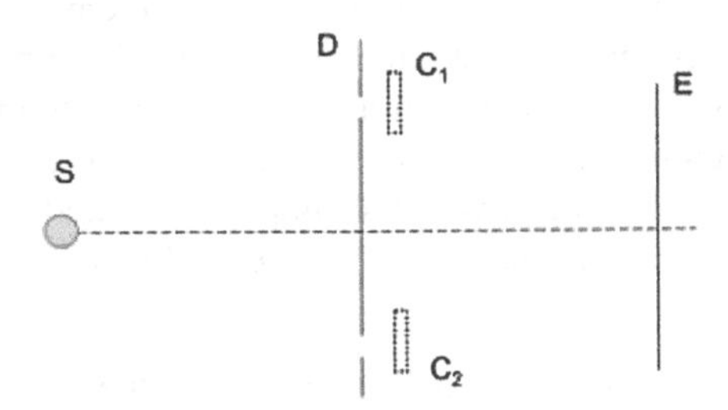

It is an experiment that allows a symmetry plane P. We have Rydberg atoms propagating from a source S, passing through two slits in a diaphragm, and eventually interfering on the points of a screen E. We have small electromagnetic cavities where these Rydberg atoms may lose a photon. We thus have a "*Welcher Weg*?" (which way?) experiment. If a photon is deposited in a cavity, we can measure it, and in practice know through which arm the particle went, so that interference disappears.

What Englert, Scully, Süssman and Walther did was to discuss this experiment and show that, if we study it in detail, dBB theory leads to what they called "surrealistic" trajectories—and, I think, they used the term pejoratively.

Let us return to the framework of standard quantum mechanics for a moment. It predicts that there are situations where the atom deposits a photon in a cavity C_1 and, after all, since it has been diffracted by the slit, it can happen that the photon is left in C_1 and the atom falls to the lower part of screen E. This can happen, perhaps not every time, but such events can occur.

Now, how is this experiment described with dBB theory.

5.3.2 The Same Experiment Seen from DBB Theory

You recall there is a no-crossing rule that precludes Bohmian trajectories from crossing the symmetry plane. It is easy enough to demonstrate: in quantum mechanics, the probability current always stays in the symmetry plane and thus never has components that are perpendicular to this plane. Thus particles that come near it can never cross the plane; they can only bounce.

What do Scully et al. say? If the predictions of standard quantum mechanics are verified, then there are situations where a particle leaves a photon in the upper cavity, but terminates its course below the symmetry plane P. Its trajectory has crossed the symmetry plane which is impossible in Bohm theory, and therefore Scully et al. have concluded that dBB theory introduces surrealistic trajectories. The argument is fitting and interesting.

However, we will see that this situation is paradoxical only if we place ourselves within a mixed theory, i.e. if we want to be both “standard” and “dBB”. Combining these incompatible elements of logic in that way cannot work.

A first comment that seems so obvious and almost trivial that I hesitate to bring it up: we have fallen precisely in the trap that Bell pointed out, i.e. the point of view whereby Bohm trajectories are real but not the wave functions, which remain probability amplitudes like in standard quantum mechanics. A Bohmian purist would retort instantly that the wave function is in fact a field endowed with reality and the simple fact that the wave function has an effect in the cavity has suppressed the existence of the symmetry plane and thus the impossibility of crossing it. It is the first answer that comes to mind, but it still shows that there is a logical incoherence in the attack.

We can go even further. Indeed, we must take into account the mechanism of “empty waves”, which as we will see makes the “surrealistic” trajectories disappear. In fact, the conceptual difficulty introduced by Scully et al. relies on the reasoning whereby, in quantum mechanics, we consider the cavity fields in a quantum manner (with the aim of performing a partial trace and showing there is no interference), whereas in dBB theory there is no variable associated with this field. However, this asymmetric treatment is an error of logic. In fact, this criticism refers only to a truncated version of Bohm theory, the one where the fields are not allowed to have a Bohmian variable.

What we must do if we want to use the real dBB theory in this experiment is to assign a variable position to the cavity fields. Then, the effect of the “empty waves” I spoke of earlier occurs, and a single wave radiated by the two interference slits plays a role. The particles are no longer stopped from crossing the symmetry plane, and the “surrealistic” trajectories disappear.

What I am trying to highlight is that if we perform a standard quantum calculation where we take into account the quantification of the cavity fields, then it is illogical to carry out a dBB treatment without also taking into account the dBB field variables. The rule is simple: each time quantum variables are introduced in standard mechanics, we must introduce the corresponding dBB variables.

Jean Michel Raimond. In concrete terms, if you excite the oscillator, you reintroduce the right to cross the symmetry plane. This gives you part of the wave function where we are allowed to cross the symmetry plane. The photon restores the possibility of crossing.

Franck Laloë. Absolutely. That is a good way of putting it.

Michel Bitbol. That said, we still have a difficulty. It is not a question of completely refuting dBB theory, of course, but this theory raises (at least) one philosophical problem. With this theory, the particle can be given two profoundly different, not to say incompatible, ontological statuses. The first status, which Bohm wanted to restore in his first theory, is the idea of a small corpuscle that moves with continuity along its trajectory and with permanence in its existence.

The second ontological status of the particle is that of a field excitation mode, which was favoured in the later developments of Bohm theory.

This duality of status cannot be perceived as problematic by Bohm supporters, but it seems to me that it weakens their initial claim whereby they re-establish the classical concept of the particle with an uninterrupted course and a spatio-temporal continuity.

Franck Laloë. I would say this differently. If we use Bohm theory, we must associate the trajectories with the usual particles. However, if we use dBB field theory, then we must allow the electromagnetic field in the cavity to have a Bohmian variable. It will no longer be a position but an electric field. We must consider this Bohmian variable of the electric field correctly. Why should the introduction of additional variables be limited to particles, to the exclusion of fields? If we widen the quantum system by restricting the Bohmian variables to only a part of the quantum system, it does not work. It is not appropriate—and we fall outside of Bohmian logic. The same incoherence occurs in standard quantum mechanics if we consider one part of the experiment classically and another part in a quantum manner.

Bernard d'Espagnat. I understand Michel Bitbol's comment in the following way: since we speak of the creation of photons, we are compelled to work within field theory. We should then do this completely, meaning we take the relativist Bohm theory that, I believe, you will speak about shortly, and we abandon the idea that particles are small corpuscles. We must abandon this and hold in its place the idea that particles are field quanta.

Franck Laloë. Exactly. If we take a cavity, we do not assign a position to each state of the cavity but a value to electric field.

Alain Aspect. We have therefore lost the "agreeable" part, if I may say so, of the initial vision.

Bernard d'Espagnat. Yes. We have lost what was agreeable. And then, aside from any question of agreeableness, we have changed ontology. We have gone from an ontology of corpuscles to an ontology of fields. Considering this is a theory whose most agreeable feature was that it was ontologically interpretable, this is disconcerting! What will be the next development? What is considered real? As far as I am concerned, it is this oddity that fuels the most my scepticism regarding this theory. Unless we can keep at the same time the intrinsic reality of particles and that of the field. However, considering the creation and annihilation phenomena, this may be difficult.

Alain Aspect. I rather agree. In that case, we can ask ourselves whether the descriptions are really more comfortable than those of usual quantum theory.

Franck Laloë. The fact that the "Welcher Weg" mechanism is due to a field or a particle makes no difference in this thought experiment. To avoid opposing field and particle, let us replace the field in each cavity by a particle that is deviated by the passage of the test particle; we thus have another mechanism allowing us to reiterate the argument in the same way. Each additional particle has obviously then a Bohmian position, and exactly the same phenomenon of "empty waves" occurs: the surrealistic trajectories disappear. Whether it is a field or a particle makes no difference. The only thing we must not do when we introduce a quantum object in standard theory is to do so without giving it a Bohmian variable. You are either not at all Bohmian or completely Bohmian, but you cannot be somewhere in between.

Another comment is that in standard theory, we always take great care to explain to students that a photon has no position. The goal of dBB theory has never been to reintroduce a position for all particles, including photons.

Jean Michel Raimond. For now, nothing in Bohmian theory is any more agreeable than in usual quantum mechanics. If I may say so, Englert's experiment is completely trivial from the point of view of quantum physics. It is nothing more than a facet of the "Welcher Weg". Whereas here, I suppose you need five pages of calculations to find that a particle can cross the axis. I have not found this, up to this point, to be any more comfortable.

Franck Laloë. I think it is sufficient to realize that the "empty waves" mechanism comes into play, and this does not require any calculations in particular.

Alain Aspect. There is still something agreeable and pleasant in all this. If it were not for the complications in calculating the trajectories mentioned by Michel Bitbol, then we would be willing to forget all that quantum mechanics says. We would consider that we have launched a particle, that it went one way or the other, but that whatever happens, it will be guided by valleys and will always end up on shining fringes. There is a palpable aspect that is rather agreeable. However, if we need to come up with calculations that are at least as complicated as those of quantum mechanics, then we lose this agreeable aspect. Is that what you meant?

Michel Bitbol. Exactly.

5.3.3 Temporal Correlation Functions

Franck Laloë. I will now come back to the hydrogen atom, even if I need to speed up through lack of time. We consider a hydrogen atom, or more simply a one dimensional harmonic oscillator. Its wave functions are real, so that the particle does not move in Bohm theory. However, standard quantum mechanics predicts that the correlation function of the position of the particle is an oscillating function of time. Is there not a contradiction? As a result, a number of articles, including that of Michele Correghi and Giovanni Morchio: "Quantum mechanics and stochastic mechanics for compatible observables at different times" [17], consider that dBB theory is not equivalent to standard quantum mechanics with regards to correlation functions.

In fact, when we calculate a correlation in dBB theory, we must point out that it is a correlation between two measurements. The first measurement necessarily introduces position variables associated with the measuring apparatus and necessarily introduces the mechanism of "empty waves". From then on, the particle sees this functioning and begins to move. The end result is that the correlation functions are exactly the same in dBB theory and in standard quantum mechanics, but we should never forget to take the measuring apparatus into account.

I would like to conclude with a different perspective, and explain why, in my opinion, this theory does not fulfil the initial programme we could have assigned to it. What we have seen up to now is that each difficulty encountered in Bohmian mechanics is also present in standard quantum mechanics. I therefore consider both points of view in the same way, without any preference. What is essential is to ask everyone, when they take one point of view, to do so in a coherent manner and not mix up the two.

So the question is now: do we achieve in dBB theory a more satisfactory point of view where "things are real" and where theory is truly ontologically interpretable? I am not convinced.

Bernard d'Espagnat. You have not spoken about the relativist generalization, i.e. Bohmian field theory. The common objection that we find in the literature against Bohm theory is that it is not relativistic. In fact, you pointed out to me some articles where the authors dealt with the problem of fields in Bohm theory, even better than Bohm did himself back in 1962. You even told me that in this theory we could consider that either the magnetic field is real, and thus the electric field is not, or vice versa.

Franck Laloë. As time was ticking away, I skipped the last slides to get to the conclusion!

Bernard d'Espagnat. It would be worth mentioning, if only to contradict those who think Bohm theory is not amenable to relativistic generalization. If I understood you correctly, their claim is not exact.

5.4 Bohmian Field Theory

Franck Laloë. Let us say a few words on dBB field theory. I will not pronounce myself on field theory in general as I am not enough of a specialist. What is clear is that we can perform an electromagnetic theory in Bohm theory relatively easily, on the condition we accept, as I said earlier, the introduction of additional variables associated with the fields.

What is unsatisfactory, or at least what does not satisfy me, is that from then on, we can choose either the electric field E of the magnetic field B, without any good reason to choose one over the other. We can see that the position of the particle is favoured over the momentum, as it is more linked to space than the momentum. Regarding E and B, which play a more symmetrical role, making a choice is more complicated. We cannot choose both at once; we must favour one or the other, or a linear combination.

Alain Aspect. Yes, because we must take one of the two variables.

Franck Laloë. Yes, we can take **E** as equivalent to **R** or **B** as equivalent to **R**—but not both. This introduces an asymmetry between the two components of the electromagnetic field which is not very satisfactory.

Bernard d'Espagnat. Someone who really wants a "pure" realist ontological theory can nevertheless decide that he prefers a theory such as this one where we choose randomly that it is, let us say, the electric field that is real rather than a theory that is restricted to human representations of reality. With this choice we have a description of nature as it is and as it would be even if we had never existed (since like classical physics, this description does not refer to measurements, instruments, etc.). I can imagine that a rational realist, namely one that wants an ontologically interpretable theory, someone like Bell, would have considered that for want of something better, this theory would do.

Franck Laloë. In any case, what is important, whatever his personal choice, is not to say that this realist position is impossible. We know it is possible. We can dislike it, it is a matter of personal taste, but this image exists and we can construct it explicitly.

Bernard d'Espagnat. If we reject it and similar images, and if we are really attached to Schrödinger's formalism (the unmodified relative quantum formalism), we never recover the realism I was speaking of. John Bell has always said this, and I think in this he was right.

5.5 Reservations. Two-Tiered Reality in DBB Theory

Franck Laloë. What seems to me to be even more difficult in this theory is that we must postulate that it is impossible to carry out a selection beforehand within the "quantum distribution". Otherwise, it would no longer be equivalent to standard quantum mechanics.

Worse than that, we can show that selecting additional variables would lead to situations where we could transmit signals faster than the speed of light [18]. That is relatively catastrophic—unless we no longer believe in relativity!

We must therefore postulate in a fundamental way that these additional position variables are and always will be completely impossible to preselect, choose and manipulate.

5.5.1 *Two Types of Quantities*

Franck Laloë. To conclude, there are two types of physical quantities in the new realist universe of dBB. The positions are directly observable but not directly manipulable. We can see, by the way, that calling them "hidden variables" is absurd since they are in fact the variables we can see in the experiments. They are observable, but human beings are not allowed to manipulate them, otherwise there would be a contradiction with relativity.

In addition, there is another type of reality that corresponds to the wave functions that have become the Bohmian pilot-waves. These waves are not directly observable. We can see them only through their effect on positions. However, they are manipulable: if we change the electric field, or if we change the Hamiltonian, we obtain a Schrödinger equation that shows how the wave function will change.

Ultimately, we had wanted to get rid of one duality, but we find ourselves with another duality imposed by the coherence of this theory. It is a two-tiered reality, with one visible but not manipulable level, and another hidden (indirectly visible) level which is the one we manipulate during experiments. I find that personally, in a way, the unification and simplification programme of dBB theory is not completely fulfilled and thus loses part of its appeal.

Alain Aspect. Are you saying that ultimately this is a lot like the quantum physics we know?

Franck Laloë. That is my pessimistic conclusion.

Jean-Pierre Gazeau. I have a technical question. In fact, we could have started with the Schrödinger equation for wave functions dependent on momentum and time. Could we not have followed the same path by introducing a momentum that would play the role of position R, with the subsequent field that combines so that when a field poses a problem, another field is there to solve it, and conversely?

Franck Laloë. We can indeed conceive a symmetrical dBB theory, but based on momenta rather than positions.

Jean-Pierre Gazeau. That's it.

Franck Laloë. However, if we want to have position *and* momentum, then a problem arises… We need to choose.

In conclusion, is this theory conceptually simpler that conventional quantum mechanics? The answer may be a matter of personal taste. I do not know the answer.

We realize that many components of dBB theory are shared with standard quantum mechanics. One thing is certain, the "empty waves" mechanism is splendid. It links decoherence and uniqueness. That is extraordinary! That being, unfortunately, the programme is not entirely fulfilled.

In any case, this theory provides an extremely useful alternative point of view. For example, without it there would not have been Bell's theorem and there would be many aspects of quantum mechanics we would still not know.

Bernard d'Espagnat. That is absolutely true. In fact, Bell first demonstrated that von Neumann's theorem did not apply; that there could be hidden variables, which comforted him in his interest in Bohm theory. Afterwards, he studied this theory and realized what we were speaking about earlier in the case of a two-particle system, namely that there was non-locality. He found that the second particle no longer obeyed its own hidden variable but what had happened "far away" to the first particle. It is within the framework of Bohm theory that he found this. And this led him to wonder whether this was not a "dirty trick" of Bohm theory or whether this was general. He studied the question and discovered it was general.

Alain Aspect. I have a completely erroneous vision of the way things unfolded. I thought it was the other way round, i.e. that he first discovered the inequalities and then he investigated Bohm theory.

Franck Laloë. In a footnote in another article on the refutation of von Neumann, he specifies that if we calculated the dynamics of the second particle, we would find exactly what Bernard d'Espagnat has said. It would be interesting to know if he already had his theorem in mind at that time, when he broached the question of non-locality [19].

Alain Aspect. I will re-examine this point in that way.

Bernard d'Espagnat. This is not of great importance, but is interesting nonetheless.
Alain Aspect. It is historical.

5.6 Discussion

Hervé Zwirn. To reiterate the conclusion and address the issue from the angle of the advantages or the interest we could find in Bohm theory, I believe there are two aspects that are put forward by those who prefer this theory. It is clear that Bohm theory does not allow us to re-establish the usual classical realism. I believe that nowadays, everyone agrees to permanently abandon the realism of usual classical mechanics. If only because non-locality is effective, that contextuality is called into play and that we can therefore give up trying to re-establish reality in the intuitive sense of the term.

The two advantages put forward by its supporters are first of all that beyond the need to expand the notion of realism to non-locality and contextuality, it is possible to express this theory without referring to an observer—which quantum mechanics does not allow. The first thing would then be to say that with Bohm theory, we can remain realists (in a different sense from classical realism) since we can forgo the observer.

The second advantage that is put forward is determinism. Supporters of Bohm theory consider it to be deterministic whereas quantum mechanics is not. Indeed, abandoning the principle of the wave packet reduction…

Jean Michel Raimond. Is it deterministic, when we have an initial probability distribution?

Hervé Zwirn. *Modulo* this distribution!

Jean Michel Raimond. Franck Laloë has insisted on the fact that we should not forget this point.

Hervé Zwirn. Of course. It is an initial distribution. It must be considered as such. I am not an advocate of dBB theory. But given this initial distribution, its supporters can consider that everything is deterministic. It obviously does not allow predictions but, from a conceptual point of view, this theory seems more satisfactory than quantum mechanics where an essential lack of determinism cannot be eliminated.

Jean Michel Raimond. He plays dice from the outset rather than during the experiment.

Hervé Zwirn. God only plays once, whereas in quantum mechanics, he plays all the time!

Édouard Brézin. You are right. We still maintain the fundamental notion in quantum mechanics that is recovered in the initial distribution: unlike anything we have known up to that point, the introduction of probabilities is irreducible. It is not a convenience used to deal with complexity. For example, as far as I know, we can perform the probabilities calculation on hydrodynamics, on a complex system, but we have no doubt that we could replace this with complex dynamics that are troublesome to solve and that we do not even need. In the same way, introducing probabilities when rolling a die is only for the convenience of calculation, it is not a necessity. In quantum mechanics, it is a necessity that subsists in the formalism that has just been described.

Bernard d'Espagnat. Where does it subsist?

Édouard Brézin. It subsists in the sense that there is no other method for reducing or eliminating the notion of probability and replacing it with a complex notion in which we introduce variables. Let me go over again the difference between quantum mechanics and rolling a die: when rolling a die, we do not doubt that we could replace the 1/6th by a description of the set of degrees of freedom where we would describe correctly the momentum of the object, the number of times it will spin in the air and its impact on the table. This would allow us to solve the motion equations. We do not do this because we do not need to. But we could classically eliminate the probability calculation. Only in quantum mechanics can it not be eliminated. It is fundamental and irreducible; it is that which subsists.

Bernard d'Espagnat. I would say that if it cannot be eliminated in quantum mechanics, then a Bohmian, believing in "hidden" variables, would reply that it is simply due to our human incapacity. We are not clever enough to do this. However, a "superphysicist" with access to these variables could use Bohm mechanics to predict with certainty, in all cases, what will be observed. We do not know these hidden variables because we are "mere animals" not refined enough to grasp them. But they exist. We could image a Laplace's demon that would be capable of knowing them.

Édouard Brézin. I do not think so.

Bernard d'Espagnat. I do not want to go too far. You are quite right to highlight the importance of the problem of the initial probability. In dBB theory, even a Laplace's demon, knowing the values of the hidden variables, would fail to predict the future with certainty if by some unfortunate event it would "comes across" a universe where the initial distribution is not defined by the squared module of the wave function of that universe.

Édouard Brézin. It is irreducible. It is in this sense that I do not believe we can do what you have just said, i.e. a being more powerful than us that would have an infinite computer who would replace the probability calculation with a non-probabilistic calculation—we can do this with a coin, but not with a quantum system, or even a Bohmian system.

Bernard d'Espagnat. If it knew the hidden variables…

Hervé Zwirn. There is still a difference. What is still probabilistic in Bohm theory is the initial state. Whereas, once we have the initial state, it becomes deterministic. Therefore, as Bernard d'Espagnat said, we could in theory calculate everything. A similar thing happens with the die. The initial state of the die should be known with infinite precision, considering its sensitivity to the initial conditions, for us to be able to make predictions.

That said, there is a significant difference between quantum mechanics and Bohm theory. Notwithstanding the initial state that remains probabilistic in Bohm theory, the rest of the process is deterministic. Whereas in quantum mechanics, it is indeterministic at all times. This is an important difference.

Alain Aspect. What do you mean by that?

Hervé Zwirn. Whatever precise knowledge we have of the state of a system in quantum mechanics, the state described by the wave function (and we think there is nothing else), there is for each measurement an indeterminism we cannot eliminate; whereas in Bohm theory, indeterminism is linked to the fact that we cannot know the initial distribution. But with a given initial distribution, there is no indeterminism. Everything is deterministic.

If we say that indeterminism is linked to the fact that we do not know the final state when we know the initial state—which can be defined as essential indeterminism (we know the initial state, we take a measurement and we cannot predict the final state)—then quantum mechanics is essentially indeterministic and Bohm theory is not.

Jean Michel Raimond. That is partially true, because each time we introduce a new measuring apparatus, we must also introduce its probability distribution and its positions. Each measurement adds its own layer of indeterminism.

Édouard Brézin. I agree.

Jean Michel Raimond. We simply push back the probability to the beginning of the calculation instead of at the end, but in my opinion, it is conceptually identical. Each time we add a measuring apparatus, it has its own probability distribution.

Hervé Zwirn. I agree. It is true. However we can go one step further and consider that if we cannot predict the outcome with Bohm theory, it is not because the process is intrinsically indeterministic but because it is impossible for us to know precisely the initial state. Whereas in quantum mechanics, the situation is a bit different: we know the initial state as precisely as is possible, since there is nothing else, but despite that, we cannot predict the final state.

Jean Michel Raimond. I forgo determinism, but I do so in a different way. We agree. I am not more comfortable with Bohm theory!

Hervé Zwirn. It is true that we are no further ahead!

Alain Aspect. I would like to mention that someone like Nicolas Gisin would say that it still changes certain things on his quantum random number generators. Security relies on the idea that it is fundamentally indeterministic since it is quantum, whereas all algorithms, as complex and sensitive to initial conditions as they are, but which have a deterministic vision in the sense you describe, could in principle be broken one day.

Jean Michel Raimond. No, since we cannot determine the initial distribution, otherwise we would violate relativistic causality.

Alain Aspect. I had forgotten that we could communicate faster than the speed of light!

Bernard d'Espagnat. The discussion we have just had on determinism is indeed interesting. In my opinion, the main interest of Bohm theory does not lie in the fact—which is debatable as you have shown—that it restores, or would restore, determinism but, I repeat, in the indisputable fact that it recovers realism in the strong traditional sense. I have not quite grasped the reasons why physicists who claim to be realists (and I believe it is the majority of us) do not follow John Bell in his idea that quantum mechanics is unsatisfactory in this regard and that, consequently, for them the last resort seems to be dBB theory, or a theory of that kind. I would like to understand the reasons better. I therefore tell myself that it would be good and above all very interesting to debate this point.

However, I do not want to impose my point of view and I would like to know what you think. Michel Bitbol, you have strong arguments against Bohm theory, but we know you are not a realist.

Michel Bitbol. Indeed. For a start, there are no arguments against the ability of dBB theory to account for all the phenomena predicted by standard quantum mechanics. I believe we agree on this point: there is no way to fault Bohm theory in its predictive concordance with standard quantum mechanics. However, we can put forward more philosophical arguments. Franck Laloë has given some excellent examples. We can add a few more. That is the only point where I think I can possibly contribute next time during a discussion on Bohm theory.

Alain Aspect. My schedule allowing, I will try to attend the next session. But I would like to start refuting now the last part of your claim. If we are realists and follow John Bell in the fact that he is uneasy with quantum physics as we know it, we are not forced to come to the conclusion that the only "escape route", if I may call it that, is dBB theory. We can imagine that there would be others. There is no demonstration of the fact that it is the only possible one.

Bernard d'Espagnat. That is true.

Franck Laloë. There are others, in particular those given by the theories with modified and stochastic Schrödinger dynamics (GRW and Pearle).

Alain Aspect. We can therefore call ourselves realists and consider this theory, which has perhaps provided the sticks with which it is beaten by being too precise and by allowing in-depth scrutiny, ultimately not that agreeable. It does not imply, however, that we give up being realists.

Bernard d'Espagnat. That is true. It is a very reasonable point of view. Besides, there are other ontologically interpretable theories, such as the theories with modified Schrödinger equations that Franck mentioned, which are very interesting.

Jean-Pierre Gazeau. By adding non-linear elements?

Franck Laloë. And/or noise. It is another point of view where the world is described by the wave function, which is real.

Édouard Brézin. What about the theory, I do not know it, of Murray Gell-Mann and James Hartle, restated by Roland Omnès?

Bernard d'Espagnat. A number of us have issued some criticism towards it. It is less agreeable now than when it was first proposed. We could also discuss it at some point.

Lena Soler. I have been listening from the beginning with a slightly external point of view, since I trained as a physicist a long time ago, and have since turned towards philosophy. You have said "not very agreeable", "not very comfortable"… I think these are feelings we can have when we have been trained, in a certain way. I wonder to what extent these feelings are not strongly determined by the scientific syllabus, and more precisely by the fact that Bohm theory is completely left out of the scientific syllabus. I am rather surprised to see that while you are all extremely experienced physicists, you are still discovering certain aspects of this theory.

I wonder if it would not be worth investigating this factor. In this respect, I would like to mention a book by James Cushing published in 1994 entitled *Quantum Mechanics: Historical Contingency and the Copenhagen Hegemony* [20]. The book discusses the main arguments in favour of Bohm theory, and puts forward the thesis that contingent historical factors that have made this theory mostly unacceptable nowadays and rejected by physicians. If we were to continue this discussion, which would interest me greatly, it would be a book to consider.

Jean Michel Raimond. I do not think we are rejecting this theory. We do not know it well, it is true. However, what we consider when we say that a theory is agreeable or not, is whether it has a different philosophical content. Apparently, if we set the initial probability, it is not obvious that we gain with determinism what we had lost. We also consider whether it allows us to reach the result we are looking for in a more succinct way. I have the feeling that it is really not the case here.

Alain Aspect. I would qualify this a bit. I agree with you on the fact that the objections you seem to raise against those who do not like Bohm theory are not acceptable if they amount to saying that people are "stubborn".

Lena Soler. That is not what I said!

Alain Aspect. I believe that all physicists are ready to accept new theories that go beyond the usual doxa. The ideal case is when it predicts new things. Even if that is not the case, we ask of it to propose an intellectual framework that allows for some creativity, meaning it allows us to imagine new and interesting situations. At this point, I cannot see much that is stimulating in Bohm theory.

I believe we are many, especially around this table, to have viewed Bohm theory with a favourable eye. However, since it does not stimulate me… Personally, what would stimulate me more would be a naive realist vision of the wave function. We do ask of new interpretations to stimulate our imagination.

Édouard Brézin. There is another point. Forgive me if I change the subject, but we are at a stage (that is far from being over) where space and time are, for many of our colleagues, emergent concepts—in the same way that temperature is an emergent concept that has no intrinsic microscopic significance.

In such a framework, where we would be in this position, it may be presumptuous to assume that we can resolve intrinsically the conceptual problems we have at the level of a quantum mechanics that completely ignores these underlying unknowns. In other words, I do not know if I am realist or anti-realist, but I am very willing to forgo realism for a lack of knowledge that does not allow us to answer this question at this time. I believe it is premature. We are in front of quantum mechanics as we were in the 19th century in front of thermodynamics. We are still searching for the underlying dynamics which are far from being understood. This could profoundly change our view of space and time and, probably, of quantum mechanics. I do not know if this will satisfy our desire for realism, since we all share this desire, but in any case, it seems to me to be premature.

Michel Bitbol. This was exactly the position Bohm adopted at the end of his life. He said that space and time were themselves emergent processes, and that we had to imagine a sub-spatio-temporal dynamics as the base level of this emergence. What he initially considered as being the real spatio-temporal trajectories of particles seemed to him, at the end of life, to be only an explicit, visible, tangible, expanded order like a superficial appearance from a *non*-spatio-temporal implicit order. In other words, trajectories were only spatio-temporal illusions of an underlying algebraic order. It is amusing to see to what extent these contemporary ideas were also at the back of Bohm's mind.

Édouard Brézin. It is clear that, at the scale of the Planck length, where there are fluctuations of the Euler-Poincaré characteristic of space-time, for example, we find it hard to speak of space-time in the rigid way we do in ordinary quantum mechanics. I therefore think the question is premature.

Alain Aspect. Coming back to the debate on Bohm and the reason why physicists are, or are not, interested in him, I would like to come back to the chronological error I made earlier, and which I think is quite revealing. When an article as simple as that of John Bell's points out something surprising in usual quantum physics, then after a while, most physicists eventually understand it because it is very simple. Bell's theorem is very simple. It points out something rather extraordinary in quantum physics. Thus, there are those like Arthur Ekert who invent applications, quantum cryptographies, etc.

Ultimately, this article is very simple. It does not call upon all of Bohm's machinery. Thus, if you like, getting all this "stuff", which seems really complicated, to come to the conclusion that perhaps we had not realized—Feynman had not either!—that entanglement is in fact more complicated than the duality in two particles, etc. Ultimately, Bell has provided a simple way to do this.

What was helpful, I think, in this series of discussions, was to say that quantum mechanics is even more surprising than we had thought. Perhaps if Bell had not come up with his simple theorem, then Bohm theory would have been of more use for pointing out the fact that standard quantum mechanics is even more earth-shattering than we could have thought at one time, after Heisenberg's microscope for instance.

I would say that because we have Bell's theorem which is so simple, it is cumbersome to resort to Bohm theory to convince ourselves… Ultimately, what does Bohm theory bring us? It brings us the fact that trying to have representations in quantum mechanics is even harder than we thought.

Jean-Pierre Gazeau. I believe it is still a matter of education. We are taught traditional quantum mechanics. We do not know Bohm's quantum mechanics. As for me, I had already come across it, but it did not really interest me. Up to now, I had no desire to go further, through laziness. But apparently, there is no contradiction between the two approaches.

Alain Aspect. That is exactly what I have just said. It allows us to realize that standard quantum mechanics, if we try to give interpretations, is extraordinarily twisted. However ultimately, that is a complicated way of arriving at this conclusion.

Jean-Pierre Gazeau. Feynman, for example, considered that we should study all possible means of access. It may be that for certain questions this approach brings something more.

Alain Aspect. History plays a part. Certain physicists, and not lesser ones, have tried to work on Bohm theory. It is not just bad physicists who have attempted to do this.

Jean-Pierre Gazeau. Let us consider classical mechanics. Sometimes, the Lagrangian point of view is absolutely impossible for certain questions. It is the same thing for the Hamiltonian point of view. However, each of these two mechanics can provide new points of view that can be equivalent up to a point.

Alain Aspect. To my knowledge, we have no example of this with Bohm theory. However, some notable physicists have tried.

Édouard Brézin. It seems to me that classical mechanics is fundamentally paradoxical and non-causal. Indeed, it results from a principle of least action. According to this principle, to find a trajectory, we must assume that light, for example, knows where it comes from and where it is going. That is absolutely preposterous. Light, when it travels, does not know where it wants to go. However, its trajectory does result from a principle of least action. It is non-causal and this non-causality is explained for the first time by quantum mechanics, which says that it goes everywhere and that what we see is the effect of all its interferences and all its trajectories. In the end, we have the illusion that there is only one trajectory at the macroscopic scale. Thus quantum mechanics resolves a staggering problem of classical mechanics that bothered no one, until Feynman or until us, which was this staggering non-causality.

Do we return to causality in Bohm's point of view, by throwing away what is one of the great achievements of quantum mechanics which was to explain why the world appears causal to us?

Alain Aspect. I am rather troubled by what you have just said. Either we speak of quantum waves that describe a single particle and what you say is correct, or we speak of light and there we have no problem because we are not forced to resort to the principle of least action. We can have local equations, i.e. Maxwell equations, local differential equations, and the field propagates itself step by step.

Édouard Brézin. That is what we usually say: in Newton's equation, it is enough to know the position and velocity to calculate the execution of the well-known principle of least action. However, that is not entirely true.

Alain Aspect. It is not true?

Édouard Brézin. Not exactly, as velocity is a derivative (which is true in Maxwell's equations), which implies that there is an infinitesimal increase of time. I am therefore not sure that this non-causality of the principle of least action is not equally present in Newton's equations as it is in Maxwell's equations. However, they disappear completely in quantum mechanics, as Feynman splendidly showed, in particular in his essay entitled QED [21], which is marvellous. He explains why and how the staggering problem of classical non-causality is resolved by quantum mechanics.

Bernard d'Espagnat. This last part of our debate is really interesting and shows that there is effectively matter to continue at a later date. It is getting late and we must stop. The next session will be devoted to the same dBB theory, and in particular to the questions you have raised which deserve to be explored in depth [22].

Alain Aspect. We would like to thank and congratulate Franck Laloë!

References

1. *J. Phys. Radium*, 8(5), 1927, 225–241.
2. Gauthier Villars, 1956.
3. Part I, *Phys. Rev.*, 85, 1952, 166–179 and Part II, *Phys. Rev.*, 85, 1952, 180–193.
4. *The Quantum Theory of Motion: An Account of the de Broglie-Bohm Causal Interpretation of Quantum Mechanics*, Cambridge University Press, 1993.
5. Sheldon Goldstein, "Bohmian Mechanics and Quantum Information", ArXiv 0907.2427; Sheldon Goldstein, Roderich Tumulka & Nino Zanghi, "Bohmian Trajectories and the Foundation of Quantum Mechanics", ArXiv 0912.2666.
6. "Surrealistic Bohm trajectories", Z. Natürforschung, 47a, 1992, 1175.
7. Richard P. Feynman, "Simulating physics with computers", *International Journal of Theoretical Physics*, 21 (6–7), 1982, p. 467–488.
8. *Quantum Mechanics*, Prentice Hall, 1951.
9. Hugh Everett III, "Relative State Formulation of Quantum Mechanics", *Review of Modern Physics*, 29, 1957, p. 454–462.
10. The Ehrenfest theorem gives the law of evolution of the expectation values of the coordinates and the momentum of a system of particles. It shows that the evolution equations of the expectation values are formally identical to the corresponding equations of classical mechanics, except that the values in the classical equations must be replaced by their expectation values.
11. The Einstein-Podolsky-Rosen-Bohm (EPRB) thought experiment, an extension of the Einstein-Podolsky-Rosen (EPR) thought experiment.
12. See session VII, "The relational interpretation of quantum mechanics and the EPR paradox".
13. See session III, "Experimental investigation of decoherence" and session IV, "Exchange of views on decoherence".
14. We designate in this way the fact that the precise location where we place the boundary between the system and the device responsible for the reduction of the wave function is arbitrary. See session III, "Experimental investigation of decoherence".
15. See session III, "Experimental investigation of decoherence".
16. *Op. cit.* 1992.
17. *Ann. Phys.*, 296, 2002, 371–389.
18. In a relatively recent article by Antony Valentini, "Signal-Locality in Hidden-Variables Theories", *Phys. Lett. Vol. A*, 297, 2002, 273–278.
19. In the aforementioned article on von Neumann's theorem (written before the 1964 article on EPR, but published after it), Bell shows the non-locality of Bohm theory but adds that we do not yet know whether the phenomenon can be generalized to all hidden variable theories. And in the footnote mentioned here by Franck Laloë (clearly added in proof), Bells adds that we now know that it is the case and cites his EPR article. This strongly suggests that it was indeed his study of Bohm theory that led him to this discovery, and not the other way around (BdE & HZ).

20. The University of Chicago Press, 1994.
21. *QED: The Strange Theory of Light and Matter*, 1985.
22. See session VI, 3The pilot-wave theory: problems and difficulties".

Author Biography

Franck Laloë is an emeritus CNRS senior research at the ENS Kastler Brossel laboratory. He is co-author with Claude Cohen-Tannoudji and Bernard Diu of the textbook *Mécanique Quantique*, which is frequently used in France and abroad. He also wrote *Comprenons-nous vraiment la mécanique quantique?* on the conceptual difficulties of quantum mechanics and its interpretations. His research has focused on quantum optics, the physics of quantum gases at low temperature, and the acoustics of musical instruments.

Chapter 6
The Pilot-Wave Theory: Problems and Difficulties

Franck Laloë

Bernard d'Espagnat. We begin the second session on the pilot-wave theory. I would first of all like to welcome Roger Balian, whom you all know and who needs no introduction.

During the previous session [1], Franck Laloë described the pilot-wave theory, focusing on the positive aspects, in particular on the fact that we can consider that it resolves the measurement problem more convincingly than any other method, including decoherence theory. However, he did not have enough time to sufficiently develop other aspects, including some of the difficulties of this theory. Consequently, in the first part of today's session, Franck Laloë will finish his presentation.

Afterwards, we will have a more general debate where those here who have reflected on the pilot-wave theory will share their thoughts with us. I hope that Michel Bitbol in particular will contribute, as well as others here today.

Franck Laloë. Thank you. Last time, with so many questions and contributions, and having spoken for an hour and a half, it seemed preferable to cut short the final part of my presentation. Bernard d'Espagnat has kindly granted me more time to revisit the conclusions of that presentation—let me say from the outset that these are my own personal conclusions.

Furthermore, Bernard d'Espagnat and I have discussed different aspects of de Broglie-Bohm theory, in particular the fact that the empty waves do not always remain empty. I would like to start with this point.

F. Laloë (✉)
ENS Kastler Brossel Laboratory, Paris, France

B. d'Espagnat and H. Zwirn (eds.), *The Quantum World*,
The Frontiers Collection, DOI 10.1007/978-3-319-55420-4_6

6.1 Addendum by Franck Laloë to His Presentation

6.1.1 Empty Waves

Franck Laloë. Let us return to Bohm's empty waves. A very simple case is that of the interference experiment we considered last time, where a particle passes through two slits of a screen and reaches a region where interference effects can be detected. In Bohm theory, when the trajectory of a particle passes through one of the two slits, it accompanies one of the wave-packets and undergoes the corresponding effects of diffraction that become the trajectory. The other wave-packet, passing through the other slit, is thus an empty wave.

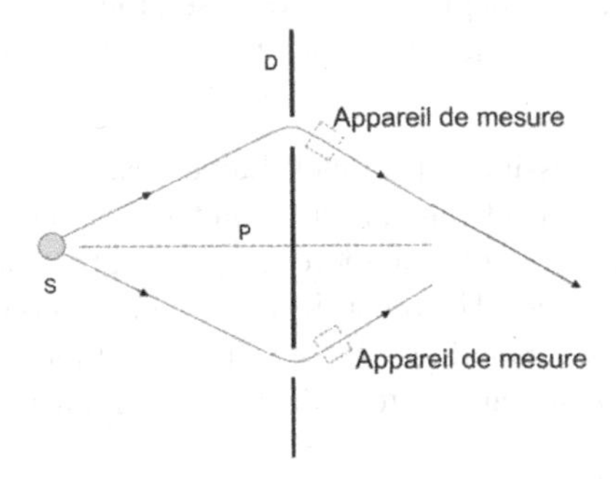

However, when the two wave-packets meet in the interference region, the empty wave-packet becomes important again as it plays a role in guiding the particle. The particle "surfs" so to speak on the waves created by the interference between the two wave-packets, and that is how the well-known interference pattern is created. Later on, after the two wave-packets have met, and because of the no-crossing rule, the particle jumps from one to the other. The wave-packets exchange the particle as would two players passing a ball. The empty wave-packet has become a pilot, and vice versa.

Thus an empty wave-packet can perfectly recapture its particle and cease to be empty. If we insert a "Welcher Weg" device on one (or both) of the possible trajectories, providing information on the path taken by the particle, a phenomenon occurs that we described last time in the thought experiment on "surrealistic trajectories" of Berthold-Georg Englert et al. [2]. We must then necessarily take into account the Bohmian position variable(s) of the device in question. If it only gives a partial indication of the trajectory, the effect of this new Bohmian variable on the trajectory of the test particle is not too important. However, the two waves from the two slits are no longer of the same intensity. The no-crossing rule no longer applies and the probability that the empty and not empty wave-packets exchange the particle is no longer 1; there is nonetheless always the possibility to "resuscitate" the empty wave. If the test particle becomes entangled with a second particle, a third particle, etc., whose Bohmian position variables are all affected, we end up with a situation where the indication given by the "Welcher Weg" device is a real measurement of the position. There is too much entanglement to be able to recover interference effects. The empty wave-packet remains empty, permanently.

Hence, the possibility of an empty wave recovering its particle and becoming a pilot wave again is closely linked to the possibility of recovering interferences. As long as the particle is not entangled with over-complex systems, it remains conceivable to bring into play the empty wave. Once entanglement has gone too far, e.g. when a measurement has been taken, the empty waves remain that way forever.

Another rule to remember, which we stressed last time, is that in de Broglie-Bohm (dBB) theory, we must take into account the position variables of all the systems. If we devise a theory that takes Bohmian position variables into account only for certain quantum systems and not others, we arrive at the same incongruities as when we consider in quantum mechanics the measuring apparatus in a classical way. It is the classic example of the recoiling screen, Einstein's objection at the Solvay conference that was raised by Bohr. Similarly, in Bohm theory, we must perform a symmetrical treatment of all the systems, failing that we arrive at a contradiction. Lack of knowledge of this point frequently leads to erroneous articles in the literature.

Roger Balian. You said this clearly during the previous session. I read the report and highlighted that point.

Franck Laloë. Yes. This time, we have added the idea that it is the Bohmian positions of particles other than the test particle that control the possibility of an empty wave-packet becoming a pilot wave-packet again. In particular, during the measurement of a path taken by the particle, it is clear that the empty wave retains this feature permanently.

Bernard d'Espagnat. Regarding the use of the term "permanently", I agree with you in *practice*, since it is radically impossible for us to build super complicated setups where coherence could manifest itself. You have shown that when there is one particle, we could still imagine ingenious systems that would allow the recovery of coherence. When there are three particles, it is already practically impossible. When there are 10 or 100, it is of course even more impossible. However, I would tend to say—to use Bell's expression—that it is impossible only *FAPP* (*for all practical purposes*), as while it is impossible for us, it does not seem to me to be ontologically impossible.

Roger Balian. If we take into account all the degrees of freedom of the measuring apparatus, decoherence is hidden somewhere. If we imagine a setup which brings back what was detected on the measuring apparatus, we could recover it…

Hervé Zwirn. It is exactly the same situation as when we consider that it is "*FAPP*" for decoherence. Of course, we will probably never be able to imagine this setup—in practice. But in theory we could imagine it.

Roger Balian. If decoherence takes place with spin and there are spin echoes, we can imagine that we are going back in time. Decoherence is therefore not total. We can finally overcome it. It is the same in Bohm theory. What is the difference? I do not see any.

Franck Laloë. I am not sure I agree with you. You seem to say that dBB theory applied to decoherence is as "*FAPP*" as standard theory. Indeed, for vector states, decoherence appears in exactly the same way in dBB theory as it does in standard theory, we agree on that point. However, the great contribution of dBB theory is the presence of a set of Bohmian positions that perform the necessary selection between all the branches of the vector state. We no longer need to say: "these branches can no longer interfere, and I decide to remove by hand all bar one". A *deus ex machina* is unnecessary in Bohm theory, since the dynamics perform the selection for us. The macroscopic uniqueness of the measurement outcomes stems naturally from dBB theory. We go uninterrupted from a situation where we can recover interferences to a situation where we no longer can, in the current state of technology. This tells us exactly how far we can go with a given technique. Personally, I do not call this "*FAPP*".

Bernard d'Espagnat. I subscribe all the more to the argument you have developed that I am, like you, extremely sensitive to the conceptual difference, necessary in the eyes of a realist, between the two theories. In dBB theory, we can believe that the world is how the theory describes it. The universe is made up of N particles (each with a well-defined position at all times and an extraordinarily complex evolution that we fail to track in all its detail) and one field, i.e. the wave function of the universe defined in configuration space. We can think that the world is really like that. In standard quantum mechanics, however, this is not possible as we do not really know what is real. We fail to distinguish unambiguously between what is real and what merely describes our perceptions and our predictions of observations. A realist would like to believe that there are certain notions in physics that correspond to reality, and consider that for this role the wave function (or the density matrix, it is equivalent in that respect) is least bad candidate; however, during a measurement, he takes the liberty to consider that all its branches bar one disappear as if they were insubstantial (i.e. only a representation of our knowledge); and with no other justification than "in practice everything presents itself as if". There is here undeniably a huge "*FAPP*" that is not present in dBB theory. When I mentioned "*FAPP*" in dBB theory, what I meant was that after a measurement, if we use a strictly ontological language (not resorting to what is "practically impossible") we cannot consider that the situation is identical to what it would be if the empty wave did not exist, since under circumstances that are infinitely improbable and impossible to generate, this empty wave *could* manifest itself. I must admit that this is a completely different meaning of "*FAPP*", which furthermore, unlike before, is clearly not an obstacle to a realistic interpretation since the theoretical possibility it refers to is also found in classical physics regarding infinitely improbable events.

Alexei Grinbaum. We have never tried to imagine what would be the spatio-temporal consequences of this Bohmian ontology (*N* particles) for space-time theory. To my knowledge, this has never been investigated by anyone.

Bernard d'Espagnat. John Bell has tried to approach this problem (without imposing that *N* be fixed). He arrived at the conclusion that ontologically, we have to separate space and time. From an observational point of view, this is not problematic since everything (length contractions, etc.) then appears as if they had combined in the way predicted by relativity. In fact, it seems that with a theory like that of Bohm's, we have to conceptually separate time and space.

Alexei Grinbaum. And favour position. This is an issue, from the point of view of space-time.

Franck Laloë. Yes, in Bohm theory we eliminate the formal symmetry between position and momentum that exists in standard theory. Is that problematic? In any case, the advantage is that we no longer need to postulate, like in standard theory, that since decoherence is such that no one can reasonably say that we will manage to recreate interferences, something magical happens that we do not try to explain, which is that all the branches of the vector state bar one disappear. We avoid this in Bohm mechanics because a mechanism automatically selects one of the branches, which seems to me to be a considerable advantage. To summarize, it would probably be correct to say that the accusation of "*FAPP*", if it exists in Bohm theory, is far less justified than in standard theory.

Bernard d'Espagnat. Exactly! To the extent that regarding the resolution of the measurement problem, I consider like you that Bohm theory is much more agreeable, if I may say so, than any other theory.

Roger Balian. Personally, I would not say "more than any other". I will perhaps be able to tell you what I have done. However, this is an aside, in anticipation.

Franck Laloë. Alain Aspect had asked a few questions about the delayed choice experiment and whether it posed a difficult conceptual problem within dBB theory. I do not think so, and I would like to come back to this point.

6.1.2 Delayed Choice

Let us take the same experiment and assume that at the last moment, after the particle has passed through one of the two slits, the experimenter inserts (or not) a lens that produces an image of the two slits. It the lens is inserted, the measurement of the position of the particle in I_1 or I_2 tells us through which slit it passed. If the lens is not inserted, interferences are observed as usual, and it is impossible to say through which slit the particle went.

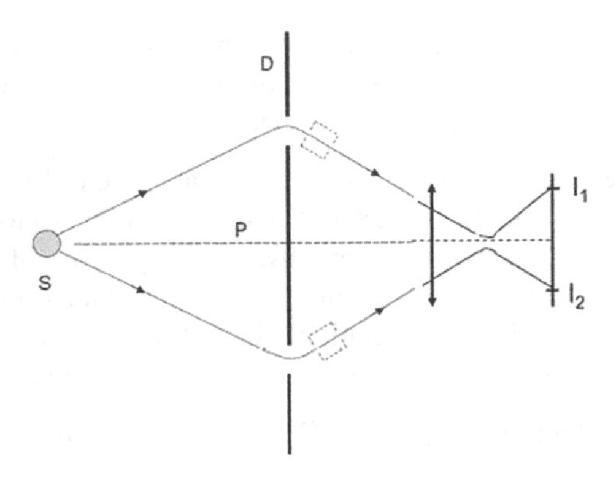

The effect of the delayed choice made by the experimenter is to modify the trajectory of the waves of the wave function. In Bohm theory, this instantly modifies their guiding effect on the position of the particle which, in the same way that waves make a cork bob up and down depending on the movement of the crests and troughs, will not follow the same trajectory in both cases. Nothing particularly mysterious occurs, the influences remaining perfectly local. However, that is not to say that nothing unexpected occurs, as we shall see!

The Mach-Zhender interferometer is a good case in point, even if in principle it is simply equivalent to the double-slit experiment. Suppose we place two detectors D1 and D2 at the exit of the interferometer: we make a yes-no decision at the last moment, while the particle is in the one of the two arms of the interferometer. If a semi-reflecting blade in inserted at the exit, the Bohmian trajectories behave as we saw last time. If the blade is removed, the detectors D_1 and D_2 become "Welcher Weg" detectors—but bear in mind—we must not reason like in classical mechanics, without waves. As a result of the no-crossing rule, D_1 clicks if the particle passes through the upper slit, and D_2 clicks if the particle passes through the lower slit, the opposite of what we would expect. The Bohmian particle did not travel through empty space in a straight line but in zigzags. The strong tendency of Bohmian particles to travel in zigzags makes this interpretation of trajectories non-trivial.

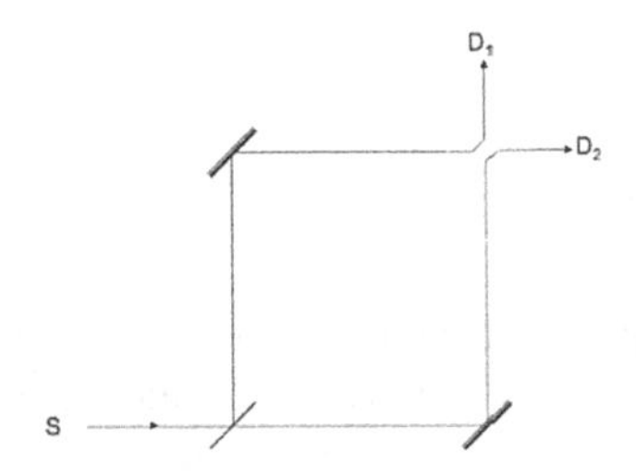

I now come to the conclusion that was cut short last time. I would like to be more critical towards dBB theory by highlighting here what, in my opinion, constitutes good reasons for considering it not completely satisfactory.

6.1.3 *Richness of Description*

DBB theory gives a richer description of quantum phenomena than standard theory. It provides a rather concrete representation, which we may or may not like. It is a matter of personal taste that is not up for debate! However, it also sometimes allows us to give a finer description of quantum phenomena. To summarize this: we have a greater richness of physical descriptions; it is often rejected, but for the wrong reasons; it brilliantly succeeds in removing the observer from measurements; the situation is more ambivalent when it comes to determinism; as for realism, the theory does not succeed in reintroducing naive realism.

6.1.4 *Rejection for the Wrong Reasons*

DBB theory has often been rejected for the wrong reasons, starting with Pauli's objections during the 1927 Solvay conference, which we can see with hindsight were not valid. Critiques of this theory have kept on being published; however, most deal with truncated versions of the theory where all Bohmian variables are not taken into account when entanglement occurs. We must not in Bohm theory make the mistakes we learned not to make in standard theory. It is interesting to see that these debates are, in a way, a revival of the historical debates of the Solvay conference.

6.1.5 *The Theory's Great Achievement: The Mechanism of Empty Waves*

Franck Laloë. DBB theory completely succeeds in eliminating any role of an observer during measurement. In other words, there is no reduction of the vector state. That is not a postulate, but something that appears naturally from a precise guiding mechanism.

6.1.6 *Determinism*

This is where problems begin. Admittedly, dBB theory provides a scheme where everything can appear deterministic, since everything derives from the initial Bohmian positions. However, as we have seen, these positions are chosen randomly and we cannot modify the initial distribution which Bohm's supporters call "quantum distribution". We are then confronted with a theory that considers that everything is deterministic, but as a function of entities we will never act on or ever know. The difference between this and an indeterministic theory becomes very subtle…

In my opinion, from a deterministic point of view, the triumph of Bohm theory is perhaps not as clear cut as some would think.

Alain Aspect. More and more do I hear very reliable people, including Nicolas Gisin and Sandu Popescu, insist on the fact that the most surprising thing with what we currently have with teleportation, etc., is the absolutely major and fundamental role played by chance and indeterminism. It allows an apparent non-locality while preventing sending information faster than the speed of light. They insist on the absolutely crucial role of what they call "fundamental chance". This is reminiscent of what you are saying.

Franck Laloë. Yes.

Alain Aspect. If we could control it, then we could send things faster than the speed of light.

Franck Laloë. Absolutely.

Alexei Grinbaum. All of cryptography is based on fundamental chance. Otherwise, quantum cryptography would not have much cryptographic interest.

Alain Aspect. It is not at all obvious. Eckert's cryptography scheme, for instance, works according to what I have just said. It is not a priori evident. Thirty years ago, Franck Laloë and I knew that the reason why we could not use quantum non-locality to transmit information faster than the speed of light came from that fundamental indeterminism. I think Mr. d'Espagnat knew it even before us. However, this was not obvious to everyone. Besides, I believe that not everyone completely agrees with it now. Nevertheless, I hear this increasingly frequently, including from extremely solid and competent sources.

Remember the debate on Bell's inequalities. At first, everyone believed that determinism was indispensable. Then, the first non-deterministic theories with hidden variables were developed—starting with Bell's theory.

Franck Laloë. Whereas I think there is a clear advantage on the side of de Broglie theory regarding macroscopic uniqueness, I do not value one over the other when it comes to determinism. Of course, if we wanted to, we could say that all the measurement outcomes detected in the universe are trivially the consequences of the initial positions of the set of all the particles in the Big Bang, namely the creation of the universe. However, insofar as we cannot have any effect on the initial distribution, we have not removed indeterminism.

Alain Aspect. Is this not significant for free will? The hypothesis that free will provides the right to arbitrarily choose the position on the measuring apparatus.

Alexei Grinbaum. There is no observer in Bohm theory. How could we formulate any kind of position of free will in such a theory?

Alain Aspect. To deny free will is a position that seems to me devoid of interest. As an experimenter, I choose the orientation of my polarizer. No one has the right to tell me that this does not exist. My question is the following: is this not a scheme where the fact that I have the right to choose the orientation of my polarizer is just an illusion?

Roger Balian. When we change the orientation of a polarizer, we completely change the trajectories.

Alain Aspect. No, because it seems that the positions are determined from the outset.

Hervé Zwirn. This is the discussion we had at the end of our last session [3]. It revolves around the subtle difference between usual quantum indeterminism, which allows free will, and Bohmian indeterminism, which does not. We can see that if we consider that the initial positions of the particles of the universe are fixed during the Big Bang period, even if we may never know them (there is therefore an epistemic uncertainty), everything remains fixed in pure physical terms. There is no longer any free will. The choice made by experimenter to fix the position was already inscribed at the time of the Big Bang. Whereas in quantum mechanics, indeterminism is played out each time—since even when completely knowing the state at a given time, there remains an uncertainty. For us, as observers or theoreticians, the outcome is the same: it remains indeterministic. Epistemically speaking, we have not made any progress. However, physically speaking the difference is considerable: in Bohm theory, everything has been played out at the outset, whereas in quantum mechanics we play all the time, which leaves a part for free will. By contrast, in Bohm theory, there is no free will.

Bernard d'Espagnat. That seems absolutely correct.

Franck Laloë. DBB theory does not succeed in reintroducing simple and naive realism, since we are forced to accept that a reality external to us does exist but must be described on two levels. If we accept the relativistic constraint that prevents us from sending a signal faster than the speed of light, we must also accept that we cannot modify the distribution of Bohmian positions. No one can manipulate these elements of reality. We must therefore accept two levels of reality, which are very different:

1. The level of reality associated with wave functions (which de Broglie and Bohm said they were as real a field as electric or magnetic fields), which we can manipulate (e.g. by changing the external potentials) but which travel not in usual space but in configuration space (six dimensions for two particles).
2. The level of reality described by Bohmian positions, which cannot be manipulated directly but which are directly observable and travel in usual three-dimensional space. It corresponds to a reality that escapes direct human

action. We can act indirectly on Bohmian positions, by acting on the wave functions (by guiding particles differently as in the delayed choice experiment), but we cannot directly take a Bohmian distribution and "squeeze" it, by flattening it one way and extending it the other way.

It seems to me that a sort of dualism reappears at this stage—which was not part of de Broglie and Bohm's initial programme. Furthermore, this dualism is asymmetric: the reality of wave functions acts on Bohmian positions, whereas the latter have no means of retroaction on the wave function. We do not achieve the type of unification we had hoped for, nor a perfectly simple vision, since once again we arrive at a conception of independent reality that is made more complex by the existence of these two levels. This is where, to my mind, dBB fails. In my opinion, the initial programme that motivated this theory is not entirely satisfied.

Let us note in passing that the tension that exists between quantum mechanics and relativity is also found in Bohmian mechanics. To criticize Bohm theory while accepting this in standard theory is not balanced. However, there is still the fact that Bohm theory does not simplify calculations, without providing a total clarification of the concepts in compensation. To conclude, in my opinion, the interpretations that propose to modify the Schrödinger equation [4] are far more promising and interesting, but that is another story. That is all, thank you for your attention.

Bernard d'Espagnat. Thank you very much. Let us now proceed to the second part of the session.

6.2 Discussion

Bernard d'Espagnat. Who among us has reflected on these issues or has read papers on this subject and would like to contribute? Michel Bitbol?

Michel Bitbol. I do not really have anything new to add on this subject; however I will remind you of a certain number of the philosophical objections that could be raised against dBB theory. Does this theory live up to its claims? Is it satisfactory when it is measured against its own ontological research programme? I have grave doubts about the latter.

The first problem, which I had the opportunity of mentioning during the previous session [5], is the multiple ontologies we must consider if we wish to carry out Bohm's realistic interpretation programme in all areas of quantum theory. There is, of course, the theory's original ontology from 1952, the ontology of particles with intrinsically defined trajectories, governed by classical fields and by a new quantum field. However, there is another radically different ontology. It is the one that has allowed us to take into account the relativistic effects within the Bohmian equivalent of quantum field theory. This ontology is no longer an ontology of material points travelling along a continuous trajectory, but an ontology of fields with instantaneous punctual levels of excitation. In this other ontology, there are no material points with continuity in space and time, but on the contrary, instantaneous events that can occur, for instance, during measurement. This representation has almost nothing in common with the initial ontology of particles. It is clear enough from reading Bohm: "the use of the descriptive term 'particle' in this quantum context is very misleading [6]." Finally, there is a third ontology that Bohm himself considered the most profound of all. It is an ontology that considers as fundamental the underlying non spatio-temporal structure, an "implicate order" *dixit* Bohm, from which would emerge an "explicate order", i.e. the order of events and phenomena we observe through our senses and experimental instruments in ordinary space and time. Thus we have three ontologies for the same reconstruction project of one ontology. This is puzzling. We could consider that Bohm finally settled on a good ontology, namely the third one. However, we cannot help being troubled by the fact that, over the course of his career, there is a form of drift of ontological images. A single researcher repeatedly casting aside previous ontologies reminds me of what the epistemologist Larry Laudan said regarding the historical physico-chemical ontologies containing archaic notions like phlogistics, caloric theory, or the ether: how do we know that our current scientific ontologies will not be discarded in the future? And what guarantees do we have that these ontologies designate real beings?

I see another problem with Bohm theory, that of the "supplementary structure". Must we add to ultra-phenomenal (i.e. that rescues phenomena but expresses nothing other than probabilities that experimental phenomena occur) standard quantum theory an additional structure whose sole aim would be to satisfy intuition and the need for concrete and visual explanations. The tradition, in philosophy of science, would be to answer negatively: we prefer to not have an additional structure, for the simple reason that it is not testable in isolation by experimental means and that no consensus can be achieved on this matter. The counter-argument that can be put forward to avoid this negative conclusion is that our perception that a structure is in surplus or not may be historically contingent. For instance, classical mechanics was born with a structure that had certain affinities (not all, it must be said) with the standard images of common sense according to which the world was made of an ensemble of material corpuscles (or "things") subjected to forces and moving through space and time in a continuous manner. During the 19th century, many other conceptions and interpretations of classical mechanics were born, such as Hertz theory (which assumes there are no forces) or energistic theories where the ultimate being is no longer the particle endowed with a certain amount of energy but energy itself. Clearly, these energistic interpretations were much closer to what we nowadays call phenomenalism than to "chosist" interpretations (as they were called at the time): within energistic theory, we refrain from imagining a sophisticated representation of what lies behind phenomena, and consider only what enables phenomena to happen inside the laboratory: energy, which is the quantitative factor for change, in the laboratory as anywhere else. The chronology of phenomenalism and the supplementary ontological structure is inverted here compared to that of quantum mechanics. The supplementary ontological structure was already present at the time of the birth of classical mechanics, and was progressively removed from the interpretation by a certain number of interpreters of this mechanics during the 19th century. However, the initial interpretation with a supplementary structure has not gone away, and that despite the arbitrary aspect that this surplus of interpretation represents. Why is that? Because this supplementary structure at least partially satisfies a continuity clause between physics theory and some common-sense beliefs (not all, of course!). The ontological surplus was consensual not because of its (problematic) capacity to resist a decisive experimental test independently from the rest of the theory, but simply because it preserved an ancient heritage of shared knowledge.

Bernard d'Espagnat. I follow your argument. Simply, it seems to me that there is something troublesome in the idea that we eliminate this structure. Indeed, that would mean that we are really at the centre of everything and that our thoughts create everything, without any substratum. Personally, I prefer to say that there is a substratum, but we cannot really get to it. This is more satisfactory to me. I have always been bothered, with absolute positivism or idealism, by this idea that we are the only existing beings. I believe this feeling is shared to some extent by Hervé Zwirn.

Michel Bitbol. You are correct. In any case, I understand your position. Rather than saying that we must absolutely discard supplementary structures, we could simply consider that we must remain agnostic regarding this supplementary structure, simply because it is under-determined. We do not have anything but phenomena at our disposal to say what this supplementary structure could be; however, these phenomena cannot completely remove the indeterminacy regarding it. We can deal with this problem in two ways: either we consider that if we are phenomenologists or positivists, we must eliminate all supplementary structures, or we consider that we have no reason to choose one over the other, and thus we abstain equally from rejecting them all or preferring one of them. In either case, I think it amounts to the same prudent epistemological attitude.

Hervé Zwirn. On this point, it seems that even if we do not want to commit to a choice, a distinction must be made between theories of strict phenomenalism (which leave open the possibility of adding supplementary structures that can be interpreted one way or another) and more strongly constrained theories which limit such liberties. In the former, we can adopt a position which amounts to saying: "I am satisfied; I do not want to know any more". We are therefore instrumentalists. We can add a suitable supplementary structure if we want to be realist, or another one if we do not want to be. We have the choice. It is a type of minimal formalism of phenomenalism. In the latter, we have formalisms that dictate that the putative supplementary structures must go in a certain direction—as is the case in quantum mechanics. Indeed, with quantum mechanics, we can easily avoid all interpretations and remain instrumentalists. However, if we want to have a realist interpretation, we "get stuck". That is why I consider that we must distinguish theories where we "get stuck" from those open to any interpretation. The former, in a way, tell us a lot more; by prohibiting certain things, they teach us something. By contrast, the latter remain neutral in a way.

From that point of view, quantum mechanics is more interesting because it prevents us from thinking certain things that we would like to think. Obviously, Bohm theory, reproducing the same phenomena, allows us to do what quantum mechanics does not. This means that in a way, on the meta-level of theories, it is likely (we have known this for a long time, it is basic epistemology) that no instrumentalist theory can prevent us from constructing an alternative instrumentalist theory that allows a radically different interpretation from the first one. The question will probably remain open forever…

Alain Aspect. I find that you are a bit categorical if we consider history. Perhaps I have misunderstood your argument, but I am thinking of historical examples where something happened that removed an indeterminacy. I am thinking, obviously, of the example of atomism. At the end of the 19th century, we could not say whether atoms were real or not. We could believe it or not, until technology settled the argument. Is this inconceivable here?

Michel Bitbol. I believe the example of 19th century atomism is different from what is happening in quantum physics.

Alain Aspect. I am sure it is!

Michel Bitbol. For a very simple reason: in atomism, there is no internal clause that stipulates that the claims of atomists are impossible to test experimentally. However, in Bohm theory, there is something of that sort. For instance, this theory claims that the Bohm trajectory is determined by a universal quantum potential, even if we do nothing. It so happens that this trajectory, through the construction of Bohm theory, is inaccessible since each time we put in place an adequate apparatus for measuring the position of a particle at one point, we change the quantum potential in its entirety, and consequently the trajectory. There is therefore a problem. Atomism never said this. It never automatically, constitutively preserved the properties it postulated from experimental validation. Even if the Bohm variables are not really "hidden", the trajectory he postulated in 1952 is clearly outside the reach of experiments.

Hervé Zwirn. I was referring to the concept of the underdetermination of theories by experiments (an idea put forward by Quine), and it goes without saying that it does not concern observable entities. Atoms have become observable and thus this indeterminacy, or underdetermination, has been eliminated. I was referring to non-observable entities (sometimes called theoretical entities as opposed to observational concepts) that are postulated to improve the explanation or our understanding but which are not directly observable elements.

Michel Bitbol. That is right.

Another objection stems from a very nice experiment conducted in London in the 1990s by Bohm theory supporters on neutron interferometry [7]. The results were such that the only way to explain them with Bohm theory was to assume that the neutron mass was dispersed over the entire space of the interferometer rather than localized at the point where the neutron was meant to be. That is very troubling philosophically, as we can ask ourselves how a particle can exist without having localized properties where it exists! This mass dispersion is also perhaps a way to lift other difficulties inherent to the theory, namely the fact that Bohm particles love doing zigzags. If they do zigzags and have localized inertia, then we must apply localized forces on them. However, the quantum potential does not really apply localized force.

Roger Balian. Yes it does.

Michel Bitbol. Quantum phenomenon is meant to act by information.

Roger Balian. It can be interpreted as a force.

Michel Bitbol. Bohm writes that "the effects of the quantum potential are independent of the force (i.e. the intensity) of the magnetic field psi, and depend only on its shape [8]". This feature does not resemble that of a standard potential.

Franck Laloë. It is not a classical potential but a quantum potential, which is very peculiar.

Michel Bitbol. That is right, the quantum potential is very peculiar.

Franck Laloë. Quantum mechanics has not transposed all the classical ideas.

Michel Bitbol. Indeed.

Bernard d'Espagnat. For the problem of the hydrogen atom in its fundamental state, we can say—and that is what Bohm says—that the quantum potential exerts an opposite force to the electrical force of attraction. Equilibrium is established between the two. In this case, the quantum potential is considered a force.

Michel Bitbol. It is equivalent to a force.

Alexei Grinbaum. If I understood correctly, this does not depend on that, in any case. Regarding the question of the properties of quantum particles, those who have conducted experiments on neutrons consider that these are quantum particles of Bohm theory, which is not tenable. Quantum particles, with delocalized properties, are strange objects in themselves. What does it mean to have a delocalized property? It does not depend on the force.

Michel Bitbol. Indeed. The objection I was referring to is of a philosophical nature, and still holds.

Hervé Zwirn. It would be interesting to look more closely if it does not fall in the trap that was mentioned.

Michel Bitbol. In any case, supporters of Bohm theory have considered that this strange characteristic of the neutron mass considerably altered their view of particles in general. If the properties of particles are no longer localized where they are found, then particles can only be *bare particulars*, naked material individuals that do not have an ounce of "clothing" property on them.

Bernard d'Espagnat. Not even mass!

Michel Bitbol. Exactly! For a neutron, the only property to consider was mass. However it was no longer localized.

Bernard d'Espagnat. Considering the philosophical notion of the atom, this is a serious step backwards.

Michel Bitbol. Yes. At least, it is an oddity. There are certainly other objections to dBB theory. I would like to mention one last one, also of a philosophical nature. It is that, in this theory, the observation is always contextual and depends on the configuration put in place to access certain properties through an apparatus. It is the whole of the apparatus' constitution that distorts the trajectory. This is a strongly non-naive way to consider knowledge theory, since knowledge no longer provides a direct access to something that would exist independently from the apparatus, but an access to a trajectory that is continually modified by the apparatus that measures it. If we consider that the relationship between knowledge and ontology is very indirect, inasmuch as knowledge does not provide instant access to things and trajectories as they are, but only through distorted quantities provided by the apparatus itself, then how can we justify from them the adherence to common-sense proclaimed by supporters of the 1952 Bohm theory? Generally, common-sense is associated with a naive vision of knowledge: things are as we see them, there is no distortion as a result of looking at what happens, the world is made up of localized material bodies. If we introduce knowledge-induced distortions, we must admit that there is no guarantee of the direct correspondence between the knowledge we have and things as we describe them. We have the feeling that in 1952, Bohm's position was rather forced. He wanted at all cost to retain some of the images of common-sense, particularly those pertaining to the trajectory of particles, while basing this ontological conservatism on a theory of knowledge that was completely non-conservative and that accepted that there were no guarantees of correspondence between objects and the knowledge we have of them. There is a lack of philosophical affinity between the desire to recover a standard ontological concept, similar to common-sense, and an extremely sophisticated theory of knowledge, far removed from common-sense.

I think I can stop there!

Bernard d'Espagnat. Thank you. We have here an extensive overview of the significant objections. Who would like to speak next?

Roger Balian. I have a certain number of questions in mind, which are not philosophical in nature, but are more those of a theoretician. For example, one aspect of Bohm theory that bothers me the most is that in this framework there is no invariance. For my part, I quite like invariances. Quantum mechanics is a unitary invariant. The fact that there is no invariance in de Broglie and Bohm bothers me a lot. Position plays a particular role: when we perform a unitary transformation in quantum mechanics, nothing is changed.

Franck Laloë. You mean that we break the symmetry of quantum mechanics between position and momentum, or between function and its Fourier transform.

Roger Balian. We break the entire unitary invariance of quantum mechanics since we choose a very particular base i.e. the base linked to position. It is the R base. Everything happens in R base. This is rather problematic for theoreticians used to invariance groups. It bothers me a lot.

Franck Laloë. All the usual quantum calculations are exactly the same in Bohm theory and in classical theory.

Roger Balian. Yes. It is the trajectories that break unitary invariance—"break" in the sense that we choose a base. They are not defined in any other base.

Franck Laloë. They favour a base. It seems to me that it is not possible to use the expression "to introduce a dissymmetry", which means something else.

Alexei Grinbaum. There are two equations in Bohm's version. One for particles and one for the wave function. For the wave function, nothing is broken.

Roger Balian. Indeed. For a wave function, everything works well, unitary invariance is there. However, particles are defined only in a particular base. It is shocking, even if it is allowed.

To mix something that is fundamentally unitary invariant with something in ordinary space… It may be for sentimental reasons, but I really dislike it. The idea itself becomes dramatic when we turn to quantum field theory, because in place of the wave function there is the functional of the field. We define the trajectory as an equation of the functional of the field rather than a gradient of the wave function—and, in the same way that we choose between R and P, we must choose between the electric field and the magnetic field. From a relativistic point of view, this is very problematic.

It is therefore not an invariant by Lorentz transformation. If even Lorentz transformations are fixed, it is problematic.

Furthermore, I would like to speak about the analogy I see between Bohm's point of view and that of Heisenberg on dynamics. I quite like Heisenberg's point of view on quantum mechanics, in that he completely dissociates dynamics from probability statistics. In the Heisenberg picture, there is an observable algebra that is defined in space and follows for an isolated system a totally reversible and deterministic unitary evolution. What is indeterministic is the choice of the wave function—or rather the density operator (as, as you know, I hate the wave function that comes from statistical mechanics, but never mind!). The entire probabilistic part is defined by the state. There is something that is somewhat similar in Bohm's representation. Indeed, the trajectories are perfectly deterministic. We dissociate, just as well as in Heisenberg's theory, the probabilistic part from the deterministic and dynamic part.

Alain Aspect. I can see your analogy, but how do you deal with the fact that in Heisenberg there is only one outcome at the end?

Roger Balian. I will send you the manuscript we are currently working on [9]. It would take too long to answer your question. I can try to do it briefly. When we place ourselves in the Heisenberg representation, there is the object and the apparatus. When we look at the evolution of all possible observables, of the system and of the apparatus, there are, among these observables, those of the object plus the apparatus that we want to measure and all those that do not commute with the former. When we consider Heisenberg dynamics, all observables that do not commute with those we are measuring disappear and will cancel out at the end of the measurement. At the end of the measurement, there remains only an abelian algebra that can be described with ordinary probabilities and not by non-commutative probabilities.

Alain Aspect. At the end, there are still probabilities.

Roger Balian. These are ordinary probabilities.

Alain Aspect. You do not therefore avoid an additional postulate.

Roger Balian. Indeed. We formulate an additional postulate where when we have ordinary probabilities it means that, like with any ordinary event, each event arrives at its own frequency. The important point is that we no longer have non-commutative quantum probabilities, which disappear since only the things that commute as a result of the dynamic interaction between object and apparatus remain.

Hervé Zwirn. Is it the same mechanism in the fact that all that does not commute is cancelled out and in the choice of the preferred base in classical decoherence theory?

Roger Balian. Overall, it is similar. However the analysis in terms of wave function or rather of density matrix is to be understood *for all practical purposes*. When the apparatus is big, it is only with a zero, or negligible probability that what I say is true. It is because the apparatus is big that we can say this.

Hervé Zwirn. This means that we take a path that is equivalent to choosing a preferred base and that we also take the one that leads to the diagonal matrix in classical decoherence theory with a partial trace…

Roger Balian. I certainly do not perform a partial trace; that would amount to cheating.

Hervé Zwirn. We end up at nearly the same point, even a bit further: when we perform a partial trace, we end up with probabilities that are not classical. It is the same path at the start, but we take a detour to get a bit further ahead without a partial trace.

Roger Balian. It is a bit more complicated than that. However, we are far from Bohm.

Bernard d'Espagnat. I have a question concerning spin. I was under the impression that Bohm theory had some difficulties with that. This was the case for some time. In the last book by Bohm and Hiley [10], it still seemed rather unclear. Franck Laloë, you talked about spin a bit too briefly last time as we were short on time. I would like to ask you the following question. Since 2000, or since 1995, has there been any progress in this area? Are the things you have told us drawn from current theories?

Franck Laloë. I cannot answer your first question as I am not familiar enough with the literature. I was not even aware that there was any great difficulty in dealing with spin in Bohm theory. As for the second question, the figures I have shown are from Peter Holland's book [11], based on simple calculations with the Pauli equation, without adding a Bohmian variable associated with spin. The only Bohmian variable is the position of the particle, whose velocity is given by the probability current in ordinary space, obtained by trace on the states of spin.

Roger Balian. The more fundamental Dirac equation is not in fact a wave equation like the Pauli equation, but concerns quantum field theory: it determines the evolution not of a wave function but of a fermion operator.

Bertrand Saint-Sernin. I may have a question. During the 19th century, Cournot claimed that he was a realist but that he did not believe that if a Laplace demon existed, it would see a deterministic universe. Chance is something fundamental in the universe, but which does not preclude the development of science or, in certain areas, the possibility of faithfully reproducing certain processes of nature, as synthetic chemistry has shown. I would like to know if the progress of physics during the 20th century leads us to consider this problem in a different way. What are we heading towards? Are we heading towards the side of Laplace, for whom chance was linked to ignorance, or towards the side of Cournot, for whom chance is a fundamental aspect inherent to nature?

Roger Balian. Did Laplace really say this? I reread his famous words. He considered it was stupid to imagine that we could know all the initial conditions. In reality, it is impossible and we are forced to use probabilities.

Alain Aspect. I agree.

Roger Balian. This quotation, from the preface of *Traité des probabilités* [12], is there precisely to introduce the notion of probability and argue that we cannot do without probabilities.

Bertrand Saint-Sernin. What would be interesting would be that both think that the laws of probability calculation are the same but that in one instance probability is mostly linked to ignorance and in the other it reflects how nature actually operates. Antoine-Augustin Cournot said a very intriguing thing in his last work *Matérialisme, vitalisme, rationalisme* published in 1875 [13]: if there are two independent sciences, even deterministic ones (in other words, if we can consider that the universe is made up of regions that develop at least for a time independently but which are likely to converge at some point), then we find chance in nature. Indeed, we find the encounter of unpredictable events that have, deep down, a type of double heritage or multiple heritage arising from these diverse origins.

Bernard d'Espagnat. It is chance "à la Cournot", meaning the encounter of two independent causal series.

Bertrand Saint-Sernin. It assumes that, for scientists, the universe can be made up of regions that we can treat independently from one another. By contrast Stoics, who thought there was a general interaction of everything in the universe, failed to develop any form of science, except perhaps a bit of logic.

Bernard d'Espagnat. Yes, the mental attitude that has been—and in most areas still is—the most favourable to scientific development has for a long time been a form of legalist, but not excessively dogmatic, realism. And the notion of chance "à la Cournot" is probably the one that best allows the integration of these two apparently contradictory requirements. Furthermore, at the time, the situation was simpler than it is today as scientifically, no theoretical discovery called into question the type of realism philosophers call "naive" and that we call "conventional".

Bertrand Saint-Sernin. Cournot's realism is not at all naive. For him, the paradox of physical science is that it stems from instruments with particularities, with a singularity and whose construction has certain contingent aspects, while having the extraordinary and possibly unattainable goal of uncovering the universal. All these aspects are perfectly understood, all the more since during the first third of the 19th century the cognitive conditions of scientific knowledge were under intense scrutiny. It is therefore not at all naive in Cournot. Furthermore, someone who left school at 15, did nothing for four years, then was "cacique" at *Normale Sciences* on his first attempt, is not just anybody!

Bernard d'Espagnat. Nevertheless, Cournot considered two real situations to be independent. He thus believed, it seems to me, that he had the right to consider situations that are what they are independently of our knowledge—which he perfectly had the right to do at the time. Nowadays, that would be more difficult.

Bertrand Saint-Sernin. I think he considered that the nature of the problem changed with the development of synthetic chemistry. If we are capable of faithfully reproducing natural objects, which nature itself has not produced, and introduce them in nature, we have an empirical criterion for saying that in certain (perhaps restricted) domains, we are capable of following natural processes. Cournot does not go any further. He does not consider that it is a perspective that could be applied universally and that would be true in all instances.

Bernard d'Espagnat. We are capable of following natural processes. That can be understood in purely phenomenological terms, I think. We are not forced to think that there is a reality per se underneath.

Bertrand Saint-Sernin. What strikes me with what you are saying is that when we shift our considerations from physics to biotechnology and we see the extraordinary resistance towards GMOs in France and in Europe, we realize that realism has a very important political dimension that is linked to fear. People ask whether the food they eat, which is an artificial product (forgetting the fact that all agriculture and livestock farming create artificial things), has the same qualities as the natural product of the same name. I believe the problem of realism is a speculative problem, but also an extremely important practical problem that sometimes leads to very strong negative reactions.

Alexei Grinbaum. I would like to make two points: one on the current state of affairs and another more historical one. Regarding your question on what is happening now, since the publication of the work by Bohm and Hiley: Hiley has continued to publish on his own, but he is not the only one. There is also Anthony Valentini, one of the main specialists of Bohm theory today. In my opinion, the problem of the introduction of spin has more or less been resolved. However, we have not resolved the fact that we do not recover the Lorentz invariance. There is no real solution.

Furthermore, regarding the dualism we are forced to consider in Bohm theory, i.e. the two realities we are forced to accept, I must say that before 1952, de Broglie's approach was very different. Like Schrödinger, he does not immediately realize that it is a description in configuration space rather than in real space. This current interpretation is the one we give a posteriori to the work of de Broglie and Schrödinger. In 1927, they are not at that stage. De Broglie was trying to change Newtonian mechanics, by refuting Newton's laws in real space (he did not yet know about configuration space). He refutes Newton's first law and proposes a dynamics based on velocity rather than on acceleration, from which the pilot-wave theory ensues. It is historically rather remarkable, even if we know that de Broglie's approach was saved by Bohm who decided to restore a dynamics based on acceleration and introduce the quantum potential—which has led to two realities. The quantum potential does not exist prior to Bohm. Historically, there were neither two realities nor the temptation to introduce some sort of *über phenomenon*. At first, it was an approach for building a mathematical model in real space and for refuting Newton's laws. The philosophical criticisms we have listed today regarding Bohm theory are not necessarily relevant if we consider de Broglie's initial endeavour.

Bernard d'Espagnat. Indeed. To pursue de Broglie's initial endeavour, however, we must do what Bohm did.

Alexei Grinbaum. I agree. But on a philosophical level, these criticisms are aimed more at Bohm's solution than at de Broglie's motivations.

Bernard d'Espagnat. Yes, but de Broglie was still a realist.

Franck Laloë. I believe that, from a historical perspective, neither Bohm nor de Broglie was aware of the impossibility they introduced to manipulate the positron. The proof of the fact that, if we changed the quantum equilibrium, if we could manipulate these variables, then a signal could travel faster than light, only dates back 10 years. De Broglie and Bohm did not know of these elements.

Alexei Grinbaum. Indeed. In that sense, the only way to save de Broglie theory is not in the manner of Bohm, but is what Valentini [14] showed in 2001 or 2002, where no signal that can travel faster than light if we adopt a certain interpretation of Bohm theory. The right interpretation is very recent. It dates back to 2002. Even Hiley did not know of it.

Franck Laloë. I would like to come back to the objections mentioned by Michel Bitbol earlier. One of these was that when we measure a particle, we necessarily perturb it. Indeed, we all agree on this point. However, the aim of dBB mechanics is not to restore all the ideas of classical mechanics where the perturbation of a position can be made arbitrarily weak. On the contrary, this mechanics says these positions are continually guided by wave functions: it is a genuine quantum mechanics. The fact that dBB mechanics allows arguments like Heisenberg's microscope is not, in my opinion, a fundamental objection. I would even say that on the contrary it seems natural to me that when a particle enters the measuring apparatus, it is perturbed by it. I do not think this is a strong argument against de Broglie theory.

Michel Bitbol. Indeed, from this point of view, Bohm's 1952 vision is similar to Heisenberg's initial interpretation: a particle has a trajectory that is unfortunately deflected when we send a photon on it to observe it. However, after Heisenberg, there was Bohr and his meta-reflections resulting from his discussion with Einstein in 1935. Bohr understood at the time that the image of a predetermined trajectory that is subsequently unfortunately perturbed by observation goes against his own vision of quantum mechanics. Bohr therefore asked physicists to no longer even imagine a trajectory that would be initially intact and subsequently perturbed, but to accept that the only thing that has any meaning is the phenomenon as it happens in the context of an apparatus. Not the existing process per se followed by the observation-induced secondary perturbation, but the global phenomenon where what there is and what the apparatus does are inextricably linked.

Overall, Bohr rejected the image of perturbation, whereas Bohm reintroduced it in his own way.

Alain Aspect. I completely agree with you. We must stop referring to Heisenberg's microscope as an answer to perturbation. Perturbation is not something so naive. It is not necessarily because we send a photon on the electron that the electron is perturbed. We know of examples of perturbation without contact, linked to entanglement.

Michel Bitbol. The idea of Heisenberg's microscope is a nice image at the start: it is an image that allows us to represent quantum novelty with the conceptual tools of classical mechanics. However, once we have benefited from this image, Bohr suggests we discard the scaffolding of this representation derived from classical mechanics and commit ourselves to understanding what he calls indivisible phenomena, i.e. phenomena that are inextricably determined by the whole formed by what is being investigated and by the instrument of investigation. Bohm knew this very well. In his 1993 book, he said that the difference between himself and Bohr was that Bohr would not represent what he did not know, whereas he considered we could represent it even if it was not experimentally accessible.

Franck Laloë. My point was not to defend the argument of Heisenberg's microscope. I agree with all you have said, and I do not say this argument explains everything. The only thing I say is simply that we cannot reject dBB theory with an argument we would not use against standard theory. That would be illogical. If we accept something in standard theory, then we must also accept it in dBB theory.

Michel Bitbol. Indeed, but what bothers me in Bohm theory is not found in standard quantum theory. In standard quantum theory, Heisenberg's microscope simply plays the secondary role of a heuristic device that allows us to become metaphorically acquainted with the novelty of quantum situations. Whereas in Bohm theory of 1952, it is more than that: it is a means for theoretical representation to distance itself from the detail of what can be demonstrated experimentally.

Roger Balian. Whether we do standard quantum mechanics or Bohm mechanics, either way we have to say that a measurement is the interaction between two quantum objects: one quantum object is the measured object and the other quantum object is the measuring apparatus. Since the measuring apparatus is big, we have a visible outcome. We must understand this phenomenon but, ultimately, it is globally quantum. The approximation whereby the object is classical and the apparatus perturbs it, and the approximation whereby the apparatus is classical, are both wrong. The whole is quantum, but the measurement outcome is not a property of the object itself. It is a property of the object as we observe it with a given apparatus.

Michel Bitbol. Exactly.

Bernard d'Espagnat. Therefore we abandon ontological realism.

Roger Balian. Yes.

Bernard d'Espagnat. It is an important step away from classical thinking which equated reality to something that exists in itself and that we can know.

Alain Aspect. I am still vehemently a realist.

Roger Balian. So am I, but reality is not the image we have of it, of a particle that would have a position and a velocity. Reality is something else that we try to apprehend.

Bernard d'Espagnat. Can we not consider that realism in the usual sense is a combination of two ideas? One being that it makes sense to say that there is a reality in itself, independent of us and our knowledge, and the other being that this reality is in principle knowable by man. Usual realism is a combination of these two ideas. What seems clear to me, in the light of the recent developments of quantum physics, is that on the contrary it seems reasonable to separate them. Once we have mentally separated them, there is no evidence that prevents us from keeping the former, which avoids the presumptuous attitude I was alluding to earlier. As for the latter, we must treat it with caution, while keeping, if deemed appropriate, the hope that it will be verified one day.

Roger Balian. No.

Bernard d'Espagnat. Whereas classical physicists thought the things they found (such as general relativity) provided information on reality per se, I think that the reasonable position nowadays is a more modest outlook which consists of considering that that is probably not the case. Even the things we are trying to interpret as being real in themselves are not really like that. It is a description of reality as we perceive it.

Roger Balian. I would like to say things in a more moderate way. First, it is obvious that reality does exist. Secondly, it is equally obvious that we cannot completely know reality in itself. But we can try to get closer to the laws that exist, with the tool that is probability. Probability is a tool that our mind has created to try to get as close as possible to this reality in itself that we will probably never fully grasp, but that we can get increasingly close to. That is how I see the world, or science.

Bernard d'Espagnat. I cannot see how probability allows us to get closer to reality in itself.

Roger Balian. When we go from classical mechanics to quantum mechanics, we get closer to this reality.

Bernard d'Espagnat. We get closer to phenomena but nothing stops us from taking this word literally, in the etymological sense: a set of appearances (in the sense of "what appears") that are the same for everyone.

Roger Balian. Our theories get closer to something all the time. It is asymptotic. In classical mechanics, there were also probabilities because we never measured things with any great accuracy. But it went unnoticed. Nowadays, we must have approximate theories. We have found the right tools to get as close as possible within the current state of science.

Bernard d'Espagnat. We can also say that, for a very long time, men thought an eclipse or the like was a mysterious or diabolical phenomenon. Then they discovered epicycles with Ptolemy's theory. So doing, they progressed significantly in that they could make *predictions* (of eclipses in particular). Then, with Newton, they had another breakthrough. If this is how you understand it, then yes, I agree. Epicycles were already a breakthrough.

Roger Balian. Yes.

Hervé Zwirn. I do not quite agree.

Roger Balian. There is something pessimistic, in a way, in that we cannot access things in themselves.

Alain Aspect. Yes we can.

Roger Balian. “In itself” I said.

Alain Aspect. But they exist.

Roger Balian. Of course they exist, but we cannot know them. The world exists. And there is something rather remarkable, which is that our brain has evolved, through biological processes, in such a way that we can get close to it. It is extraordinary! However, we can only get close, we never access it in itself. In itself, from my point of view, does not exist. That is my personal philosophical point of view.

Hervé Zwirn. You say we have progressed because probabilities have allowed us to move forward and get closer.

Roger Balian. We have to use them.

Hervé Zwirn. I agree with that with one *proviso*: we must discriminate between different levels, which we have not done. Probability is a tool that allows us to get closer to phenomenal reality, i.e. what manifests itself during our observations. And there, science has progressed. That is undeniable. And obvious. However, this has nothing to do with the question we were asking earlier. It is funny, this deviation always happens when we discuss realism in this way. We try to determine whether things exist or not, independently of the observer. We answer in the positive, then we go on to describe the things in question, by means of scientific explanations, and come to the realization that science has progressed and that, using tools such as probability, we can describe with increasing accuracy not the things we postulated as existing independently of any observer, but of the phenomenal image we have of it. The fact that we have significantly progressed in describing phenomena, which is undeniable, says nothing about whether we are any closer to what we have postulated.

Roger Balian. Yes it does.

Hervé Zwirn. No, because the divide between the two can remain as large as we want it to be. Let us imagine that something exists, that “it” exists and that this “it exists” manifests itself to us through a certain phenomenal image that is at a certain distance from the “thing” (in the sense of a metric I cannot describe of course and whose very nature prevents it from being described). Imagine that this phenomenal image was hard to describe and understand in Antiquity, and that, bit by bit, we kept getting closer. Nevertheless, the distance I postulated between the "thing" and this phenomenal image has not decreased at all. We have not progressed at all in that respect. Therefore, we have not answered the initial question which was to know whether we were describing reality in itself better. We know simply that we describe the reality of phenomena better.

Roger Balian. I would formulate this differently by saying that we make images that are closer and closer to reality.

Hervé Zwirn. No. They are closer and closer to phenomena.

Roger Balian. They are closer and closer because we are capable of acting better.

Alain Aspect. If, ultimately, we are capable of acting, we cannot but think that we are a bit closer.

Roger Balian. Yes, there is progress.

Alain Aspect. I must leave the meeting.
(*Alain Aspect leaves the meeting*).

Hervé Zwirn. We act in our world of phenomena, on which we have increasing control, undeniably. However, the logical step that goes from that conclusion to the one where we are getting closer to reality in itself (however we define it) is not allowed.

Roger Balian. We are not capable of doing better than an image of reality in itself, but we are capable of constructing images that are probably closer, "closer" meaning nothing to me, since reality in itself may exist for me, as an animal endowed with reason, but will always remain inaccessible since I am a part of it and I am an animal like any other.

Bernard d'Espagnat. It seems to me that our points of view are converging.

To come back to epicycles, it is true that we have progressed when we discovered that eclipses could be explained by Ptolemy's theory, by epicycles. However, we then realized that in a way, the description was completely inaccurate. In other words, we have progressed when we discovered epicycles and Ptolemy's theory, but it was more a breakthrough in the knowledge of phenomena that a breakthrough leading to a better understanding of reality in itself—since, ultimately, reality in itself as Newton then Einstein claimed to describe it is really completely different.

Roger Balian. I would not say that.

Bernard d'Espagnat. It seems to me in any case that we cannot say there was a continuous improvement in the image we have of the world. It was not all plain sailing.

Roger Balian. I would not say it like that. A pencil drawing is less informative that a monochrome photograph, which is itself less informative than a coloured photograph: we get closer to the object we want to represent.

Bernard d'Espagnat. For the metaphor to be pertinent, we must still accept that in the pencil drawing, much erasing had to be done.

Roger Balian. Of course, the pencil drawing is a lot more imperfect that the coloured photograph, but that does not stop it from being a representation of reality. In the same way, epicycles were a way to get closer. This image was not as good as that of Newton, which itself was not as good as that of Einstein, etc.

Bernard d'Espagnat. The road has been chaotic to say the least, however I readily admit that in our power of synthesis and prediction (in other words, in our "knowledge of empirical reality"), there has been a truly considerable advancement, with a strong underlying element of continuity due mostly to the use of mathematics.

Roger Balian. That is how I see scientific progress.

Michel Bitbol. I still think that Mr. d'Espagnat is right. As Hervé Zwirn reminded us, we often confuse two things: reality in the empirical sense and reality in the sense of what there is prior to the empirical and what causes the empirical (we could say the empirical and the metempirical, using the terms of Vladimir Jankélévitch). They are very different. We can perfectly get as close as we want to empirical reality without scratching the surface of the metempirical domain. For example, if we add an extraordinarily large number of epicycles, we will manage to reproduce the planetary trajectories as well as Newton's theory or even Einstein's theory. However, all things considered, we have done nothing more than reproduce phenomena. We will have learned nothing about the nature of planets, gravitation, etc.

Roger Balian. I would not say that. Increasing the number of epicycles is complicating the theory. New theories, like Newton's, are simpler. And it is very likely, since it works.

Michel Bitbol. Indeed, there is also the simplicity criterion.

Roger Balian. And the "it works" criterion.

Michel Bitbol. We need both "it works" and "it is simple". "It works" in itself is not sufficient.

Roger Balian. If we are capable of acting, that means that if we had added all the epicycles, it would have been very difficult to succeed. There, it is easier. We are therefore closer to reality.

Michel Bitbol. In truth, the criterion of empirical adequacy is not sufficient to convince ourselves that a certain representation is a good image of an external reality "in itself". We need more than that. We need additional criteria that Larry Laudan called ampliative criteria (as they amplify the domain on which a theory is based by going beyond the sole experimental findings). Among these criteria, we find unity, simplicity, beauty... However, these criteria are not completely convincing. We can ask ourselves whether it is not our mind that is at stake when we put forward these criteria, whether these criteria do not depend on us, on our aesthetic and intellectual preferences rather than on a reality external to us that we claim we want to resolve.

Hervé Zwirn. I would like to push the argument a bit further. There is a paradox in saying that there is at the same time something independent of us (of our mind) and that a theory that seems simple, for us human beings, must be true (meaning that it corresponds to something) because it is simple. Either reality exists independent of us and the notion of simplicity, which is completely linked to human beings, has nothing to do with the truth. Or what we want to describe is a creation of the human mind and we can, in that case, retain the simplicity criterion linked to our mind. If what exists is completely independent of the human brain, it is not because something is simple that it corresponds to it.

Roger Balian. Of course. I said it was simpler *and* more effective.

Hervé Zwirn. Effectiveness is a criterion that we could accept, but simplicity… It is true it is often invoked in scientific theories. If a theory is simpler, it is more practical, more attractive. But to say that is to judge in terms of our mind. It is not acceptable to judge in terms of our mind when we speak of the existence of something that is completely independent of that mind. That is an error in reasoning. Perhaps if there is something that is completely independent of our mind, that something would function in a completely different way from what appears simple to us. Why is it that because it is simple, it is truer? There is no reason.

Alexei Grinbaum. In the philosophy of quantum mechanics, there are only two ways to be realistic with regards to entities: Everett's [15] and Bohm's. Often, physicists tend to reject both. This does not affect their right to believe that "it exists", or "this gives the impression of being an entity". However, spontaneous realism, if it is developed systematically, must always be confronted with this choice between Everett and Bohm. Otherwise, it is not a realism of entities as a philosophical position but simply as an impression. Physicists usually find this hard to accept. Indeed, they want to believe in entities, but the philosophically defendable realism is different from the realism of entities. It is something else. It is a philosophical point of view that you add to a physics theory. If we want to be realistic regarding quantum mechanics, there are only the two solutions I mentioned. Otherwise, we must adopt a much more graphic realism, like structural realism.

Franck Laloë. When I spoke about dBB theory I tried to defend it because that was the remit I was given. I would like to specify that, in my opinion, the most interesting way to alter quantum mechanics is not it. I find the theories that modify the Schrödinger equation, like that of Bohm and Bub, and that of Ghirardi, Rimini and Weber (GRW) [16], far more constructive and interesting.

Roger Balian. And they are testable.

Bernard d'Espagnat. We could, of course, also discuss these. From my point of view, I was under the impression that GRW theory was probably not relativistic (Ghirardi confirmed this himself to me a long time ago). Have things improved since?

Alexei Grinbaum. The initial version of GRW theory was not relativistic. However, that of Pearle [17], who is another specialist in this kind of approach, does respect the necessary constraints.

Bernard d'Espagnat. It is getting late. I would suggest drawing the session to a close. The next meeting will take place on the 16th of January. Matteo Smerlak will tell us about Carlo Rovelli's theory and its application in solving the EPR problem. He will show us how, with this theory, to recover locality—while making compromises with regards to realism.

References

1. See session V "The pilot-wave theory of Louis de Broglie and David Bohm".
2. "Surrealistic Bohm trajectories", *Z. Natürforschung*, 47a, 1992, 1175.
3. See session V "The pilot-wave theory of Louis de Broglie and David Bohm".
4. GRW, Pearle. See session V "The pilot-wave theory of Louis de Broglie and David Bohm".
5. See session V "The pilot-wave theory of Louis de Broglie and David Bohm".
6. David Bohm, *Wholeness and Implicate Order*, Ark Paperbacks, 1980.
7. H.R. Brown, C. Dewdney & G. Horton, "Bohm particles and their detection in the light of neutron interferometry", *Foundations of Physics*, 25, 1995, p. 329–347.
8. David Bohm & Basil J Hiley, *The Undivided Universe: An Ontological Interpretation of Quantum Theory*, Routledge, 1993, p. 31.
9. Armen Allahverdyan, Roger Balian & Theo Nieuwenhuizen, "Understanding quantum measurement from the solution of dynamical models", *Physics Reports*, 2012. arXiv: 1107.2138v2.
10. *Op. cit.*, 1993.
11. *The Quantum Theory of Motion: An Account of the de Broglie-Bohm Causal Interpretation of Quantum Mechanics*. Cambridge University Press, 1993.
12. *Essai philosophique sur les probabilités*. 1814.
13. *Matérialisme, vitalisme, rationalisme. Études sur l'emploi des données de la science en philosophie*. Hachette. 1875.
14. Antony Valentini, «Signal-locality in hidden-variables theories», *Phys. Lett. Vol. A*, 297, 2002, 273–278.
15. Hugh Everett III, "'Relative State' Formulation of Quantum Mechanics", *Review of Modern Physics*, 29, 1957, p. 454–462.
16. David Bohm & Jeffrey Bub, "A proposed solution of the measurement problem in quantum mechanics by a hidden variable theory", *Review of Modern Physics*, 38, 1966, p. 453–469 and Giancarlo C. Ghirardi, Alberto Rimini & Tullio Weber, "Unified dynamics for microscopic and macroscopic systems", *Phys. Rev. D*, 34, 1986, p. 470-491.
17. Philip Pearle "Combining stochastic dynamical state-vector reduction with spontaneous localization", *Phys. Rev. A*, 39, 1989, p. 2277–2289.

Author Biography

Franck Laloë is an emeritus CNRS senior research at the ENS Kastler Brossel laboratory. He is co-author with Claude Cohen-Tannoudji and Bernard Diu of the textbook *Mécanique Quantique*, which is frequently used in France and abroad. He also wrote *Comprenons-nous vraiment la mécanique quantique?* on the conceptual difficulties of quantum mechanics and its interpretations. His research has focused on quantum optics, the physics of quantum gases at low temperature, and the acoustics of musical instruments.

Chapter 7
The Relational Interpretation of Quantum Mechanics and the EPR Paradox

Matteo Smerlak

Bernad d'Espagnat. Most of us, of course, have heard of the new and daring interpretation of quantum mechanics proposed by Carlo Rovelli, who is here today, called *Relational Quantum Mechanics*.

After introducing its central idea, Matteo Smerlak will tell us how Carlo Rovelli and he propose to apply this interpretation of quantum mechanics to the resolution of the EPR paradox in its current form. Thanks to the work of J. S. Bell, Alain Aspect and others, we now know that we must abandon either realism or locality. Most physicists prefer to abandon locality, but there are exceptions. Rovelli and Smerlak's approach is one of those, since, as Matteo will show us, it preserves locality at the cost of considerably weakening reality.

Matteo Smerlak, the floor is yours.

7.1 Presentation by Matteo Smerlak

Matteo Smerlak. Thank you, Mr. d'Espagnat, for giving me the opportunity to speak before this audience about ideas that are not fundamentally my own but that of Carlo Rovelli. Let things be clear, I had the opportunity to collaborate with him on one of his projects, but the basis of the relational interpretation is proposed by him, in an article published in 1996 entitled "Relational Quantum Mechanics". The relational interpretation claims that the object of quantum mechanics is not the entangled state of physical systems, as we often say implicitly or explicitly when we teach quantum mechanics, but their relations [1]. In that, it claims to extend and

M. Smerlak (✉)
Perimeter Institute for Theoretical Physics, Waterloo, Canada

B. d'Espagnat and H. Zwirn (eds.), *The Quantum World*,
The Frontiers Collection, DOI 10.1007/978-3-319-55420-4_7

deepen the notion of Einsteinian relativity, and thus dissolve some of the persistent paradoxes of quantum mechanics, for example the measurement problem (the reason why Carlo Rovelli proposed this interpretation) but also the EPR paradox (of which I will also speak here).

7.1.1 The Relational Interpretation

The observed observer

Let us start by describing the relational interpretation itself. For that, let us consider the following situation: a quantum system S (represented below by the Feynman diagram) and two observers O and P who simultaneously observe S. Let me specify straight away that despite this drawing, the observers must be considered as measuring apparatus, i.e. observers in the sense of quantum mechanics, and not necessarily conscious beings.

O and P are two measuring apparatus capable of coupling with the degrees of freedom of the system S to measure a certain observable.

We can imagine that at the start, the state φ of the system S as quantum mechanics prescribes it is a superposition of two particular states, 0 and 1, corresponding to two specific values of an observable I will not specify, with normalized coefficients α and β.

Quantum measurement: a contradiction?

Let us consider first of all the measurement of system S as described by O. We can find this type of description in the original works of von Neumann and Wigner [2]. From the point of view of O, the measurement has induced a transformation of the state of the system, which we call a collapse or a projection, that leads the initial state towards one of two values, 0 or 1, with respective probabilities $|\alpha^2|$ and $|\beta^2|$. In both cases, after measurement, the observable takes on a defined value: either 0 or 1. There is no doubt on that point.

Let us now consider the point of view of P, the second observer, who does not interact with S but observes the measurement of S by O (in other words, the coupling of S and O). From P's point of view, the initial state of the system is a product state of φ for system S and an initial state for O. At the end of the measurement, the degrees of freedom of O are correlated with those of S and the system results in a superposition between two coupled states for S and O, without projection. There is a unitary evolution of the system from its initial state to its final state. The important point of this observation is that, from P's point of view, the value of the observable on S is not determined. *It remains indeterminate.*

We have an apparent contradiction. From O's point of view, the observable has taken on a defined value. From P's point of view, that is not the case. That is the problem of quantum measurement.

Two fundamental postulates

Commentators of this problem mention incompleteness (Einstein), a form of contextuality (Bohr) or the existence of a break between the quantum system and the classical observer (Heisenberg). Such are the classic commentaries on the measurement problem.

By contrast, Rovelli postulates that first of all (*hypothesis 1*), quantum mechanics provides a complete description of the physical world, adapted to the current level of experimental observations; and secondly (*hypothesis 2*), all systems of nature are equivalent. There is no fundamental difference between quantum systems and classical systems. Notably, macroscopic systems are quantum systems of a particular type.

Furthermore, he claims to want to deduce the interpretation of quantum mechanics from its formalism and not add metaphysical preconceptions that could lead to a reformulation of the formalism of quantum mechanics. From that point of view, the attitude of Einstein (who wanted a realist description of the world) resembles a metaphysical preconception. Rovelli claims to get rid of this type of preconception in order to take the formalism of quantum mechanics seriously.

Given these two postulates, we must accept that the points of view of O and P are equally valid. Indeed, O and P being quantum systems, there is no reason to favour one point of view over the other.

This leads us to the conclusion that the value and even the actuality of an observable of S are relative to the system with which S interacts. Indeed, I said that for O, the value of the observable of S is determined after measurement, whereas it is not for P. This means that the quantum states are relative to the system that measures them. We must therefore speak of the quantum state of system S *relative to* O. In other words, we must index quantum states no only by the system they target, but also by the observer that observes them.

Revisiting the measurement problem

Let us return to the problem of the observed observer and describe it from this point of view.

From the point of view of O who interacts with S, the state of S passes, after collapse, from superposition φ to one of the two possibilities, let us say 0, which is the final state of system S relative to O. From the point of view of P—and I note that P does not interact with S—the product state of the state of S and the initial state evolves towards the famous superposition we mentioned earlier. The important point is that 0 on the one hand and the superposition on the other hand do not correspond to the same quantum states because they are not indexed by the same observer. In the first instance, we have the state of S relative to O. In the second instance, we have the state of S and O relative to P. Different observers, different systems: there is therefore no contradiction in the fact that the states are different.

Let us note that the violation of unitarity from O's point of view results from an incomplete description of the measurement system. Indeed, S is correlated with something that evades O: O itself. That is why I have stressed that, from the point of view of O and from the point of P, there is an asymmetry: O is coupled with S, whereas P is not coupled with S. There is a physical interaction between O and S, but not in this case between P, S and O.

This could potentially explain the difference in behaviour between unitary evolution in the latter case, and non-unitary evolution in the former.

A historic leitmotiv

Rovelli proposes to clarify the interpretation of quantum mechanics, as we have seen, not by introducing new concepts but by abandoning an old concept: namely, that of a system's intrinsic state. For him, it is *one hypothesis too many*.

This reminds us of the elucidation of relativistic kinetics by Einstein, who derived Lorentz transformations not by postulating new microscopic principles, but by purely and simply abandoning the idea of absolute time. It is this analogy that Rovelli wants to push with his interpretation of quantum mechanics.

On a historical level, it seems that physics has progressed towards more and more relativity. It would therefore not be fundamentally surprising to discover that in quantum mechanics as well, we are progressing towards a new stage in relativization.

Relation and information

The relational perspective is naturally compatible with the notion of information, which we know is found nowadays in all areas of physics. Indeed, information quantifies the amplitude of (cor)relations between two systems or between two variables. If two systems S_1 and S_2 are strongly correlated, we would say that S_1 contains a lot of information on S_2 and vice versa. Formally, the information contained by an object O on another object S is the number of binary questions on S for which we can predict the answer by measuring O.

I will come back to how Rovelli uses the information concept in his relational interpretation.

Ideas for reconstruction

To go even further in the elucidation of the formalism of quantum mechanics, a number of authors have suggested a *programme* of reconstruction of the formalism of quantum mechanics, namely Hilbert space, the algebra of observables or unitary evolution from clear physical postulates. I can cite, among the earliest attempts, von Neumann himself, then Birkhoff and Mackey, who embarked on a programme of quantum logic with the aim of reconstructing Hilbert space from physical principles. There are more recent approaches, including that of Rovelli. I would like to mention on this matter Alexei Grinbaum's review paper on the reconstruction of quantum theory and on Rovelli's contribution to this reconstruction [3].

Rovelli's informational postulates

What are the clear physical postulates that Rovelli claims can allow us, in time, to reconstruct entirely the formalism of quantum mechanics? At this stage, I would say there are two postulates, even if three postulates were suggested in the original article [4]—the third one seems to me more conditional and remains, I think, to be formalized.

Rovelli's first informational postulate is that there is an upper limit to the quantity of information we can extract from a system.

The second postulate is that it is always possible to acquire new information on a system.

We are confronted with an apparent contradiction: the first postulate tells us there is limited information and the second postulate tells us that information is unlimited. In truth, these two postulates are not incompatible or incoherent, simply because in quantum mechanics there are incompatible observables and we can obtain new information on a system by measuring a new observable that does not commute with the previous one. In this way, we can obtain new information. The price to pay is that we degrade the information we had previously.

These two seemingly contradictory postulates may actually be a seed from which we can grow back the formalism of quantum mechanics, with its incompatible observables and its mechanism of information degradation that is specific to quantum mechanics and is the source of its indeterminism.

You will see in the original article that it is possible to derive certain aspects of the formalism from these principles. However, progress needs to be made. I consider, for my part—and I believe Alexei Grinbaum agrees—that the notion of entropy will certainly play a role in the reconstruction programmes of quantum mechanics.

So much for the relational interpretation in general and how I can summarize it. I now come to the EPR argument, which is so problematic in quantum mechanics, as we know, and yet is so productive—from both a theoretical and an experimental level.

7.1.2 The EPR Argument

The relational interpretation—as I will suggest—discards the concept of quantum non-locality in the EPR argument, according to an analysis described in the paper "Relational EPR" [5].

Let me remind you that similar arguments have been described previously by Michel Bitbol himself in 1983 [6] and by Federico Laudisa in 2001 [7].

The EPR-Bohm argument

Let me remind you of the EPR argument in its simplified version as given by David Bohm. We consider a source producing two quantum particles α and β that are sent in two arms of an interferometer—two arms of an experiment—in such a way that they meet two observers A and B—I can call them Alice and Bob—separated by a spatial distance such that there is no direct connection possible between A and B.

In other words, in terms of special relativity, A and B are in two regions separated by a space-like interval.

We hypothesize that the source produces the particles α and β that are in an entangled state. We could, to make things clear, consider the observable spin and say that it is a singlet spin state,

$$\begin{aligned} |\psi > &= 1/\sqrt{2}(|\downarrow >_\alpha|\uparrow >_\beta - |\uparrow >_\alpha|\downarrow >_\beta)| \\ &= |\, 1/\sqrt{2}(|\leftarrow >_\beta - |\leftarrow >_\alpha|\rightarrow >_\beta)| \end{aligned}$$

I wrote this state in two different ways, following the decomposition in two possible bases of spin space: the vertical spin and the spin in the other direction which I consider horizontal. The EPR argument is the following: the measurement of the observable spin in the vertical direction y or the horizontal direction x would actualize, because of entanglement, a defined value of the spin of β with no causal relation between α and β. If Alice measures α and finds a spin value, then instantly, according to the formalism of quantum mechanics, this produces a defined value for the spin of β as it would be measured by Bob.

This is problematic for relativistic causality, as we have said that Alice and Bob in this set up are separated by a space-like interval.

A counterfactual difficulty: *EPR and Bell's inequalities*

I repeat this argument: the counterfactual possibility of measuring incompatible observables on α, along with the previous observation, leads to tension between quantum mechanics (such that predictions are formulated by the step described above), realism (for a reason I will explain) and locality (Alice and Bob are separated by a space-like interval, without any possibility of communicating). It is as if information was communicated without being causally transferred.

As Mr. d'Espagnat said, this tension can be quantified with Bell's inequalities. This was a considerable breakthrough regarding this argument. It so happens that these inequalities can be tested experimentally and that they are violated, which rules in favour of quantum mechanics as opposed to "realism and locality". In this three-sided argument between quantum mechanics, realism and locality, it seems that it is quantum mechanics that must be maintained, and therefore the concept of local realism that must be altered.

Confusion reigns …

A more in-depth interpretation of these results seems to generate a certain amount of confusion. In the literature, we find that the majority of physicists accept the image of a "strange [quantum] non-locality"—as Christopher Isham puts it—strange because, without directly threatening relativistic causality (I have in mind the theorem where there is no possible instantaneous signalling in quantum mechanics), this non-locality undermines the foundation of our knowledge of space and time, of which relativistic causality is the fundamental expression.

Other authors speak of "non-separability", like Alain Aspect or Mr. d'Espagnat who writes: "*If the notion of reality independent to man, but accessible to its knowledge is to have any meaning at all, then such reality is necessarily non-separable.*" This quotation introduces the concept of non-separability, of which we can say, like non-locality, that it is strange. Indeed, let me remind you that there is a complete set of observables that commute for the system α and β, the pair of particles, that is only accessible by measurements on α and only on β. Is this criterion not precisely separability as we can conceive of it? In any case, the least we can say is that the concept of non-separability is unsatisfactory [8].

Critical reappraisal of the EPR argument

I do not claim to provide a reasoned criticism of these arguments, but simply show that there is a certain confusion surrounding the concepts of non-locality and non-separability.

Let us return to the EPR argument, from the point of view of the relational interpretation. Let me remind you that according to the relational interpretation of quantum mechanics, there is a redundant hypothesis in the traditional formalism: that of the entangled state of a system. However, the EPR argument is fundamentally based on this redundant hypothesis.

Indeed, when Alice measures the spin of α, the value she obtains is instantly actualized for Bob—for the state of the couple α and β, without having to specify relative to which observer this state is defined. If we asked the following question: "Relative to which observer should this non-local actualization take place?", we immediately realize that this observer should himself be non-local, since he would be simultaneously correlated with Alice and Bob. We can thus postulate that it is the existence of a super-observer, capable of knowing simultaneously the measurement outcomes of Alice and Bob, which violates locality and not the quantum probabilities themselves.

According to the relational interpretation, there is no correlation between the spin of α relative to Alice and the spin of β relative to Bob. These two states are defined in relation to two different observers. It makes no sense a priori to compare the measurement values obtained by different observers, except when considering a new observer. However, in the EPR experiment, this new observer needs to be non-local to instantiate these correlations—and therefore, does not exist.

To sum up …

If we return to the arguments and introduce very clearly the observers involved, we realize that the individual measurements are decidedly not in violation with causality.

The spin measurement taken by Alice on α in the vertical direction gives a value written as $\Sigma\alpha/A$. Let us say that afterwards, Alice decided to measure the spin of β, at a later time, and finds the value $\Sigma\beta/A$. According to the prediction of quantum mechanics, the correlation is such that these two values are necessarily opposite. The spin of α in relation to Alice will be the opposite of the spin of β in relation to

Alice. The important point is that the events associated with this comparison are not in violation with relative causality: the spin measurement of β by Alice will necessarily take place after the spin measurement of α in relation to Alice.

Can we go further and test the coherence of these different relational accounts of the EPR experiment? We can, for instance, ask ourselves whether the accounts of Alice and Bob are compatible. Let us imagine that Alice wants to compare the outcome of her own measurement with that of Bob. Alice will measure the state of the system <α and β and Bob> and will find, according to the standard formalism of quantum mechanics, that the value of the observable measured by Bob is such that as seen by Alice it is equal to the spin value of β as it is observed by Alice. There is thus compatibility between what Alice has seen herself regarding the spin of β and what Alice sees of B regarding the spin of β. In other words, Alice never exposes any conflict between her own description of the state and the description B has of it.

If we introduce a third observer C who compares the values obtained by Alice and Bob, then C would measure a system that would be α and β, the two particles, and the two operators, Alice and Bob, write this state for this compound system and, depending on the state, instantly deduce that the spin value measured by Alice and the spin value measured by Bob are opposite. There again, there is no contradiction. There is not, within the framework of quantum formalism, the possibility of making the different relational observations by the different observers contradict.

In other words, the quantum formalism, although fragmented into partial descriptions relative to different observers, is perfectly coherent. From this point of view, we can see that the relational interpretation frees the EPR argument from the problem of non-locality, which has caused so much debate.

7.1.3 Some Philosophical Correlates

I now come to the third part of my presentation, where I would like to mention certain philosophical correlates raised by this interpretation, in the form of three questions. I do not pretend to have unequivocal answers; for that I defer to the philosophers present at this table.

Is the relational interpretation solipsistic?

Let me quote the comment of one referee of my joint article with Carlo Rovelli [9]: "*If the authors are actually comfortable advocating for ridiculous philosophical views like Berkelian Idealism or outright solipsism (in the name of making sense of Quantic Mechanics) let them say so openly and clearly.*"

Confronted with such philosophical aggressiveness, it is appropriate to defend ourselves and give a rebuttal. I would personally reply that I do not see any solipsism in saying that physical properties are sometimes only defined relative to operators. Is it solipsistic to say that Alice's eye colour, for example, is given by a wave length of 463 nanometres? I do not think anyone would take this accusation seriously. And yet, due to the Doppler effect, this is a relative claim. Let me remind you that if you move relative to Alice, if you begin to accelerate very quickly in Alice's direction, the eye colour will change. "Alice has blue eyes" is a typical relational proposition of relativistic physics. There are many relative claims in current physics, and yet we do not speak of solipsism, I believe.

Is the relational interpretation antirealist?

It is undeniable that the relational interpretation strips the object of its substantial attributes: no more intrinsic state, no more intrinsic property, nothing resembling the attributes of traditional realism. We can therefore ask ourselves if the relational interpretation is a form of antirealism. I have two comments to make on this point.

The first is the idea that the relational interpretation, namely the relativity of quantum observables, is not an ontological commitment. To modify what I call *objectivity* (i.e. the nature of what being an object is within the framework of physical theories) is not equivalent to taking sides in the realism/antirealism debate. The two metaphysical options are compatible with the relational interpretation. It simply consists of updating the concept of object. We have seen many examples of this type of update over the course of the history of physics, and none are strictly speaking equivalent to an ontological commitment.

The second point is that the relational interpretation is monist (let me remind you that all quantum systems are thought to be equivalent in the relational interpretation) and that by contrast, the question of the external nature of reality, which implies the question of realism, is fundamentally dualist. Indeed, to choose between realism and antirealism is tantamount to defining a hierarchy between me and the world. These concepts are not admissible in the relational interpretation. I would like to stress that to introduce the realism/antirealism debate in the relational interpretation is to introduce concepts that are not in it—precisely because of the property of monism.

Is the relational interpretation realist?

We can ask ourselves whether there is any sense in ontologizing the relations themselves, i.e. to say that, ultimately, relations and not objects make up the metaphysical fabric of reality. I personally consider that this is a relatively pressing issue since from the outset Rovelli has argued that quantum mechanics is complete. If we must content ourselves with qualifying relations between systems, it is perhaps because ultimately the ontology is made up of these relations. It is a difficult question. There again, I have two comments, two entry points to this question of the ontology of relations.

Firstly, because relativity is involved in the relational interpretation of all descriptions, the objects and relations are in turn naturalized and transcendental. In its relation with S, the observer is transcendental: he enables the description of S. By contrast, from the point of view of P, O belongs to the world of objects and is described by the formalism of quantum mechanics. Thus, objects and relations are not defined in themselves. Depending on our perspective, objects and relations appear either naturalized or transcendental.

Secondly, I observe that the objects of quantum mechanics, if they must have an ontological status, are "bare" since they no longer have any intrinsic property. The possible values of observables, on the contrary—e.g. the possible spin values in the EPR experiment—remain absolute. In relational quantum mechanics, we do not make the observer depend on the possible values of a measurement, but only on the actual values at the end of that measurement.

There may be here a philosophical difficulty in expressing potentialities that seem absolute, and actualities that are only relative.

I conclude this very superficial philosophical discussion by reminding you that the realism of relations is not a particularly revolutionary idea in philosophy. We find it in numerous authors, from Heraclitus to Nietzche, Bachelard and Simondon. In addition, it is an idea that has many points in common with structural realism, which is much discussed currently in philosophy. It is therefore not necessarily as radical a metaphysical option as it may initially seem.

Comments on the relational interpretation

More, and better, comments on the philosophy of the relational interpretation can be found in Bas van Fraasen's article [10] and in one of the last chapters of Michel Bitbol's book [11]. I recommend these two to go further in the philosophical discussion of the relational interpretation of quantum mechanics.

I would like to end my presentation with a comment on two paintings by Kandinsky. The first painting (*Bleu du ciel*, 1940) provides a view of what I consider to be pre-relational physics. We see objects floating on a blue background. There is no particular relation between them. They are there, on their own, in themselves.

By contrast, in the painting entitled *Composition X* (1939) we see that colour, shape, geometry, everything is determined relationally. None of these objects has any meaning without the others. It is an image I keep in mind as representing a more relational metaphysics.

7.2 Discussion

Bernard d'Espagnat. Thank you Matteo for this outstanding presentation. We will now move onto the discussion.

Matteo Smerlak. Let me remind you that Carlo Rovelli is here and can also answer on certain points.

Catherine Pépin. I would like start the discussion with a naive question. Actually, I have not at all understood in what way Carlo Rovelli's hypothesis provides a position that preserves locality. I really have not understood in what way it can solve this paradox.

Roger Balian. I have a similar question. Alice measures what is close to her, i.e. she measures α, but she also measures β.

Catherine Pépin. Actually, we cannot see what has been added.

Matteo Smerlak. The problem with the traditional arguments is the fact that the measurement of α by Alice leads to a defined value for β as it will be observed by Bob. In other words, it is as if Alice's actions influenced Bob's future, when in fact they are separated by a space-like interval and that there is no possible communication between the two observers.

From a relational point of view, this cannot even be formulated, because to say that there is an influence between one and the other is already to say that there are correlations between A and B. And to instantiate these correlations, to make them real, we must say in relation to which observer they would be visible.

The measurement made by Alice on α only produces an outcome for Alice. This outcome is independent of the measurement made by Bob on β. Once again, all that matters are the measurements relative to a given observer. There is no reason to compare or relate Alice's measurement on α and Bob's measurement on β. If we want to compare these two measurements, we must introduce an observer that would take the two measurements consecutively. It can be Alice herself. Let us imagine, for example, that α and β are brought back to the same point in space at the end of the experiment and that this time Alice measures the spin of β. She will find an opposite value to the spin of α, there is no doubt about that. However, that observable will be the spin of β in relation to Alice, and will have nothing to do with the spin of β in relation to Bob. The two measurements made by Alice will be perfectly causal. They will take place one after the other. We will have two correlated measurements, but one after the other. There is no difficulty, a priori. What is outside Alice's light cone is Bob's measurement. And Bob's measurement bears no relation to Alice's description.

Is this any clearer?

Catherine Pépin. It seems to me that Alice, when she measures the spin on β, as this takes place in the future, is no longer really Alice. What is an observer? If it is in the future, it is not the same observer.

Matteo Smerlak. Why? Generally, when we speak of observers, we speak of a physical system and we imagine that it is re-identifiable over time.

Catherine Pépin. In that case, if Bob does the same thing as Alice, with the same number of particles and the same type of macroscopic observer… I mean that a future Alice or a Bob-type Alice is the same observer. Either it is the same thing or it is not the same thing when we consider the future. We have to agree at the start. It seems to me that if we perform statistics on the number of observables, we can correlate the observations in A and in B. This is how I understand Bell's paradox, regarding spin.

Carlo Rovelli. I can give a *poor man's version* of this argument. Two measurements have been made at points A and B of space-time. We note and compare the outcomes. But for that, we must have the outcomes of the two measurements at the same point C of space-time (we do not compare at a distance!). The comparison will therefore be carried out at point C which is in the future of A and of B. That is where, and only where, we carry out the comparison. We can imagine that the outcome of the measurement made (for example) in B is written down on a piece of paper and kept secret up to C. For that, the piece of paper must be considered a classical object and not a quantum object (which could be in superposed states). However, in relational quantum mechanics we consider that everything is quantum. Thus classical pieces of paper do not exist. This means that from the point of view of the observer making measurement A, the piece of paper is still in a quantum superposition of the two possible outcomes of measurement B. In relation to observer A, no measurement has been made at point B. Collapse occurs only at time C. The paradoxical EPR non-locality is only an appearance since it results from forgetting the fact that the piece of paper on which the measurement outcome in B is written is also a quantum object.

Only when observer A observes does the measurement outcome of B become well-defined. The idea that something happening here can affect something happening there in an *observer-independent* manner is wrong. If we remove this independence in relation to an observer and if we maintain the quantum nature of all elements of reality (there is reduction of the wave function because we place an indeterminate state in a determined state), we cannot speak in terms of two events separated by a space-like distance.

Jean-Pierre Gazeau. I asked myself a question when I was looking at the slide on the notion of relation. It felt like the object and the observer were playing perfectly symmetrical roles. Which is the observer and which is the object? It seems that they are interchangeable, according to this formalism. Is there a complete symmetry of S in relation to O and of O in relation to S?

Roger Balian. I can equally be an observer or a measured object.

Jean-Pierre Gazeau. However, there is a difference, because the observer is capable of producing a well-defined outcome, of saying what he detected. This means that there are other states in the formalism. It is complex. There is something else, in the way it is described, and that is what we strive for.

Matteo Smerlak. In concrete terms, a measuring apparatus in a laboratory raises a microscopic correlation to the macroscopic level, following a type of printing process on a piece of paper. But, once again, this is a physical process. In principle, nothing stops us from considering that a simple atom is a measuring apparatus for another atom, two atoms that would then interact.

Roger Balian. No, because a measurement supposes a well-defined recording and outcome. This is what characterizes a measuring apparatus that discriminates between objects.

Matteo Smerlak. What is a recording?

Roger Balian. A recording is the fabrication of a verified macroscopic outcome.

Matteo Smerlak. As we have seen in Wigner's argument on the observed observer, this recording does not take place in an absolute way. For some observers, there is a recording and for others, if we simply describe the unitary interaction between the measuring apparatus and the quantum system of interest, there is no actual recording. We can see that at the end of the coupling between S and O …

Roger Balian. … this argument supposes that the first small observer has not made a recording. If a recording was made by this small observer, then the irreversible process of fabrication of a well-defined macroscopic outcome, i.e. a measurement outcome, has taken place.

Carlo Rovelli. If the irreversible process of fabrication of a well-defined outcome has taken place, then the second hypothesis of the relational interpretation, i.e. the hypothesis that all systems are equally quantum, is wrong. If quantum mechanics is strictly correct, there cannot be any process that is perfectly irreversible.

Roger Balian. No. The systems are equivalent, perhaps. In statistical mechanics nothing distinguishes a priori a system of two molecules enclosed in a ring from a system of n particles where n is 10^{23}. And yet, one is irreversible and the other is not. It is the same thing.

Carlo Rovelli. Let us imagine that I make a spin measurement in a laboratory and I obtain spin +. Do you think it is possible in principle to measure the interferences between—to use this language—the branch where I measured the outcome *up*, and the branch with outcome *down*, or not? If you say "No, it is impossible in principle, even by measuring the state of the world, no observable can see an interference between the two", that would mean—if all systems are equivalent—that quantum mechanics is sometimes violated and that the calculation is not valid. It you tell me that is it possible…

Roger Balian. … I would say that it is possible, but extraordinarily difficult, therefore practically impossible.

Carlo Rovelli. We agree on this point.

Roger Balian. It is therefore exactly the same thing as the irreversibility paradox in statistical mechanics. We have macroscopic systems, and given that they are macroscopic, they do not behave qualitatively like microscopic systems—even though there is nothing more and they strictly obey the same laws.

Carlo Rovelli. Yes. I think that here as well, we agree. This means that the distinction between something recorded and something unrecorded…

Roger Balian. … is relative. It is relative to our possibilities. It is contingent.

Carlo Rovelli. So it has to do with the number of degrees of freedom we have lost.

Roger Balian. Exactly.

Alexei Grinbaum. I would like to say a few things about observers, since I have just written a paper on this topic. The issues that have been raised here appeared immediately, or rather reappeared, after the publication of Carlo Rovelli's article in 1996 [12]. For example, Asher Peres, referring to the work by Rovelli as well as to David Mermin's position, countered that it would be absurd to say that two electrons in state S in an atom, thus perfectly correlated, are one an observer of the other.

Let me remind you that in the history of the notion of observer in physics, there has been one contribution whose importance is comparable to that of Wigner's discussions on consciousness. It is the definition given by Hugh Everett in his famous article [13], where he introduced the idea of multiple branches of the wave function. Everett asked himself the question of knowing what an observer is. His answer was that it is a quantum system endowed with memory. But what is memory, he asked. For Everett, it is something that stores information of past correlations between our observer and the systems it interacted with in the past. So, what is the correct physical definition of an observer? What are its defining elements: it is memory? Consciousness? Neither? It is not clear.

The debate is still widely open today regarding the foundations of physics. It may even be the most controversial question. Indeed, there is a theoretical vacuum that physics has not been able to fill between the hypothesis of universal observers like that proposed by Carlo Rovelli and Matteo Smerlak, and that of singular observers, endowed with memory, consciousness, or something else. Perhaps, as Bohr said, endowed with a macroscopic description. But I wonder if this still unresolved problem does not stem from the fact that we have not understood, or that we have not pushed it to its logical conclusion, the relativization approach mentioned by Rovelli and Smerlak. More concretely, I am surprised by the opening statement of Matteo Smerlak's presentation: "There are two observers, O and P, and a system S." How do they know? How can O know that there is S? There are indeed correlated degrees of freedom between O and S, but does O have a special memory that stores this information? Has someone told him? Does this knowledge come from prior observation? How has O identified S? I therefore wonder whether the controversy we speak of does not stem from the fact that we have not yet pushed to its logical conclusion the idea that we should define, relative to observers, not only physical states but the very constitution of systems, i.e. if S is seen as S by O, perhaps P can see differently the degrees of freedom implicated in the interaction between it and S.

I wonder if we are not simply confronted with a problem of coherence in our approach, linked to the fact that we have assumed that systems belong to the fabric of the world, whereas systems should be defined in relation to the observer.

Bernard d'Espagnat. That seems to me to be true. I think that in "orthodox" quantum mechanics (with no hidden variables), physical systems, no matter how big, must be regarded as relative and not as existing "in themselves". Furthermore, this is what happens in quantum field theory. Systems do not have an existence "in themselves" since quantum field theory predicts that particles, or even systems with multiple particles, can be created or destroyed. And—a non-negligible point!—as we all know, these phenomena predicted by theory are verified experimentally…

Alexei Grinbaum. Yes.

Carlo Rovelli. I think you have touched upon what I consider to be two central problems. I agree with nearly all that you have said, except with the relationship between the two problems, which I consider to be separate. Let me start with the second, which is that of the definition of the notion of system.

All that has been done within the framework of relational quantum mechanics presupposes having a well-defined notion of what a system is. I believe that was singled out by Michel Bitbol when he said that something was missing there for understanding the world—if I understood correctly. And I agree with that. I would be happier if, for our understanding of the world, there was no need to break up the world into sub-systems. However, I do not consider this problem to be linked to the first problem, that of the definition of the observer.

I consider that defining the observer has always been the most opaque, most obscure part of all the constructions of quantum mechanics. Because, ultimately, there are many positions. The only clear position is that of Wigner, who said that the observer is consciousness. As for me, I prefer to start from another point of view.

Of course, we can consider that an observational apparatus is something that is thermodynamic. However, I find that, on the one hand, defining the observer is essential in quantum mechanics—otherwise I do not understand what the physical values, and what the specific values of the operator, are. But on the other hand, the only way I can understand an observer (and that was for me an a priori), is to say that it is any physical system. Therefore, what relational quantum mechanics tries to do is, in a way, use of the notion of observer in the largest possible sense.

Obviously, the question is not terminological: we can say "general observer", "type A observer", "type B observer", … The real question is knowing which of these notions is necessary in quantum mechanics: is it necessary to speak of consciousness, is it necessary to speak of an animal system, is it necessary to speak of a system that captures information, etc.?

Relational quantum mechanics is based on one hypothesis: it is possible to account for quantum mechanics and make it coherent even with hypothesis 2, according to which all systems are equivalent. An observer is therefore any physical system, in the sense of an observer in special relativity. It is a coincidence if we use the same term. Speed is a relational quantity in classical mechanics since Galilei. We can thus only speak of speed in relation to an observer, in relation to a frame of reference. The frame of reference may be the bottle in front of me. My hand goes at a certain speed, relative to the bottle. But the bottle does not need to observe my hand. The bottle registers nothing and is not conscious. It is only the speed of my hand that is relative to the bottle. The bottle is an observer only a very basic sense. In the same way, in quantum mechanics I consider that any pair of physical systems can function in the interaction where one is the observer of the other. Is it possible, from this starting point, to reconstruct quantum mechanics without using a more limited notion of the observer? I think so. Obviously, we cannot then use just any measuring apparatus. We need to write down on a piece of paper the properties of a system which are true in relational quantum mechanics. If we make observations as one would in a laboratory, we would use the appropriate "observer" systems.

Michel Bitbol. I would like to comment on the question of the observer. I was struck by Matteo Smerlak's choice of words—perhaps it was not intentional, but let me remind you. He said: "From observer O's point of view, there is collapse" and "From observer P's point of view, there is not collapse but the superposition remains". What is the point of view an observer that is precisely not an intentional observer in the philosophical sense of the mind i.e. where the observer is capable of representation? Would you say that a macromolecule linked to an electron has a point of view on that electron?

Carlo Rovelli. I use it in the following sense: from the Earth's point of view, my speed is 30 km/hr. From the point of view of a train, it is less. Obviously the train has no point of view.

Michel Bitbol. Could we not say, in this purely kinetic configuration, that "the relative speed of A in relation to the train is *x*"?

Carlo Rovelli. Yes we could.

Michel Bitbol. That would be sufficient. You would not even need to use the term "point of view" which implies that someone, on the train, is capable of seeing and realizing the measurement outcomes. The problem is that we cannot get off so easily. Let us suppose that instead of saying "from the point of view of the train, the speed is *v*", we say "the relative speed of A in relation to the train is *v*". The latter statement tacitly implies another real "point of view" of an intentional observer: that of the observer who is at the same time external to A and to the train, but who could compare their speeds. The observer endowed with intent is still there, underneath it all.

Alexei Grinbaum. I believe I have discussed elsewhere the link between the two problems, which you separate and I do not. I do not want to go into the details of my answer to this question, but I would like to highlight once again the statement with which hypothesis 2 begins: "All systems are equivalent". In your reply, Carlo, you have spoken of physical equivalence: "All systems are physically equivalent". When I read this statement, I see in it—including in the vocabulary used in the 1996 article—the fact that all systems are physically equivalent but not necessarily informationally equivalent. The idea, which comes from Everett, is that of an informational characterization of observation. Physically, any system can be an observer. But can the informational requirements and constraints for being an observer be assigned to any physical system?

In my opinion, there is in this statement ambivalence in the meaning of the word "equivalent" that allows us to bridge the two questions you have separated.

Carlo Rovelli. It all depends on what information you speak of. The information I was speaking of in the article is information in the sense of Claude Shannon. It is simply something that quantifies the number of possibilities. In this context, we can speak of the information a system has on another. It is the existing relationship between the states of one system and another. System A has information on system B if a measurement on A predicts something about a measurement on B.

Once again, although this basic notion of information is necessary to be able to speak of relative states, it is not sufficient for building realistic measuring apparatus. An atom is not a measuring apparatus in relation to another neighbouring atom. However, it has information (in that sense) about the other atom if something (classical or quantum) correlates them.

I would like a fundamental theory of the world to speak of physical systems, or relative information, and not of more complicated things. I will not satisfy myself with a fundamental interpretation of the world that is based on the fact that we need sentient beings, machines, or event pieces of paper! If the fundamental notions of quantum mechanics can be interpreted in the physical or informational sense, sensu Shannon, I would be happy!

Hervé Zwirn. I believe this question touches upon the most difficult problem of all: the question of knowing to which entities we are prepared to assign the role of observer. We would like to say that we can do without the notion of consciousness to describe the world. But do all these problems of interpretation not mean that, in essence, we place ourselves from an intentional point of view, as Michel Bitbol said earlier? Meaning that in reality, the problems we face are not problems of the evolution of the world independently of us—because the equation of the universe evolves…

Roger Balian. … there is no equation of the universe.

Hervé Zwirn. I will be prudent on this matter! We are speaking about interpretation problems. However, when we say interpretation, we speak about someone who asks the question of knowing how something is coherent in relation to him. It is very difficult, even if we would all like to, to avoid invoking an observer in the sense of someone with an intentional point of view. And it is in relation to this intentional point of view that we will then try to re-establish coherence.

I see, from the first equation presented by Matteo, that there is collapse. Therefore, from the point of view of O who interacts with S, there is indeed collapse. If we place ourselves from O's point of view, why is there a collapse? In quantum mechanics, without the principle of reduction of the wave packet there is no collapse. Thus the measurement problem remains unresolved. Presented in that way, since we have explicitly introduced a collapse for a given observer, whatever it may be, we need to explain this collapse. The Schrödinger equation does not provide an explanation. The unitary evolution equation does not either. Thus already, by writing down "collapse" as it appears on the slide, we say that there are two evolution principles—and thus the measurement problem remains unresolved.

I entirely agree with saying that there is a relativity of states or descriptions in relation to the observer. But we must explain why a collapse occurs here for observer O. And for it to happen there must be a mechanism that triggers it, and that is not given in the relational interpretation.

That it is necessary to call on—as Roger [Balian] said earlier—the fact that there are too many degrees of freedom and that it evades us, is just another way of saying that it is because our means are limited that things appear as they do. If our means were not limited, for example for an omnipotent super-observer, the world would remain quantum and there would never be any collapse since we would always be capable of putting in place the means to measure observables that would allow us to detect correlations. Therefore, even if we wanted to, I struggle to see how we can completely eliminate the need, at some point, to bring everything back to the level of the mind—and therefore to an observer who is, in a way, what we are. And therefore the idea that we can eliminate the observer by giving this role to any atomic system does not seem a good one to me. What we are trying to do is not explain how a macromolecule sees the world—and anyway, it is hard to see what that would mean -, but explain how, we, through quantum mechanics, manage to account for what we observe. We are not going to eliminate ourselves as observers!

A long time ago I suggested an interpretation that was very similar to the one presented today: the convivial solipsism. It has many points in common with the relational interpretation in that it also proposes that states be relative to the observers. It is in fact an extension of Everett's theory that I proposed in a book published in 2000 [14]—following the ideas of Bernard d'Espagnat. It is very similar, but there are two small differences. The first is that I describe the collapse as being a principle that I call "the hanging-up mechanism"—and I preserve the notion of a conscious observer: at a certain time, a given observer is presented with a choice and he, in a way, hangs himself up to one of the possible branches—knowing that, like for Everett, the wave function is never reduced. The second thing is that this is so relational, (i.e. relative to the observer) that two observers, with their own points of view, can observe totally different things—thus this resolves the problem for the EPR paradox since two observers can have seen things which will not be coherent from the point of view of a super-observer who does not exist. However, the principles of quantum mechanics prohibit, when two observers meet, that they realize their differences. I therefore completely agree with what you have just said about the piece of paper: we tend to think that if the piece of paper has something written on it, it was that way in the past. In fact, that is not the case. Everything remains superposed, except for the observer, who at some point hangs himself up to one the branches and remains there forever in such a way that what he will be able to control, compared to other observers, will never reveal any incoherence.

It is true that this conception is rather strange and it has consequences that bring into question the usual notion of realism: it means that each of us sees a world that is specific to him and that can be different from what others see, but we cannot make this difference visible. The problem of collapse remains complete. I have not solved it either. For obvious reasons, due to the images that are evoked, I call this the hanging-up mechanism but the collapse is not eliminated. In fact, the solution I propose consists in making the hanging-up mechanism rely on decoherence which itself is caused by the limits of our capabilities and thus, as a last resort, of our consciousness.

The point I would like to emphasize is that it seems difficult to me to completely forgo the notion of a minimal observer with intentions or consciousness.

Jean Petitot. Is the hanging-up mechanism a physical process?

Hervé Zwirn. It is not a physical process. It is a process that, at a given point, consists of an interpretation linked to the observer, of making a choice that eludes us—we do not control it—and which leads us to hang ourselves up to one of the branches of the superposed wave function, which itself continues along its path but for us disappears. The alternatives disappear.

Jean Petitot. So there is something else in addition to physics.

Hervé Zwirn. Perhaps … this needs more work.

Carlo Rovelli. I realize that I should have made more references to and comparisons with the convivial solipsism!

This question allows me to come back to one of the most important points for understanding what I wanted to do with the relational interpretation. There are obviously many ways of classifying the possible interpretations of quantum mechanics. I have learned about many of them with Bernard d'Espagnat. One of these consists in distinguishing between two radically different ways of thinking about quantum mechanics: that of Heisenberg and that of Schrödinger.

One possible point of view (we can call it Schrödinger's, but that is too vague) is to say that quantum mechanics is essentially the Schrödinger equation for the state of a system, a unitary evolution—therefore we must understand what happens when a collapse occurs. This opens up a number of discussion points.

A different point of view is the one where Heisenberg recalls how he came up with quantum mechanics. He has told that one night he was in a park in Copenhagen. He saw someone walking in the dark—but only seeing him when he walked under a streetlight. He therefore sees him appear and disappear, then reappear and disappear again. He deduced from it that this is what happens to an electron: it exists at different points, but we do not know what happens between them. Thus, the other way to think about quantum mechanics is to consider quantum events as high-order elements of the reality in question. When we see something going on, the result is a collapse.

Thus, the reconstruction effort in my article consists in discarding the state as a fundamental object, discarding the image of an "evolving state", and think of reality as these quantum events. I can see you here, not in superposition. I can see the electron, the measuring apparatus. From a standard "Copenhagen" point of view, it is very simple. There is a classical world and the quantum system manifests itself in relation to the quantum world in a series of discrete events which result from measurement. This series is the reality of the system. I tried to reformulate this by replacing "classical world" with any system in relation to another. From this point of view, the collapse is not a problem: it is a reality. The measurement outcomes are a reality. The problem is understanding how they are linked to one another and how, in a strange way, they seem real in relation to one system but not in relation to another. However, if we start to think that quantum mechanics is a state that evolves under the Schrödinger equation, we are in a completely different interpretation.

Roger Balian. I absolutely agree with the point of view *à la* Heisenberg you have just described. Anyway, it is the only one I understand! I would change just one small thing: I would not say that the successive flashes you speak of are reality, but images of reality. In the same way that a measurement is an image that we have of reality, an image created on an apparatus. A created image of reality, which eludes us since it concerns variables that are not intelligible—and of which we have no intuition mathematically. And with good reason, since our intuition is based on our world, which commutes. I would see things more in this way, as an image.

In the same way, regarding information: we would like to include it in quantum mechanics—that is true—but we face a difficulty which is that the unit of information is the bit. And we know that in quantum mechanics, the unit of information is the qubit. And we have no intuition of what a qubit is.

I believe we must take a leap and say that we have an image of the reality of quantum information in the form of classical information—but that quantum information is the qubit, which is what it is.

Alexei Grinbaum. I would like to ask Matteo Smerlak about a statement he made earlier in passing. He said that there was a collapse in relation to O because O itself is not observed. Thus, self-observation is never mentioned in his approach. Since I believe the question of the collapse in the relational interpretation is fundamental, it is necessarily linked to the question of self-observation or, more generally, self-reference.

What is the status of this prohibition of any self-reference in the relational interpretation? What motivates it? What is the current line of thought on self-observation: is it necessarily considered an additional postulate, or is it a theorem, hence something that can be deduced? If it is a postulate, it must be made explicit and added to the list of postulates.

Matteo Smerlak. There are several mysteries in quantum mechanics! One of these consists in understanding why O and P give different descriptions—the word "description" being taken with a pinch of salt after what Michel Bitbol said. The other consists in understanding the origin of indeterminism in quantum mechanics. These are two different problems. Relational quantum mechanics operates on the first problem. It does not provide any fundamental explanation regarding the second problem, namely the origin of indeterminism. It simply allows us to point out something that is not in traditional quantum mechanics, namely that the collapse is linked to an element of self-reference. The asymmetry in the relational interpretation between the interpretation of O and that of P is that O interacts with S, and P does not interact with S. There is an element of externality in P that is not found in O.

We can therefore postulate that the indeterminism of the measurement made by O—the fact there is a collapse with no possibility of knowing which value will emerge from this measurement—is linked to the intrinsic limitation of O to describe himself as interacting with S. That is to say that, from this point of view, there is nothing more than the following observation: there may be link between collapse and self-reference.

I am not aware of any recent advances on self-reference in relational quantum mechanics. All I can say is that—as already mentioned in the 1996 article—this question has been investigated by the Italian researcher Marisa Dalla Chiara.

Alexei Grinbaum. In addition to Marisa Dalla Chiara's work, there is the Breuer theorem, which is one of the most fundamental theorems on self-reference.

Matteo Smerlak. That as well. These are the two references I can cite. I do not know of anything more recent.

Catherine Pépin. I will adopt the point of view of a physicist, by going along the same line as Roger Balian. Some things remain unclear. I am thinking of three points in particular.

First of all, concerning the EPR paradox and Bell's inequalities, it seems to me that if observers cannot communicate with each other, if everything is always in relation to an observer, then we cannot even speak of a correlation between what is observed in A and what is observed in B. We can define observers as large systems that make wave functions collapse—these are classical systems. Once the classical system has interacted, we have a result, as Roger Balian said. If we make a measurement on α and on β, these classical systems provide correlations only if α and β are initially in the same quantum state. Now, if α and β are initially in different quantum states, these correlations do not take place. I really do not see how introducing something relative to the observer can overlook this.

Matteo Smerlak. You have said that it is not possible to compare the results of Alice and Bob, hence to demonstrate that there are correlations, when all measurements are relative to the observers. It is not the case. It is perfectly possible for Alice, once she has measured the spin of α, to correlate herself with Bob. It is possible to bring back β in the presence of Alice and that Alice then measures the spin of β. She will have measured, over time, the spin of α then the spin of β. She will thus observe a correlation between these two outcomes. She will find that these values are opposite.

Catherine Pépin. Fine. So how do you explain that the correlation between these two measurements is different in EPR paradox where particles are initially in the same quantum state compared to a situation where we measure particles with initially different quantum states? The fact that the initial quantum state is the same must induce a correlation in the observation, at a later time, of A or B.

Matteo Smerlak. It induces a correlation.

Jean Petitot. Yes, but at a later time.

Matteo Smerlak. Yes. This correlation is the following: the spin value of β in relation to Alice is the opposite of the spin value of α in relation to Alice. This correlation exists. However, it is causal. It happens for two events where one occurs after the other.

Catherine Pépin. If it exists, then it means there is non-locality. That is what I do not understand. If this correlation exists, causal or not…

Matteo Smerlak. … demonstrating this correlation does not involve two events separated by a space-like interval. It involves two events separated by a time-like interval.

Hervé Zwirn. I think it would be good to describe what happens step by step, over time. The particles separate. Alice measures the spin of α at a given time, and finds a certain value. There is no causal influence in the space-like interval, thus there is no violation of locality. Which means that strictly speaking the spin of the particle that is far away is still not determined.

Jean Petitot. For Alice.

Hervé Zwirn. Yes, for Alice it is still not determined. To clarify things and find a consensual explanation, perhaps we should explain how, when Alice and Bob meet again (Bob has measured independently and obtained something), if the outcome for Alice is not determined, then how come when Alice and Bob get together, that Alice finds a posteriori that Bob has indeed found the opposite of what she has measured. I think that is the point that bothers you. I think we should explain this step by step, so that it is no longer troublesome.

Catherine Pépin. Let us be clear, it is the same problem for Bell's inequalities.

Hervé Zwirn. Concerning Bell's inequalities, there is something more. Here, however, I think it is very simple and is easier to explain. To follow it step by step would enable us to understand it better.

Matteo Smerlak. I would not do any better than what you have just done.

Hervé Zwirn. I left something to one side, namely the precise explanation of the mechanism. Let me go over it. The particles separate and at some point, Alice measures the spin. Let us say she finds + along Z. But for Alice, the spin of the other particle is not—along Z. It is still indeterminate. Bob measures the spin of the other particle; and he finds what he finds.

Carlo Rovelli. And what happens for Alice?

Hervé Zwirn. For Alice, when Bob measures, nothing happens.

Carlo Rovelli. Yes, something does happen!

Matteo Smerlak. During my postdoc and when trying to write the article, I got stuck on this question for a long time. I wrote the article after finding the answer.

Hervé Zwirn. I think this is precise point that needs to be explained.

Catherine Pépin. So what happens for Alice when Bob makes a measurement?

Carlo Rovelli. It is not collapse but entanglement that occurs between the state of the particle and the state of Bob. Bob himself enters into a situation of entanglement with the particle, but the two remain in a quantum superposition. More precisely, this means that Alice knows that from that point, the measurement outcomes on Bob or the particle will be correlated.

Hervé Zwirn. This means that for Alice, there no longer are simply the two particles, but there is one particle plus Bob.

Catherine Pépin. There is thus completely non-locality. When Bob measures, as Alice interferes on Bob's measurement, it is completely non-local—since they are not in the same place.

Alexei Grinbaum. It is a process of redefining the systems.

Hervé Zwirn. It is an entanglement of the systems.

Alexei Grinbaum. The systems are redefined, of course.

Hervé Zwirn. The two particles were already entangled, Bob becomes entangled.

Catherine Pépin. Bob and Alice are not in the same place.

Hervé Zwirn. No.

Catherine Pépin. So a non-locality appears.

Hervé Zwirn. In the entanglement.

Catherine Pépin. It must reappear somewhere…

Alexei Grinbaum. It is not a change of state. It is a change of what they consider as systems.

Carlo Rovelli. Let us return to the previous question. The element of reality here is not the state, which is simply the mathematical system that allows us to calculate the probabilities of future events—based on previous observations. For Alice, when Bob makes a measurement, nothing happens other than the possibility of calculating future events. What is, for Alice, the significant future event? She will receive a call from Bob. She will then collapse the state of Bob and see if the predicted correlation is realized or not.

Catherine Pépin. Fine. I understand much better now.

Roger Balian. I would not quite say it like that. I would say that Alice carries out a series of measurements of her spin, and sifts through them. She shows that measure no. 1 has this value, measure no. 7 this value, etc. She writes down the outcomes on a piece of paper that she sends to Bob by post, asking him to select such and such measurement and to note that all the spins obtained will be prepared in a given position. It is therefore a preparation, a directive for selecting different successive measurements to be carried out, rather than a collapse. There is no collapse. There is selection. And nothing else. At each stage, for Alice, there may have been her own collapse—corresponding to the fact that her apparatus has given her a certain outcome. This collapse, in fact, is linked to the activity of the apparatus that allows her to operate a selection and to select such and such measurement in her message. What is shocking about that?

Carlo Rovelli. May I reply to your previous observation? You said that you liked the idea of linking reality not to states but to quantum events. But you added that you preferred thinking in terms of an image of reality rather than of reality itself.

Roger Balian. Because a measurement provides an image.

Carlo Rovelli. Personally, I prefer using a term other than image, by calling this the information a system can have on another system. That is how the term information is used. When an electron strikes the screen and leaves a mark, the screen has information on the fact that the electron is there. Therefore, all a system can know of another system are quantum events. Why are we able to think of them in terms of information? Because they are physical events. What the history of physics tells us is that if we know the quantum events, they allow us to predict (as physical laws) the quantum events that will follow. In this sense, if we are clever enough, if we have enough paper, we can predict the events to come on the basis of these quantum events. They are thus information on the system. Quantum mechanics tells us this information is of a particular kind, because it is lost when we look at another information.

Personally, I would like to reduce quantum mechanics to this: quantum events that are always between two systems—by nature. If we choose a system and another system, we have a set of events and we can, if there are enough, calculate the next probability. A state is only the set of future events that allows us to calculate the events to follow. This is, to my mind, what relational quantum mechanics is.

Hervé Zwirn. I would like to ask Carlo [Rovelli] again a question that was briefly addressed earlier but received no clear-cut answer—if that is even possible. Can we, yes or no, definitely abandon the notion of intentionality or consciousness—or whatever we wish to call it—for an observer?

We know that Bernard d'Espagnat, in many of his articles, insisted on the fact that quantum mechanics, unlike classical mechanics, seems to not be describable without referring to the point of view of an observer. This is one of the main points he insists on. When the decoherence phenomenon was beginning to be described and better known—we have already discussed this during our two sessions dedicated to decoherence so I will not go over it in detail [15]—in Wojciech Zurek's early descriptions, we had the impression that he presented it by saying that with decoherence, we no longer needed an observer and that the world becomes classical spontaneously. Later on, we know this point of view was disputed. Nowadays, there is still a debate although we cannot say that the world becomes classical independently of any observer since its classical "aspect" stems from the impossibility for us human beings to make measurements that would show that "in reality" it is not classical.

Can we say that the relational interpretation discards completely the notion of consciousness, hence the intentional point of view? Or are we in a "for all practical purposes" situation by specifying that even if the correlations, recurrences, etc. are inaccessible to us, they are still there. And therefore, in that case, from a philosophical point of view, we cannot say that we have eliminated the need for consciousness or an intentional point of view.

Do you think you can do without it? Or do we still retain, from a relational point of view—and it seems to me that it would be difficult to do otherwise, but I would like to know your exact position—the fact that an observer is an entity with limited means who makes cuts in relation to the degrees of freedom he cannot observe, and that because of that the world appears as it does? And if we were super-observers with unlimited means, the world would appear quantum because the world remains quantum in its very essence. It never becomes classical—even if to us, that makes no difference. If that is the case, that means that all explanations or interpretations designed to account for what we observe as human beings must refer at some point to these limits—human limits—and therefore to consciousness.

Jean Petitot. Finiteness does not imply consciousness.

Hervé Zwirn. In any case, finiteness is the cause of what appears to our consciousness and we can only use the verb "appear" in relation to consciousness.

Matteo Smerlak. I understand the relational interpretation as an effort to separate the problem of consciousness—which appears in all epistemologies, all attempts to account for the state of science—from the inherent difficulties of quantum mechanics. At all stages in the evolution of science, there will be a Michel Bitbol who will say "what about transcendence, experimental conditions, conditions of possibility, the intentional observer?". This will still be the case probably even after quantum mechanics. In my opinion, the relational interpretation operates within quantum mechanics an important clarification: it does not discard the problem of consciousness, but reduces it to something even more external—i.e. to disentangle quantum mechanics, strictly speaking, from the problem of consciousness.

Michel Bitbol. Your project aims to sufficiently disentangle the problem of the interpretation of quantum mechanics from the human intentionality of consciousness. Then let me ask you this: who benefits from this project of disentanglement?

To which intentional subject, which conscious being does it matter to free science from the intentionality of consciousness? You see, I am playing my part!

Matteo Smerlak. I do not have to answer!

Carlo Rovelli. I completely agree with what Matteo has said. I do not want to ignore the problem, or deny the fact that it differs from quantum mechanics. Our effort consists in doing everything possible to separate them. But the question remains. Do I think the effort has a modicum of success? I would like to say yes, in the sense that the effort reduces quantum mechanics to something where we only speak of physical systems, quantum events, values of quantum observables. But success comes at a price, a hefty price in this case, and reopens the connection with the problem you mentioned. That price is that the notion of the reality we speak of is weakened, as Mr. d'Espagnat said earlier, much more than the notion of speed when Descartes, Galilei or Copernicus realized that the notion of speed does not mean anything other than in relation to another object. It is true that physical properties (i.e. spin up or down) are relative to our physical system. But it is much more than that. It is the very fact that certain events are realized or not that is relative. I must therefore say that in relation to a certain physical system, such and such thing will happen—but it will not happen in relation to another system.

The notion of reality is thus weakened. It is more difficult to describe reality as a series of unambiguous events. The answer is therefore positive, as much as possible: let us try to separate the problems. But the price to pay is what I have just described.

Hervé Zwirn. I completely agree. I give the example—albeit exaggerated—of two people talking to each other and one thinks he is talking about relational quantum mechanics and the other thinks he is talking about his skiing holiday in the Alps. In fact, neither is right or wrong. Simply, they cannot, either of them, realize that there is an incoherence. Reality in itself has no meaning, since it is not collapsed. It is a superposition of everything. It must be expressed only in relation to each of the observers. This means that the notion of usual realism is completely swept aside.

Jean Petitot. You have often made the parallel with classical relativity—a parallel that Vladimir Fock was already making in the 1930s. He explicitly said that the problem of measurement and of the measuring apparatus in quantum mechanics should be considered as a strict generalization of the principle of relativity. Indeed, in all theories with relativity there is a loss of realism. In general, we define at the start the entities which, we think, possess a certain reality, then we realize that they are relative, that there is a group of relativistic invariance, and thus they lose their reality. This was the case, for instance, for speed with Galilean relativity. At the time, it was extremely traumatic to consider that speed was not a property of bodies. But beyond the principles of relativity, we can still recover some invariants. My question is therefore the following. You introduce a new relativity in your relational point of view that weakens or even dissolves the reality of the quantum state.

However, you should recover, somewhere, new invariants since there cannot be relativity without invariants. What is the purpose of these new invariants?

Jean-Pierre Gazeau. Spectra.

Roger Balian. It is the algebra of observables.

Jean Petitot. These invariants would therefore be C* algebras, or something like that. This leads us formally towards a set of mathematical considerations.

Carlo Rovelli. It is a very mathematically-inclined way of thinking. I think that these invariants are first of all spectra, i.e. a set of possibilities and transitional amplitudes. Given a certain sequence of quantum events, the probability to have one of them is a defined number which is fixed.

Jean Petitot. But these invariants are already present in non-relational theories. They are not new. The question I ask you is this: how do you recover these things that are already known in your approach which claims to be more fundamental? It is a bit like when we try to find informational axioms that allow us to recover what we already know.

Alexei Grinbaum. The history of the reconstruction of quantum mechanics (Matteo Smerlak has already mentioned this in his introduction) tries to answer this question. Attempts at reconstruction, which can in no way be considered to have been successful, consist of mathematically deriving the formal content of quantum theory.

Jean Petitot. That was my question.

Roger Balian. Has this not been achieved? Is it not the algebra of observables?

Alexei Grinbaum. No.

Carlo Rovelli. We know the starting point. We know the end point. Because we know the invariant part of quantum mechanics. If I measure a spin along Oz, then along Ox, the probability of obtaining outcome + in the second measurement is 50%. The relation between these two measurements is a fixed number, which tells us what the physical processes that are possible in the world and their intrinsic characteristics are. It therefore provides something objective.

Roger Balian. It is in the algebra.

Carlo Rovelli. But something is missing. The result we would like to have is the following. Considering this interpretation, considering this way of apprehending quantum mechanics—the clear and simple postulate according to which there are quantum elements that are relative to systems—it is a fact of nature that I take to be true that, given a certain number of quantum events, we can calculate the probability of events to come. I take that as a postulate. Can we, based on this postulate, reconstruct the whole of quantum mechanics? The answer is “nearly”. Or “yes”, by adding certain technical details. I we could do without these technical details, we would be very happy. But they are still there. And they tell us that in this probability we think we have found, in truth, something is missing.

Alexei Grinbaum. Along the way, you discover that these technical details are not all that technical, because they are very profound.

Carlo Rovelli. Yes.

Catherine Pépin. Could you describe these details to me?

Alexei Grinbaum. A first example would be the existence of continuous and irreversible transformations between pure states—i.e. the origin of the notion of continuity which is part of the formalism of quantum mechanics. During Matteo Smerlak's presentation, at no point did you hear the term "continuity". The postulate that leads to this continuity must be added, whatever the attempt at logical or informational axiomization. You cannot do without it. And the approach which consists in trying to understand the origin of this postulate—what can we replaced it with?—is a subject in itself.

Bernard d'Espagnat. Thank you. We have had a fascinating debate where we have touched upon (but by no means exhausted!) the main questions facing us. Should we continue with another session dedicated to this topic? The answer is not obvious. I believe we should have a second session; it would be good if someone would reopen the debate.

Alexei Grinbaum. Michel [Bitbol] and I have worked on rather different approaches: Michel has taken a more philosophical angle, whereas I have focused on more technical aspects. Perhaps we could each speak for 10 min.

Michel Bitbol. OK.

Bernard d'Espagnat. Problem solved. We will meet again on the 12th of March at 4:30.

References

1. *Cf.* Carlo Rovelli, "Relational quantum mechanics", *International Journal of Theoretical Physics*, 35, 1996, p. 1637–1678.
2. John von Neumann, *Mathematical Foundations of Quantum Mechanics* [1932], Princeton University Press, 1996; Eugene Wigner; *Group Theory and its Application to the Quantum Mechanics of Atomic Spectra* [1931], New York, Academic Press, 1959.
3. Alexei Grinbaum, "Reconstruction of quantum theory", *Brit. J. Phil. Sci.*, 58, 2007, p. 387–408.
4. See (1)
5. Matteo Smerlak & Carlo Rovelli, "Relational EPR", *Found. Phys.*, 37, 2007, p. 427–445.
6. Michel Bitbol, "An analysis of the Einstein-Podolsky-Rosen correlations in terms of events", *Phys. Lett. A*, 96(2), 1983, p. 66–70.
7. Federico Laudisa, "The EPR argument in a relational interpretation of quantum mechanics", *Found. Phys. Lett.*, 14, 2001, p. 119–132.

8. See Bernard d'Espagnat's comment at the start of session VIII, "Exchange of views on the relational interpretation".
9. See (5)
10. Bas van Fraassen, "Rovelli's world", *Found. Phys.*, 40(4), 2010, p. 390–417.
11. "The tentative advance of "immanent relativity" in contemporary interpretations of quantum mechanics. II) Relational interpretations", in *De l'intérieur du monde*, Flammarion, 2010, p. 137.
12. See (1)
13. Hugh Everett III, « 'Relative state' formulation of quantum mechanics » , *Rev. Mod. Phys.*, 29, 1957, p. 454–462.
14. Hervé Zwirn, *Les Limites de la connaissance*, Paris, Odile Jacob, 2000.
15. See sessions III « Experimental approaches on decoherence » and IV "Exchange of views on decoherence".

Author Biography

Matteo Smerlak is a post-doctoral researcher at the Perimeter Institute for Theoretical Physics (Canada). His research area is at the crossroads between general relativity, quantum mechanics and statistical physics. His research interests include the "thermodynamics of the vacuum" and the different quantum phenomena related to it (Hawking effect, dynamic Casimir effect). He has also worked with Carlo Rovelli on the aforementioned relational interpretation of quantum mechanics.

Chapter 8
Exchange of Views on the Relational Interpretation and Bell's Theorem

Round Table

Bernard d'Espagnat.

Michel Bitbol and Alexei Grinbaum have kindly volunteered to start off the discussion. However, before that, I am unfortunately obliged, quite in spite of myself, to give you some complementary information. And that because of a passage in Matteo Smerlak's excellent presentation, which you have read in the report of our session from the 16th of January 2012. Here is the passage in question:

> Other authors speak of "non-separability", like Alain Aspect or Mr. d'Espagnat in one of his famous works: "*If the notion of reality independent to man, but accessible to its knowledge is to have any meaning at all, then such reality is necessarily non-separable.*" This quotation by Mr. d'Espagnat introduces the concept of non-separability, of which we can say, like non-locality, that it is strange. Indeed, let me remind you that there is a complete set of observables that commute for the system α and β, the pair of particles, that is only accessible by measurements on α and only on β. Is this criterion not precisely separability as we can conceive of it?

From my point of view, this passage needs to be commented on and expanded. Indeed, he claims that, with this quotation of mine, I *introduce* the concept of non-separability. Doing so, he ignores the fact that I had given a previous definition. And what is more troublesome, is that he interprets this quote according to his *own* definition of separability, the one we have just read, where the outcome would apparently be that the particles α and β of EPR are separable; that, equally, after impact, the two quantum systems are still separable … and that consequently this sentence of mine that he quotes makes no sense …

Without any hard feelings at all towards Matteo who has probably not done it on purpose, I would like to point out that this substitution—involuntary, of course!—of one definition with another is very misleading. Because in my book *Conceptual Foundations of Quantum Mechanics* (published in 1971 and still available in bookshops) [1], I had—of course!—taken great care to define non—separability ("in one's mind"). I had done so by pointing out that when two quantum systems,

B. d'Espagnat and H. Zwirn (eds.), *The Quantum World*,
The Frontiers Collection, DOI 10.1007/978-3-319-55420-4_8

each represented by a wave function, collide then separate, even after separation, they generally lose the property to each have their own wave function. There is only the shared wave function that defines them, thus we can no longer think them as having their own complete set of commuting observables with well-defined values, and as I have said "that is the essence of what is sometimes called the non-separability of systems" (page 79 of the aforementioned book, 1989 edition). We can like this definition or not. But at least it has the merit of existing and of being coherent with standard quantum mechanics. And considering the estimated frequency of collisions occurring between particles, it fully justifies the sentence that Matteo was implicitly criticizing.

Of course, the sentence in question supposes the validity of standard quantum mechanics, which takes the notion of quantum state seriously, and is valid only within its framework. What I have just said does not directly impact Carlo's theory, which sees in "quantum states" only a tool for predicting observations.

However, I would like to raise two additional points, which do call this theory into question. But first, we must hear the contributions by Michel Bitbol and Alexei Grinbaum. I will therefore come back to these points later on.

Roger Balian. May I comment briefly on what you have just said? Non-separability is the existence of a correlation in classical mechanics. That would also be the case in quantum mechanics.

Bernard d'Espagnat. It is entanglement, which is more than a correlation.

Roger Balian. It is more than a correlation, but it is of the same nature as a classic correlation in classical statistical mechanics.

Bernard d'Espagnat. No. If we take quantum mechanics seriously—I am not speaking here of large systems but simply of particles—we cannot equate entanglement with a simple correlation.

Roger Balian. I do not equate it. I simply make a comparison, by pointing out that it is the quantum equivalent of a correlation that is created if we take two independent particles, in the sense of classical statistical mechanics. We take two beams, make them interact and they become correlated.

Bernard d'Espagnat. Yes, indeed.

Roger Balian. And it is equivalent to entanglement.

Bernard d'Espagnat. No, precisely not. The difference is that in classical mechanics, we can think the two particles, once they have separated, as having as many properties as before. Whereas in quantum mechanics, that is not the case. As long as we have not perturbed them by making measurements on a complete set of the type of commuting observables mentioned by Matteo, we will not be able to *think* particles as having all the properties corresponding to these measurements, or others of the same quantity; whereas before impact, the formalism would have allowed it. It is therefore more difficult to think the particles in questions as "correlated but separated" than in the case of billiard balls where, on the contrary, it is natural to do so. It was in order to characterize this new, specifically quantum event, that I used the term "non-separability" (I believe, in fact, that I was not the first to use this neologism).

Roger Balian. My point was that it was similar not to classical mechanics but to classical statistical mechanics.

Bernard d'Espagnat. Perhaps in certain ways, but that is a different point.

Matteo Smerlak. I am certain we are many across the world to have read your works—and misinterpreted them. It is a great privilege to know you in person, and to be able to benefit from your corrections of my interpretation of this passage!

Bernard d'Espagnat. By the way, where did you find this quote (that I am not denying of course!)?

Matteo Smerlak. I believe it was quoted in an article. I quoted it from a secondary source. I could find it again, but I cannot recall where it came from just now. And, also, I did not have access to the book itself at the time, which explains my mistake. Thank you for this clarification.

Bernard d'Espagnat. Let us move on to the scheduled presentations. Michel Bitbol has the floor.

8.1 Presentation by Michel Bitbol

Michel Bitbol. You have gone back to the origins of your book *Conceptual Foundations of Quantum Mechanics*, which I read with much interest in 1982, and which was one of the first comprehensive works published on the philosophy of quantum mechanics. As for me, I will presently speak to you about my first article on the philosophy of physics, from 1983 [2]—a bit before Matteo was born! This article, which you have cited in some of your works, was spurred by a result obtained by John Bell in 1981 that is quite remarkable but has not been commented on much. This result was published in a rather well-known article, which was very clear and informative [3]. To explain it, I will start by recalling certain elementary notions of Bell's theorem, to make you grasp the difference between Bell's original theorem from 1964 and the later version of Bell's theorem from 1981.

You all know that Bell's theorem from 1964 shows that no local hidden variables theory can be compatible with the predictions of quantum mechanics regarding Einstein-Podolsky-Rosen type correlations. In other words, quantum mechanics rules out the conjunction of two hypotheses: realism regarding the properties of particles and locality of each of these properties (the fact that there are no signals making them instantly communicate with each other). Let us note that both excluded hypotheses, realism and locality, deal explicitly with *properties*.

However, upon reflection, Bell significantly increased the scope of his conclusions by showing that quantum mechanics ruled out the conjunction of two more wide-ranging hypotheses: realism regarding *experimental events* and locality of these *events*; thus a major qualitative leap between the 1964 and the 1981 theorem.

From there, it remains to be determined which of the two hypotheses, whose *conjunction* is ruled out by quantum predictions, must be rejected. As long as the hypotheses deal with the properties of particles, rejecting either remains possible. We can abandon knowing the locality of properties, but we can also abandon knowing the realism pertaining to these properties. It is not implausible that we cannot describe the intrinsic properties of physical systems, but only the relation between these systems and a measuring apparatus. However, when the hypotheses in question deal with macroscopic experimental events, observable with the naked eye in the laboratory, only one choice seems possible. The only way to resolve this seems to be to reject the locality of experimental events. The other option, i.e. rejecting the realism of experimental events, seems much too far-fetched to be retained. Who would deny the reality of what is visible and tangible at the scale of the experimenter? John Bell counted on the unthinkable nature of the latter option to make the rejection of locality necessary.

Rejecting the realism of experimental events goes against the most minimal realism we can imagine, namely that of empiricists: empiricists do not believe in any reality except that of visible events. They do not believe, in particular, in the entities created by theoretical physics. They differentiate between what they consider possible to attribute a reality to (macroscopic objects and events), and what they do not consider possible to attribute a reality to (theoretical entities). Is it permissible to be even more anti-realist than the empiricists? Is it permissible to be even more sceptical than the empiricists, by doubting the proven, tangible, visible macroscopic reality of experimental events observable in the laboratory? Can we go as far as to distrust the experimenters Alice and Bob who claim that at the initial time of their local measurement, they had already observed a certain spin value of the particle they were measuring, and that the correlation had already occurred at that time, prior to the moment when they compare their measurements? However, in order for this correlation to occur at that time, we must invoke some form of non-locality! Here is the problem. I am pushing this problem to its limits, so that we can clearly see why Bell thought it was impossible to ignore non-locality.

Nonetheless, as daring (or fool-hardy) as it may seem, I have been trying since 1983 to explore an alternative possibility consisting of rejecting the hypothesis of an intrinsic reality of experimental events, including events proven in retrospect by observers, and on the contrary maintaining the hypothesis of locality. I was influenced by Mr. d'Espagnat, who carefully read through the manuscript before its publication and who pointed out that it seemed to juggle two diametrically opposed philosophies: one that was completely empiricist, even anti-realist, and the other that was ultra-realist, that of Hugh Everett. This judicious comment by Mr. d'Espagnat (for which I am thankful to this day) lingered in my mind for years, and I realized that there was indeed something surprising in the conjunction of these two radically different philosophies, but that I could reconcile them in a way. To achieve this, I identified two seemingly diametrically opposed strategies that could allow for the construction of a theory that does not suppose the intrinsic reality of experimental events, and showed that they had one major but generally overlooked point in common. The two strategies are the following:

1. The first one could be called "radical empiricism", with reference to William James [4]. It restricts the ontological involvement to what is contained within immediate experience, here and now (i.e. we declare that only immediate experience, in the present, *exists*). From this point of view, it is clear that past experimental events have no reality "in themselves", but only relative to immediate experience. Still from this point of view, it is clear that events described a posteriori cannot be considered as having existed intrinsically in the past, but only in *the present* as exact translations of a state of the observers' memory. The events depend at all times on a process of reconstruction and comparison, from data of the actual experiment. We cannot say they "exist" but only that they are (re)constructed *ex post facto*. In my 1983 article, this strategy was formulated with more caution, by trying to avoid sounding too openly idealist. I contented myself with mentioning that the comparison between two experimental events which occurred at a distance from one another in the past supposed itself an event in the present. It is only within the context of this present event of confrontation that the two past events are retrospectively *constructed* as being comparable to each other. Another way of presenting this would be to replace the present experience of the event with a present intersubjective agreement regarding the event. We replace an extreme quasi-solipsistic idealism with a language pragmatics, and thus with a speech community.
2. The other strategy seems to be the polar opposite of the first. It relies on the "radical scientific realism" of entities of theoretical physics. In quantum mechanics, this amounts to acknowledging the realism of vector states. We are not talking about discrete events (the famous "reductions of wave packets"), but about a continuous evolution, in accordance with the Schrödinger equation, of these vector states. However, we must find a way to link such an event-less formalism to something that is *recognizable* by an observer *as* an experimental event. A fascinating way to achieve this link was proposed with the "interpretation of relative states" by Everett in 1957 [5]. Everett attributes a vector state to the observer and introduces a notation for the content of the latter's memory. His memory contains, among other things, the recollection of *having been a witness* of such or such experimental event. From this "radically realist" interpretation of quantum mechanics, we arrive at the surprising conclusion that experimental events do not *exist* "in themselves"; but that relative to each counterpart of the observer, from his partial point of view, a well-defined experimental event *appears* to have occurred.

Strangely enough, consequently, when taken to extremes, empiricism and realism meet up on a crucial point. Both suggest that an experimental event does not *exist* in absolute terms, but relative to the point of view of who bears witness to it. In radical empiricism, the event "is" relative to the point of view of the ultimate observer who represents the "present transcendental consciousness"; and with the radical realism of vector states in quantum mechanics, the event "is" relative to a counterpart of a naturalized observer, himself associated with a vector state. Bell clearly sensed this convergence, when he described what was novel in Everett's interpretation: "a repudiation of the concept of the 'past', which could be considered in the same liberating tradition as Einstein's repudiation of absolute simultaneity [6]." More precisely, in Everett's theory, "there is no association of the particular present with any particular past [… and] this does not matter at all. For we have no access to the past. We have only our 'memories and 'records' [7]." Bell considered this description of Everett's interpretation as its "*reductio ad absurdum*" (when he concludes that it amounts to a radical solipsism of the present). But nothing stopped him from supporting it philosophically!

The recent revival of the option where realism rather locality is rejected in the treatment of EPR correlations shows that in any case dismissing it in the name of common sense is no longer a self-evident position. We have heard the relational version of this option, with Matteo [Smerlak]'s and Carlo [Rovelli]'s presentation. Others are emerging, for example in a very recent article, Andrei Khrennikov [8, 9] claims that we can do without non-locality if we fully accept all the consequences of contextuality in quantum mechanics, if we consider "everything contextually", if we accept that any determination, whatever it may be (on its own or comparing two singular determinations), is relative to a context. If we accept this, we can then perhaps do without non-locality. This is what Khrennikov proposed anew in his 2011 article, after Matteo and Carlo.

8.2 Discussion

Roger Balian. I was surprised by this story of experimental realism. In reality, in Bell's experiments, we put together experiments that cannot be carried out at the same time, that are incompatible. Consequently, we find things that are contrary to common sense, even though they are perfectly reasonable in quantum mechanics—but which depend on the experimental context.

Michel Bitbol. Absolutely. I think the way you phrased this—perhaps you will disagree with me—is extraordinarily Bohrian. Indeed, you say we cannot put in the same formula experiments that were not carried out simultaneously.

Roger Balian. I do not know if it is Bohrian, I am not much of a philosopher! But it is down to earth.

Michel Bitbol. It is somewhat similar to what Bohr replied to Einstein, when he said it was too much to ask of quantum mechanics to describe situations that would require, to be demonstrated, two incompatible sets of measurements.

Carlo Rovelli. Michel [Bitbol], in the first part of your presentation you formulate the problem as a tension between locality and realism. In the second part, you show that two extreme solutions are possible, where the notion of reality is called into question. It seems to me that, even with these two solutions, locality is weakened in a way. In the radical empiricism solution, there still is a form of locality, but it is somewhat superfluous: from the moment the experiment is linked to an observer, it is necessarily local—which is fine with me. In the radical realism solution, currently in vogue in England [10], locality seems to me to be lost in a brutal way: a type of physics that becomes realist in the sense of a wave vector is not physics in space-time. There is therefore a ruthless non-locality. From this point of view, if the problem is a tension between the two, the second solution seems far worse than the first.

Michel Bitbol. First of all, we must come back to a crucial point in my presentation. Ultimately, the two solutions (radical empiricism and radical realism) have one point in common: they are anti-realist regarding events. Thus, radical realism is realist with regards to mathematical entities (*e.g.* vector states), but not events. An event, for an Everettian, is simply an account by the counterpart of an observer represented by a certain vector state within the superposition. The event does not exist, in the same way it does not exist in radical empiricism. That is the point the two possibilities have in common. Besides, you say that this formal realism, the realism of the mathematical entities of quantum mechanics, maintains a form of extreme non-locality. However in my opinion, I believe that it does not have a set idea regarding locality. It is a-local. It is not concerned with locality. But at the same time, it allows us to attribute spatial coordinates to its objects from the values the various counterparts of the observers measured for observables X, Y or Z. If we accept this version of radical realism, this formal realism of mathematical entities of the vector state-type, we can say that is supposes a-locality on the one hand (the a-locality of the vector state entity occurs in Hilbert space rather than ordinary space) and on the other hand the absence of necessity to lose locality in the traditional sense of the term, i.e. in the sense of instant non-communication between the different positions as they are measured when we determine the observable spatial coordinates. With this type of realism, the two are true at the same time: a formal a-locality and a compatibility with locality in the usual spatial sense of the term.

Bernard d'Espagnat. I believe there is another way to recover radical realism. I am thinking here of the discussion between on one side John Bell and on the other Abner Shimony, John Clauser and others, the latter claiming that we could avoid violating Bell's inequalities if we assumed that, contrarily to what we think intuitively, the positioning of the different instruments does not stem from a truly free choice on our part. If we postulate absolute realism and absolute determinism, we must say that everything is determined, including our actions that we consider a manifestation of free will, therefore including the positioning of the instruments in question. These authors have shown that given this hypothesis, Bell's demonstration does not work.

Bell could only defend himself, more or less, by providing very good arguments of plausibility to reject this hypothesis of absolute determinism applied to the positioning of instruments.

Michel Bitbol. That is perfectly true. On this topic, let me point out that Asher Peres and Wojciech Zurek wrote an excellent article in 1982 (entitled "Is quantum theory universally valid" [11]) on the fact that we must choose between two options: absolute determinism or the possibility to give meaning to science. To give meaning to scientific data, one must assume that the observer is free to choose the experimental conditions and that he is not himself strictly determined.

Alexei Grinbaum. Nevertheless, absolute determinism is still alive today, notably with superdeterminism [12] which the physicist Gerard 't Hooft has been defending for years. It is very coherent, even if I do not know if anyone apart from him believes in it.

Michel Bitbol. This shows that by pushing the limits of metaphysical hypotheses, we can achieve interpretations that are radically different from Bell's own interpretation. Bell did not allow himself certain options, like absolute determinism or solipsism of the present—he even criticized the latter, as you saw. But if we do not have the same restrictions as Bell, there are a fairly large number of "exotic" solutions to the EPR paradox. Often, the range of possible solutions in physics is artificially restricted by the excessive importance of the "metaphysical superego".

Roger Balian. I would like digress one moment on Bell's inequalities. Among the multiple models we have studied, there is one that, in a way, allows us to have an image of the object in the apparatus. In ideal measurements, the images are totally correlated: when the apparatus says *up*, it is *up*, when it says *down*, it is *down*. With this model, there is a way to build a dynamic device that makes measurements where we obtain a probabilistic image of what the apparatus says compared to what the object says. What is amusing is that when we look at the correlations between effective measurements (what the pointer shows), Bell's inequalities are not violated simply because the classical image on the pointer of the existing correlations in quantum mechanics becomes squashed to the point that Bell's inequalities become satisfied regarding the measurement outcome. However, what we infer from these measurement outcomes on the apparatus when we apply the theory of dynamic object/apparatus interaction leads us to recover Bell's inequalities. It may be a rather basic comment, but it shows that macroscopic objects do not follow the same laws as microscopic objects, and yet they are perfectly correlated within a completely quantum framework.

Alexei Grinbaum. If I may add a few words to what you have said, the theory of quantum discord—which is not the same thing as entanglement—studies these correlations, which do not violate Bell's inequalities but are nonetheless a valuable resource.

Bernard d'Espagnat. I think it is time to let you start your presentation.

8.3 Presentation by Alexei Grinbaum

Alexei Grinbaum. I have the feeling that this discussion on radical empiricism could lead us to speak of counterfactuals, but I would like to speak today about something completely different—and I have the privilege of mentioning articles published after Matteo was born!

I will broach two subjects. The first is a historical overview. In his 1995 article on relational quantum mechanics, Carlo Rovelli [13] tried to propose a formal, mathematical development of his ideas, and still within the framework of relational philosophy, to derive from informational principles the elements of quantum mechanical formalism. This approach was outlined in Carlo's article, and I attempted to develop it myself eight or ten years ago. At the time, we used the ideas of quantum logic. I do not wish to go into any detail, but simply note that it allowed us to understand the limitations of this attempt.

In his 1995 article, Carlo Rovelli proposed two axioms. The first is that there is an upper limit concerning the amount of relevant information to a system. The second is that we can always obtain new information about a system. From this starting point, I tried to use quantum logic to derive the structure of Hilbert spaces and elements of quantum mechanical formalism. Doing so, I realized we needed to add many supplementary principles to these axioms. Indeed, even if they capture something essential by considering that there is an upper limit but also that there is a possibility to renew our "hidden" information, our memory, we need to add other postulates to arrive at quantum mechanics. In particular, we need to add a structure that introduces continuity in the transformations between pure states as an axiom directly postulating the real, complex or quaternionic numerical body. We therefore need many elements to get Rovelli's programme to lead to a Hilbert space. From that point on, I asked myself what the problem was. By needing a certain number of additional axioms, the mathematics were clearly indicating that the philosophy on which was based Rovelli's approach, put in place using the formalism of quantum logic, was perhaps not sufficient. It is possible that it is not yet sufficiently understood, since the "cost" of the additional assumptions is high. We do not have in the 1995 version a natural correspondence between the conceptual postulates and their mathematical expression.

I therefore asked myself what was this conceptual flaw that I could, perhaps, help elucidate. It seemed to me that this flaw could be found in the absolute—and thus non-relational—use of the notion of system. Here are two quotes. The first is by Carlo Rovelli: "Any system can be an observer for another system. Information is nothing else than correlation." The second, taking the opposite view, is by Asher Peres: "No one would say that two electrons in the fundamental state of a helium atom are observers for one another or measure each other."

Carlo Rovelli. This claim is obviously wrong! He says "no one would say that…", and yet here is precisely someone who does …

Alexei Grinbaum. As I have had the opportunity to say in the past, I consider that the question of the observer and his choice of observed system is still unresolved. Is the observer a universal notion, i.e. can any system be an observer in quantum mechanics, or is it only under certain conditions, and if so, which ones? To speak of information and observers, I have chosen to consider two ideas. The first, even if it was formulated prior to this, was defended by Léon Brillouin in the 1940's: we need to exclude human elements, such as consciousness, from the information analysis [14]. The second was expressed by Hugh Everett, and says that what matters with the observer is memory.

As for me, I would like to briefly introduce a research programme—which currently is unlikely to be funded—divided into three components: a mathematical component, an experimental component and a logical/computer science component.

Is an ice cube a system that could be described by quantum mechanics? It is clearly a physical system. But when it starts to melt, at what point is it still a system—not necessarily quantum, but simply identified as an "ice cube" system? It is clear that within a given field, for example the mechanics of solids, it ceases to be a system at a certain point. Why? Because the degrees of freedom that were associated with an "ice cube" have disappeared. In the liquid state after melting, the degrees of freedom are completely different. Consequently, it is essential for an observer within a physical theory to retain the identification of the system under observation. In other words, the observer retains the information regarding the relevant degrees of freedom. This can be succinctly formulated as: to identify a system is to retain the identification of the relevant degrees of freedom of this system.

What is a relevant degree of freedom? To answer this question, I propose a simulator that is as abstract as the Turing machine, the abstract simulator of a computer. I propose to consider that the observer is an algorithm. Whatever its physical format, it is this algorithmic aspect that allows us to distinguish the observer from any other heap of physical degrees of freedom. At an abstract level, the observer has a sort of strip of tape in front of him, with many degrees of freedom—and he puts a cross in front of all the relevant degrees of freedom. Obviously, this is not a physical empirical realization, in the same way the Turing machine is not a real computer. But all observers follow this abstract model, exactly like all real computers are abstract Turing machines. The definition of observer as an algorithm that identifies systems provides a general model of observation.

If the observer is an algorithm, the mathematical characteristics of this algorithm must be independent from the practical physical realization because, physically speaking, an observer can be a butterfly, or a gas molecule, etc. This abstract observer, this algorithm, has an invariant characteristic of its material format which is its Kolmogorov complexity. It characterizes not the empirical physical system, but the algorithm itself. When the abstract observer identifies a system, he writes a sequence of 0 and 1 that indicates the relevant degrees of freedom. If the complexity of this sequence is less than the capacity of identification of the observer, i.e. the complexity of the observer as an algorithm, we would say that the observed system is by definition a quantum system.

Catherine Pépin. Could you define the Kolmogorov complexity?

Alexei Grinbaum. It is a notion used in computer sciences and in the theory of complexity. Algorithmic complexity, or Kolmogorov complexity, is the length of the shortest programme that reproduces a given sequence. One can show that this length is an invariant quantity, up to a constant, and does not depend on the machine that is executing the programme. The Kolmogorov complexity is the only characteristic of an algorithm that is independent of its physical realization.

I would like to present three results of this research programme, starting with the mathematical result, and then presenting the result from experimental physics.

First of all, the mathematical result. I tried to replace Carlo Rovelli's axiom stating that any system can be an observer, with something more nuanced. To observe a system as a quantum system, i.e. to retain all the degrees of freedom through measurement and the evolution of the system seen by observer, the observer must be more complex than the system. But does he need to be much more complex? Then, if the first observer is very complex but the second is only slightly more complex than the sequence of symbols that describes the system, will they identify it in the same way? Can we provide an objectivity criterion, namely the criterion of an identification agreement between two observers observing a quantum system? I propose a mathematical criterion: if we have a sequence of different observers observing different systems (systems S observed by observers O), it will be necessary that the addition of any new observer does not change the Shannon entropy of this process of multiple observations. There is a very interesting link between the Shannon entropy of this process and the Kolmogorov complexity of these observers as identification algorithms. The objectivity condition I set mathematically is that when we add observers, the rate of Shannon entropy for the process is equal to zero. I derive from this a condition on the Kolmogorov complexity of the observers that agree among themselves, i.e. observers that give an "objective" description of the system in question. Concretely I show, from Brudno's theorem as conjectured by Alexandre Zvonkine and Leonid Levin, that the complexities of the sequence of identification produced by these observers must increase less than linearly when we add systems to the observation process.

Roger Balian. Initially, the observer defines the relevant variables of what he observes. But do these observers have the same relevant variables, or different ones?

Alexei Grinbaum. The objectivity criterion answers the question of knowing when a correspondence between two identifications is possible. If we take two real physical observers and describe them as producing sequences of 0 and 1, these sequences are always defined. From then on, the condition I have just mentioned becomes simply an identity condition between two sequences of 0 and 1. However, since we do not know the empirical reality underlying this notion of observer, we have to allow the possibility of infinite sequences. And consequently the objectivity condition, or the existence of a concordance between observers, is that the rate of Shannon entropy is equal to zero.

The second component of my research programme is an experiment that has never been carried out to this day, but for which I can predict the outcome from past experiments. Take a very fine calorimeter and insert it in a fullerene (C_{60})—a single one. Let us see if this C_{60} will work as an observer. For that, let us send photons on it. It will “observe” (in the language of quantum mechanics) or “absorb” (in the language of experimental physics)—these two terms meaning there will be a correlation between the degrees of freedom of the fullerene and the degrees of freedom of the photons. In the vibrational, mechanical, etc. degrees of freedom of this highly complex molecule, there will be information on the state of the photon that has arrived. From the outside we do not have detailed access to these conditions, but is it even possible to know that they exist, i.e. to know whether an observation took place? To find out, let us send multiple photons one by one. Each time, the fullerene will observe the photon and store the information. When it will have reached its algorithmic complexity as an observer, it will no longer save this information and begin to delete it—which will generate heat. Indeed, according to the Rolf Landauer principle, to delete information is to produce heat. We can measure this spike in the calorimeter and we can measure the amount of heat. I am not suggesting any new physics, but a totally new interpretation of a physical process. This process, according to measurements taken around twenty years ago, shows that a fullerene can “observe” up to ten photons (depending on wave length and other elements). In my opinion, this new interpretation of the absorption of photons by a fullerene can allow us to understand what happens in terms of observation in quantum mechanics.

Finally, one last word on the consequences for logic and quantum information science, bearing in mind this work is ongoing and far from being finished. The approach where the observer would have limited capacity (without really knowing its physical realization) allows us to see, in a way, where the Tsirelson [15] bound in Bell’s inequalities comes from. I propose a very simple, even simplistic, scheme. Take Bell’s inequality of the form CHSH, with four correlators, and imagine that a processor that calculates the result of this formula does not have four registers to save the numbers. Imagine, to begin with, that there is only one. Then it can only perform $A + A + A - A = 2A$. If the value of A is between -1 and 1, then the maximum will be 2. Now imagine there are two registers but that these are qubits (we can save a vector quantity, but not read it twice because once it is read it disappears). With two vectors, we could first calculate the sum and difference of A and B and store them in the registers. Then, we would use them to calculate CHSH. In a way, you obtain the Pythagorean theorem: the length of A and B is 1, but the maximal length of $A + B$ and $A - B$ is equal to the square root of 2. And when you have $2\sqrt{2}$, you have the Tsirelson bound. It is simple, even simplistic, but we can see that the idea of a limited capacity of the observer can have consequences allowing us to understand the foundations of quantum mechanics.

8.4 Discussion

Bernard d'Espagnat. I am both seduced and perplexed. Seduced by the component of your programme that aims to evaluate the minimal complexity that an observer must have: that seems justified to me even from an antirealist point of view, considering that the act of observing always requires the use of instruments (eyes, for instance, for a human observer). However, I am perplexed because you seem to rally to the realist position ultimately defended by Carlo, as in this context in my opinion two difficulties (both specific to quantum mechanics) remain. However, I will come back to them later, as I have said.

Jean Petitot. I think Alexei [Grinbaum] is fundamentally right, and in fact, all our observations in the most naive sense correspond to what he said from the moment we take into account what we are: how we see, how we calculate with our minds. Take a sensor like a photoreceptor in the retina; it observes a photon exactly as Alexei said a fullerene does. A cortical neuron does the same thing, but in a more complicated way. Therefore I think that not only is Alexei right, but that ourselves, as observing minds, are informational machines according to his definition. Simply, a fullerene is much simpler that our sensors and neurons.

I am always surprised to see that with this type of discussion on observation, we constantly speak about the sight of the human observer who "reads" the instruments he uses, but we never speak of these observers in terms of biological machines made up of photoreceptors, neurons, etc. However, a retinal photoreceptor is an absolutely extraordinary quantum photon detector that satisfies the principles mentioned by Alexei. In the dark, it is capable of detecting one or two photons. Under normal conditions, it is capable of measuring fluxes and wave lengths, etc. But it is only capable of that. Its photon-sensitive outer segment is a stack of thousands of membranous disks where rhodopsin macromolecules, made up of opsin with a retinal molecule at the 11 cis position, are inserted. Photons trigger a cis-trans isomerization of the retinal, which induces a hyperpolarization of the membrane and a neurophysiological chain reaction in the layers of the retina. That is how ganglion cells are activated by light and send information through the optic nerve. All perception begins with this retinal isomerization, which is in the strictest sense an observer.

In short, a photoreceptor is capable of coding certain degrees of freedom and it is thus typically an observer. But as soon as there is too much information to code, it deletes it—thankfully! I would like to point out that if photoreceptors did not constantly delete the information they code, I do not know what our perception would be like.

Alexei Grinbaum. A problem was raised at the start and I still do not know how to answer it: how is all this quantum mechanics? This story of limits in observation can be something more general.

Jean Petitot. When you have a protein and all you need is a photon in order to have a stereo-chemical phenomenon that changes its configuration, then we have a typical quantum phenomenon.

Alexei Grinbaum. Yes, in the sense of Carlo Rovelli's axioms which limit the information available. On the one hand, I can see the analogy. But on the other hand, I also see a challenge: to my mind, it precedes quantum mechanics since it allows us to speak of a notion that is not defined by it—that of the observer. It is a meta-theoretical notion in quantum mechanics. We need to show where the Hilbert space comes from. At present, I do not know how to.

I think that with this approach, we will be able to understand the important elements of this quantum formalism, which are structural like the amount of non-locality, rather than derive unitary evolution over time.

Hervé Zwirn. Since we are here to discuss things, I think we can disagree! As for me, I disagree on a number of points regarding what has just been said. Admittedly, the fact that a photoreceptor is capable of detecting a photon is undeniable. But so what? What does that have to do with what we are discussing? It is not because something is true that it contributes something to a particular question. It can be completely irrelevant to the problem we are considering. And this is where I think lies the ambiguity in our discussion. I do not disagree with everything that has just been said, but I consider it to be beside the point. The fact that a photoreceptor in the eye is capable of detecting a photon, which no one will deny, is completely disconnected from the questions we are discussing.

As I had the opportunity to mention previously, my main is problem is primarily with Alexei Grinbaum's use of the Kolmogorov complexity: first of all because it is not applicable to algorithms, but only to strings of characters. The Kolmogorov complexity of a finite string of characters is defined as the length of the shortest programme that allows a given universal Turing machine to write this string on its tape from an initial blank tape. The interest of the concept is that it is possible to prove (via an invariance theorem) that this definition depends on the universal Turing machine considered only in the following way: given two universal Turing machines U and V, there is a constant C dependent only on U and V such that, for any string s, the Kolmogorov complexity of s according to U and of s according to V differs in absolute value by an amount less than C. Once we have chosen a universal Turing machine, we can calculate the complexity of any string and be certain that the complexity calculated by another machine will not differ by more than a fixed constant that depends only on the choice of the machine and not of the string. There are a number of problems with what Alexei Grinbaum has said. The first problem is that the Kolmogorov complexity of an algorithm is not a defined notion. There is confusion between algorithm and string. An algorithm is indeed a programme, hence a string of characters, but researchers in the field know that an algorithm can be implemented in different ways and that depending on the chosen language and the way of implementing this language this can lead to totally different programmes, therefore to different strings, and therefore to completely different complexities. As in this case there is no equivalent to the invariance theorem, then the complexity of the algorithm (without further precautions) is an undefined concept. But beyond that, the algorithm itself seems to me to also be undefined. What does "the algorithm of the observer" mean? What meaning can we give to "the algorithmic complexity of the system"? I know that Charles Bennett frequently uses this notion of likening an object to a string, and allows himself in all his articles, albeit in a very hypothetical manner, to speak of the algorithmic complexity of an object, even of the depth of the programme required to produce this object. But just as these notions have rigour in mathematics or computer sciences, when we transpose them to physics, they become extremely problematic, poorly defined, very hypothetical and ambiguous. And so I cannot see what this algorithm is and consequently what the inequality $K(S) < K(O)$ [16] means.

Besides, I am very surprised that we speak of an observer or a measurement without having defined precisely what we mean by that, which is different from their usual sense. What do we mean when we claim that a fullerene makes a measurement as an observer? The least we could do would be to specify the meaning we have in mind since it can obviously not be the usual intuitive sense according to which a defined value is determined. It seems that in the philosophical problems we consider, there is a notion of interpretation of the formalisms and measurement outcomes, especially in quantum physics. If we were fullerenes and did not have a personal viewpoint, a feeling for the world we are trying to understand, we would not be asking ourselves this question of interpretation in quantum mechanics. The question of knowing how come the world appears as it does, i.e. classical, with objects having properties with well-defined values, when in fact it is fundamentally quantum, is the main question, or at least one of the main questions. However, a fullerene does not think about that sort of thing! But I cannot see what definition of the measurement concept Alexei uses.

Alexei Grinbaum. I will answer by starting with the Kolmogorov complexity. It is defined for strings, sequences of 0 and 1. Theoretical computer science establishes the possibility, when we speak of the Kolmogorov complexity, to consider the following notions, going easily from one to the other: the complexity of whole numbers, the complexity of partial recursive functions, the complexity of a sequence of symbols. All these notions can be called "Kolmogorov complexity". When I take an image of a tape, it is to represent the sequence I am speaking of in the sense of the Kolmogorov complexity. The inequality $K(S) < K(O)$ is an expression of "relevance", from the fact that the observer retains the identification of the relevant degrees of freedom. When S is observed, $K(S)$ depends on observer O identifying S.

Hervé Zwirn. How do we define $K(S)$?

Alexei Grinbaum. Let me come back one moment to Bennett and Zurek, who tried in a way to introduce the notion of Kolmogorov complexity in quantum mechanics. Zurek, at the end of the 1980s, and Bennett later on (although more vaguely to my mind), said that when an observer identifies a quantum state (but note I have never myself at any point spoken about quantum states), it is information—thus it is saved in his memory. There is a Kolmogorov complexity linked to the fact that the observer has stored this state as information. If the state changes, not only does the entropy of the system change, but also the Kolmogorov entropy, which is linked to the fact that we have rewritten the information in the observer's memory. For Zurek, the physical description contains two entropies, and so to calculate the total physical entropy, we need to add the Kolmogorov entropy to the Shannon entropy. For my part, the act of identifying the state of the system and the act of identifying the system is in fact one and the same thing. I tend to think there is no clear distinction, informational or physical, between the identification of a system and that of its state. My proposal would therefore be a development of Zurek's idea, leaving open the question of entropy. What is clear is that in this act of the observer who looks at a system and who says that since he is looking at it, there is information (a string of 0 and 1), there is a Kolmogorov complexity. It is essential.

I would also like to comment on the physical interpretation. In 1905, the idea that the notion of time could be defined for a photon travelling at the speed of light and that its specific time was not at all the time spent by an object travelling much more slowly, seemed very strange. How can something that is not a human being have a specific temporality which is completely different from ours? In my opinion, it is the same thing with the idea that a fullerene can be a quantum observer. Since the fullerene does indeed observe photons—as we do! But, obviously, it cannot communicate this to us. There is no a priori reason to consider that man is special and that the notion of observation is restricted to him; in the same way that the notion of time is not restricted to him.

Hervé Zwirn. We must then define what observation is. If we reduce it to the notion of interaction, why not? That does not bother me, if it is a definition. However, that is not what we usually call an observation, where there is a notion of interpretation, because it raises a number of questions for us—who are not fullerenes—questions that a fullerene does not ask. Now, if we wish to change the vocabulary and call observation a simple interaction, why not? But, then, why retain the two words? In "observation", there is something more than in "interaction"—and this "something more" is very difficult to assign to a fullerene. As, in all likelihood, the fullerene does not have this "something more".

Jean Petitot. It is the same with the needle of a macroscopic apparatus.

Roger Balian. No, because we can programme a computer which will filter the results.

Michel Bitbol. Einstein would have never said that a clock was an observer. He did not confuse observer and clock, which is a measuring apparatus.

Carlo Rovelli. The interesting question is not that of the definition of the notions of information, observation or measurement. We can define these as we wish. The real "observers" in laboratories have all sorts of properties that we can include, or not, in the definition of "observer". The interesting question is rather the following: since quantum mechanics uses the notion of observer in its formalism, what is the minimal notion of observer necessary to understand this formalism? The question is finding the minimal characteristics of an observer for quantum mechanical formalism to have any sense.

In the context of special relativity, we often speak of observers, but no one would say that an observer needs to be conscious. In this context, "observer" means a reference point—and can thus be defined as any inanimate object. We say that the velocity of an object is defined only in relation to an observer, which means "in relation to another object".

I think that the minimal notion of observer in quantum mechanics is the following. Imagine two systems, S1 and S2, and two variables of the systems, A and B. The two systems together can be in a state such that when we measure variable A and variable B, we always find the same outcome (if 1 in A then 1 in B; if 2 in A then 2 in B). We thus speak of "correlation" between the two. It can happen through a quantum state or through a classical statistical state, but we have a very clear idea of what a correlation is: a well-defined notion which concerns only the *relative* properties in relation to the measurement outcomes on two system. Shannon introduced the possibility of quantifying these correlations. According to Shannon, we can say that in this case "system A has information on system B". It is important to insist on the fact that this notion of information is perfectly well-defined for any physical system with variables that can take on different values. It does not require either the capacity to "store" information, or consciousness. It can be defined in an operational manner: A has information on B if a measurement of variable A provides outcomes that are correlated with the measurement outcomes of B.

However, we know that each time a quantum measurement is carried out the measured system and the system making the measurement evolve into in a state where there is a correlation (I do not consider the case where the measured system is destroyed). In that very precise sense, the two systems have information on each other that is quantifiable sensu Shannon, with a correlation. It seems to me therefore that the question is the following: is the notion of information sufficient to understand the use of the notion of "observer" in quantum mechanics? Is it enough to say that an "observer" is a physical system that has obtained information (sensu Shannon) on another? The crux of relational quantum mechanics is the hypothesis that this notion of "observer" is sufficient to speak of quantum observation.

Hervé Zwirn. I completely agree up to that point. But the question we ask when we want to understand quantum mechanics (and that is probably the difference with relativity where considering that an inanimate object can be an observer—in the sense of a frame of reference—does not pose any conceptual problem) is the following. If we say that a measurement is only a correlation, this poses no problem in classical mechanics. However quantum formalism says it is not possible to think like that. The correlation that takes place during an interaction results in entanglement, and thus most of the time in superposition. However, during a measurement (in the usual sense of the term in quantum mechanics), one of the possible values of the superposition becomes determined. That is the fundamental question. It is one of the mysteries of the measurement problem. It is also what has attracted a lot of interest, leading to many articles and so many different potential solutions. This question does not arise in classical mechanics because classical statistical correlations do not pose any problem of indetermination. This problem of indetermination can obviously not be considered by a fullerene. The fullerene becomes entangled with the photon, and both are in a state that is indeed correlated but indeterminate. Whereas an observer is not indeterminate. We do not have the impression of being indeterminate when we make a measurement. Explaining all this is where the difficulty lies.

Alexei Grinbaum. The fullerene, if it could sense anything, would perhaps have, like us, the impression of not being superposed. We cannot know this. I see two elements in the measurement problem. The first is the following. For a given observer and an observed system, does the observer have the impression of being superposed? And the second is: for a given class of observers (humans, fullerenes, etc.), how come we agree when we make different observations and measurements? My answer to the first question consists in proposing that the observer who is correlated with the system produces a sequence of 0–1 that identifies this system and perhaps even its state. We are fortunate enough to be able to say that "I have the impression of not being in superposition" because we know what we know, meaning we have this sequence of symbols (encoded in a language other than 0 and 1. But perhaps the fullerene is in the same situation!). We know nothing of the self-referential capacity of fullerenes, but there is no reason to think that we are exceptional.

Hervé Zwirn. I disagree with the idea that a fullerene can have an impression. A fullerene has no impression. I cannot go as far as accepting that a fullerene as any impression whatsoever. It is not possible.

Alexei Grinbaum. I reduce the notion of impression to something that is a matter of fact, formulated in a certain language. If this language is human, it will contain self-referential elements, so a human being can know and say that he is not in superposition. But what do we know a priori of the constraints on language? As Carlo said, the question is to know what the minimal requirement is to consider that an observation has taken place. And the answer is the correlation of the degrees of freedom, relevant or not, 0 or 1.

Hervé Zwirn. I disagree. It amounts to saying that any interaction is a measurement, which is clearly not desirable.

Alexei Grinbaum. I now come to the second question: why do we all have the impression that we agree when we carry out experiments in physics? No one mistakes an electron for an elephant. The way we identify the relevant degrees of freedom for a given system coincides. I believe I can provide a criterion that shows how, for a class of observers whose complexities obey such conditions, inter-subjectivity is possible, i.e. why similar observers can identify systems in the same way.

Roger Balian. It is true that in the context of quantum mechanics, we can correlate objects totally—which does not at all mean that it is a measurement. Indeed, correlation is not a property of a particular object that is being observed, but a statistical property of all the experiments we could carry out starting from an initial state of the object and the apparatus—and, in the end, we find a correlated outcome. That is not sufficient. A complementary element needs to be added: although it is a statistical property that is relative to a set of a very large number of measurement repeats, each individual measurement is such that the object is macroscopic with a particular property, which is a property of statistical mechanics which says there is a probability that is practically equal to one to be either up or down by losing its coherences and being in a thermal equilibrium in either case. There is therefore a property of macroscopic equilibrium. We have a model that has succeeded in measuring this property, allowing us to say that each unique experiment we will carry out will lead to such or such outcome. The correlation that has been established therefore says something about the object.

It is an additional element that we must add to the correlation. Six months ago, we made the mistake of forgetting this crucial point. We can solve it within the framework of a model of pure statistical mechanics, with nothing else.

Alexei Grinbaum. I am thinking of another point, Hervé. Imagine an observer. Call it God or a Martian or something from the far end of the Universe. He does not speak human language and does not know how to interact with us. He wonders whether man has, or not, the impression of being in superposition—since he is also able to observe photons and knows quantum mechanics. How will he determine whether man has the impression of being in a superposition or not? He cannot interact with us in any way. Here is what he will do. He will tell himself that man has his memory, thus he can store information. Then he will study all the consequences that are accessible from the outside of this process of information storage, in order to know whether the information has been deleted (i.e. if some heat is emitted). Only objective data, i.e. not linked to our internal and self-referential view, are accessible to him. He therefore cannot know if we have impressions or not. He can only see whether heat is emitted or not. It is the same thing with men and fullerenes. The question of knowing whether a fullerene has impressions or not, is undecidable from the outside. We can only look to see whether there are physical consequences of storing information in the form of heat, nothing else.

Hervé Zwirn. What does this prove, with respect to our previous discussion?

Alexei Grinbaum. It is the minimal denominator necessary to conduct physics.

Hervé Zwirn. We are asking a precise question. How does this provide an answer? In what way does saying that there is heat at a given point, when we carry out this theoretical experiment (supposing it is feasible), provide an answer to the question of knowing how we resolve the measurement problem, which is the problem of all the founding fathers of quantum mechanics? And by which chain of reasoning, if it does provide an answer?

Alexei Grinbaum. The answer is the following: the measurement problem does not exist. Indeed, we have said that an observer is not necessarily a human being and any system can be an observer. The question of knowing why men see spin up or spin down is not a question of physics. Questions of physics must be asked at the level of what is observed by an observer, who is not human but informational.

Hervé Zwirn. It is a bit simple to claim that the measurement problem does not exist. This question has been much debated. Besides, I do not see any proof of the fact that an observer is not necessarily human.

Alexei Grinbaum. It is like the aether problem.

Hervé Zwirn. You think that the measurement problem is like the aether problem? We will not agree on this point! If what you claim amounts to saying that the measurement problem is like the aether problem, then we profoundly disagree, since I reject this analogy. We were able to get rid of aether convincingly. When Einstein came up with special relativity, he no longer needed aether because he provided a convincing and coherent answer to certain questions that previously required the hypothesis of the existence of aether. This is rather different, as I understand it, or else I missed something: you do not provide a different and coherent answer to the measurement problem, even if you claim with no supporting evidence that it is not a problem. This approach bothers me.

Alexei Grinbaum. It is in fact exactly the same thing, by removing from physics a notion that has nothing to do with physics and by proposing a different approach. There is no absolute objectivity in measurements. It is not true that the spin is *up* for all observers.

8.5 Comment by Bernard d'Espagnat and discussion

Bernard d'Espagnat. I would like to make two comments that, I think, directly concern the theory of *relational quantum mechanics*.

The first is on Carlo [Rovelli]'s thesis which states that any system, conscious or not, can play the role of observer in quantum mechanics. According to Carlo, this point of view would be compatible with the idea that all predictions of observables in quantum mechanics are correct as long as we abandon the notion of quantum state of a system and that the quantum formalism is only considered as a simple tool that allows us to make these predictions. For a reason I will explain, I find it tricky to reconcile this point of view with a special aspect of quantum mechanics. Let me explain. Consider the thought experiment proposed by Wigner and known as the *beam recombination experiment*. A beam of silver atoms travelling along Oy with a spin oriented along +Ox passes through a Stern-Gerlach device oriented along Oz, which splits the beam in two; further on and with the appropriate set-up (including a second Stern-Gerlach device also oriented along Oz), the beams spatially recombine into a single beam, prolonging the original beam. If—following the simplifying convention used by Bohm in his analysis of the Stern-Gerlach experiment (see his book from 1951 [17])—we equate the Sz component of the spin of the atom to the measured observable and the z coordinate of the atom to the pointer's position, this leads us to equate, somewhat bizarrely it is true, the spin of the atom to the system which in his 1996 article Carlo called S, the atom itself to the observer he called O, and the experimenter who prepared the experiment to a person he called P. Let *t* be the time when the two beams are separated and *T* a later time when the two beams have recombined. Suppose that O, knowing the set-up, measures Sz at *t* and obtains the outcome +. He will come the conclusion, as we would, that from this value taken as the initial value and his knowledge of the set-up, he can predict, by applying quantum laws at least in terms of probability, the measurement outcomes of Sz that himself and others will find at any later time. However, he is clearly mistaken on that point! This is because the fact that the second Stern-Gerlach device is also oriented along Oz *forces* him to predict that Sz will retain its value in the final state at time *T*, when at time *T*, because the beams have recombined, the spin of S is actually oriented along Ox and hence the probability of O obtaining the outcome + at *T* will not be 100% but only 50%. This shows that because O is below P in the von Neumann hierarchy, his "objectivity level" is, so to speak, lower than that of P. I am a bit troubled to not find an equivalent "gradation of objectivity" in Carlo's conception.

By equating S to the spin of the atom and O to the atom itself, I have, it is true, in the manner of Bohm, separated conceptually what is physically inseparable. It is at this cost that I can claim to have followed Carlo's law according to which systems do not have properties in themselves. For those who believe in the existence of systems "in themselves", this cost may seem too steep but we must note that the inseparability in question is not clearly derived from the formalism [18].

Carlo Rovellib. This is a very elegant example. I should have thought of it myself. First of all, in the context of the relational interpretation, the identification of systems S, O and P with, respectively the spin, the atom (its position in particular) and the experimenter is perfectly legitimate. More than that: it is an exemplary case. Secondly, the fact that the atom (its position) "has measured" the spin during the time between separation and recombination of the beam is perfectly correct in this context. Remember that in the relational context this claim means that, if we measure the position of the atom and the spin, we find a correlation. It is certainly the case in this example: if we measure the position of the atom and the spin between separation and recombination, we find they are indeed correlated: according to Shannon's definition, we can therefore claim that "the position of the atom has information on the spin". What a great example of the way the relational interpretation works! Where I think there is a problem is when you say "and retains it in the final state". You have implicitly hypothesized that relative information is kept forever. Nothing in the relational context allows us to say this. After recombination, the position of the atom loses the information it had on the spin during that brief period. Information can be lost, as we all know when we cannot find our glasses.

You point out that a measurement outcome cannot be considered as scientifically valid unless it is made public, and more precisely reproducible, at least "in theory", by another party. The notion of "public" equates a set of observers to a privileged super-observer. The approach is not at all incompatible with the relational interpretation. Besides, it seems that this approach is not fully understood: in what sense do these "communal observers", or even better these "convivial observers" differ from the rest of the Universe, which is quantum and can exist in superposed quantum states? The relational answer is simple: this "communal observer" is also quantum and, in principle, could be perceived by another external observer as part of a quantum correlation. The set of information we have could be, in principle and on another scale, like the information that the position of the atom has on the spin. There is no contradiction to consider that all these systems are quantum.

Bernard d'Espagnat. My second comment deals with the "weakening of reality", which, during our previous session, Carlo admitted was a feature of his theory. My problem is this: is this weakening sufficient? I consider that it is not. It is not sufficient to make the *properties* of systems relative. That by conserving the notion of existence "per se" of systems, his theory does not go far enough. That quantum physics forces us to abandon this notion as well. Last time [19], I had already mentioned the main argument that led me to think that way, but I would like to come back to it. It is based on high-energy physics, and more precisely on experiments of creation and destruction that clearly show that no system, not even the wrongly called "elementary" particles, is something "in itself". And that, consequently, in our attempt to understand human experience, ordinary quantum mechanics is just a stage that we must surpass to get to quantum field theory, where the "number of particles" is no longer a "c number" but a "q number"—in other words an observable, just like the properties Carlo wants to make relative. Consequently, the "systems" being made up of particles, they must also be relative.

I think the information provided here by experimental physics is philosophically very important. That it "sparks Carlo's interest" of course, but it goes beyond that. It tells us that the existence of particles, therefore of objects, is also relative, but relative to what? I am of those who consider that at this point we should call upon the well-known argument that given that we belong to the world, it would be conceited on our part to claim that we can draw from it the correct, undistorted view, supposing it even exists, that only a being which is outside and looking down on it could have. There is no point in trying to know, by experience and reasoning, reality as it is "per se". What we produce, as scientists, and that is far from being pointless, is only a synthetic account of our collective experience. Unless we accept solipsism (possibly of the "convivial" type!), we must consider that as a "matter of fact of consciousness" this collective experience does exist (it is not an illusion) and is unique (it is the same for all). This is what we would like to, wrongly, consider as reality "per se", external to us, whereas, I repeat, it is relative to us, to a sort of collective consciousness sometimes called the "epistemic subject" by philosophers, and that consequently we can only call "empirical reality"... and because this epistemic subject is clearly "more real" than the images it forms, it is the one that, for me, provisionally plays the role of thing "in itself" that Carlo believes he can retain in "systems".

I said "provisionally". I will explain this at a later date when I have more time (I have already done so in my books), as we are discussing Carlo's ideas and this would be off-topic.

Carlo Rovelli. I agree that within the framework of quantum field theory, which is clearly closer to Nature than non-relativistic quantum mechanics, we cannot equate a system to a particle or a set of particles. We can nonetheless equate a system to a region of space-time. The region has variables, described by a corresponding local algebra, which can be correlated with that of other regions. A particle detector at CERN is a machine that correlates electronic bits on a magnetic tape with field variables in the region around the point of impact of the beams.

In addition, we already knew this: a human being is not a set of particles: it is made up of atoms that change over time. Recently, we have discovered that the same thing applies to a galaxy: a large amount of matter comes in and out of a galaxy at all times. A galaxy, like a human being, like a sea wave, is not a set of particles: all are phenomena like clouds on a mountain top; the air passes and the vapour briefly condenses into a cloud when it rises to go over the summit. The cloud is the same, but its elementary constituents change all the time. It takes nothing away from the value or usefulness of the notions of "cloud", "human being" or "galaxy". We can still consider these ephemeral entities as "systems", because they are indeed described by variables. All reality is ephemeral in that sense.

That said, I share the sentiment also found in the criticisms formulated by Michel Bitbol, that the notion of system is weak. When I think of a "system", I do not think of a *concrete* object that is the basis of reality: I do not think of the atoms of Democritus. I think of a subjective idealization that allows us to better apprehend and conceptualize a part of reality with which I am in relation.

The problem becomes even more difficult in quantum gravity [20], were regions of space-time are themselves dynamic. There are *two* relational aspects, which I believe are linked: in quantum mechanics, the reality of an object is realized in relation to another. At the heart of general relativity, localization of an object can only be done in relation to another object or another field, like a gravitational field.

Bernard d'Espagnat. It was already the case in Galilean mechanics.

Carlo Rovelli. Indeed. But it becomes a lot stronger in general relativity, because of the absence of a non-dynamic fixed structure of space-time compared to which certain aspects of motion remain nonetheless absolute. I have often wondered if there was a link between the two relationships at the heart of physics: the relationship between the observer and the observed underlying quantum mechanics and the relationship between the objects that define localization. Ultimately, all interaction requires spatial vicinity. And conversely, what does spatial vicinity mean other than the fact of being in relation, in contact? And what does the relation of being in contact, which has been the basis of space since Descartes and even more so with Einstein, have to do with the relation of interaction which, I think is the basis of quantum mechanics? I see in it a great unresolved problem. But because I have a train to catch, and I am already late, I will not go into this any further!

Jean Petitot. I suggest you reread Leibniz. Reality cannot be spatio-temporal.

Bernard d'Espagnat. Reality is really well hidden! We have agreed, Roger Balian and I, that we would dedicate a session to the questions of interpretation on the 14th of May.

Roger Balian. I will send you our latest 180-page-long article, even if I will mostly speak about sections X and XI.

References

1. Bernard d'Espagnat, Conceptual Foundations of Quantum Mechanics, New York, Benjamin, 1971; third edition, Boulder, Westview Press, 1999.
2. Michel Bitbol, "An analysis of the Einstein-Podolsky-Rosen correlations in terms of events", Phys. LettL A, 96, 1983, p. 66–70.
3. John S. Bell, "Bertlmann Socks and the nature of reality", J Physique, colloque C2, suppl. 42, C2, 1981, p. 41–62. Re-worked in John S. Bell, Speakable and Unspeakable in Quantum Mechanics, Cambridge University Press, 1987.
4. William James, Essais d'empirisme radiacal [1912], Marseille, Agone, 2005.
5. Hugh Everett, « 'Relative state' formulation of quantum mechanics » , Reviews of Modern Physics, 29(3°), 1957, p. 454–462.
6. John S. Bell, Speakable and Unspeakable in Quantum Mechanics, Cambridge University Press, 1987, p. 118.
7. *Ibid.*, p. 136.
8. See session VII, "The relational interpretation in quantum mechanics and the EPR paradox".
9. Andrei Khrennikov, "Bell argument: locality of realism? Time to make the choice", ArXiv, 1108 0001v3, December 23rd, 2011.
10. This is mainly the pilot-wave theory. Cf. for example John S. Bell, "Quantum Mechanics for Cosmologists", in John S. Bell, Speakable and Unspeakable in Quantum Mechanics, Cambridge University Press, 1987, chapter 15, and Antony Vanetini, "De Broglie-Bohm Pilot-Wave Theory: Many-Words in Denial", in Simon W. Saunders et al. (eds), Everett and his Critics, Oxford, Oxford University Press, 2009.
11. American Journal of Physics, 50(9), 1982, p. 807–810.
12. No human free will.
13. "Half-way Through the Woods", in John Earmen & John D; Norton (eds.), The Cosmos of Science, University of Pittsburgh Press, 1995.
14. For Brillouin's ideas, see Jérôme Segal, Le Zéro et le un. Histoire de la notion d'information au XXe siècle [2003]; Paros, Éditions Matériologiques, 2011, chapter 5: « Une réinterprétation de la physique: les travaux de Léon Brillouin » , p. 421–486.
15. The inequality discovered by Clauser, Horne, Shimony and Holt, CHSH in short (see for example Bell, Speakable …, op. cit., 1987, chapter 16), generalizes Bell's inequality. It is written $|S| \leq 2$, S being a certain quantity calculable from experimental data, and expresses a limit that, in all local realist theories, the S value cannot exceed. In certain experiments on

quantum particles, this limit is exceeded: quantum mechanics is thus not realist and local. Besides, Tsirelson has shown (B.D. Cirel'son, "Quantum generalizations of Bell's unequality", Lett. Lath. Phys., 4, 93, 1980) that, in any case, if quantum theory is true the bound $|S| \leq 2$ cannot be exceeded. It is the "Tsirelson bound".

16. K denotes the algorithmic complexity or Kolmogorov complexity. S is the system and O the observer.
17. David Bohm, Quantum Theory, New York, Prentice Hall, 1951.
18. *Note added by Bernard d'Espagnat.* Generally, a measurement outcome cannot be considered as scientifically valid unless it is made public, and is reproducible, at least in theory, by another party. And it is this feature which I think is missing from the observation carried out, either by a system Carlo calls O or by Grinbaum's fullerene. See Bernard d'Espagnat, " A tentative new approach to the Schrödinger cat problem", arXiv; Quant-ph/0201160v1).
19. Session VII, "The relational interpretation in quantum mechanics and the EPR paradox".
20. See session X, "Loop quantum gravity".

Chapter 9
The Theory of Measurement

Roger Balian

Roger Balian. I would like to present a small part of a study we started around ten years ago with two colleagues—one being in Armenia and the other in the Netherlands and both travelling frequently, this has slowed down our work somewhat. This work is now finished and is published in detail in *Physics Reports* [1]. Our goal consisted in trying to understand how ideal measurements behaved without going beyond conventional quantum mechanics. In order to specify in which framework we placed ourselves, I will start by recalling the statistical interpretation that I consider to be the minimal interpretation of quantum mechanics, in which one places the least number of things possible. I will then proceed to the presentation of the model and its resolution, with the problems it raises.

R. Balian (✉)
Institut de Physique Théorique, Saclay, France

R. Balian
Académie Des Sciences, Paris, France

R. Balian
École Polytechnique, Palaiseau, France

R. Balian
Summer School of Theoretical Physics, Les Houches, France

R. Balian
Société Française de Physique, Paris, France

B. d'Espagnat and H. Zwirn (eds.), *The Quantum World*,
The Frontiers Collection, DOI 10.1007/978-3-319-55420-4_9

9.1 Statistical Interpretation

9.1.1 The Notion of Observable

The way I see it, the statistical interpretation is a dualist interpretation, with on the one hand the object, and on the other the information we have of it. The object itself concerns reality, whereas the information calls upon the knowledge we have of the object, which means that it calls upon the observer. Initially, what pertains to the object is represented mathematically by observables (which represent physical quantities associated with a given system in unspecified circumstances). We can translate this mathematical representation by the fact that the observables are elements of a C*-algebra, namely a non-commutative algebra in which there is correspondence between the different mathematical objects and their adjoints. The observables are the self-adjoint objects of this algebra. The observables play a somewhat analogous role to that of random variables, but present the peculiarity of being non-commutative. They are therefore both very concrete, since they concern the system itself—being real—but also abstract (as is the case each time we use mathematics). If these objects were to commute, then all that would follow would lead back to classical statistical mechanics.

It is necessary to introduce a second mathematical object: dynamics. The dynamics of an isolated system are determined by the Heseinberg equation according to which observables, not states, evolve over time. In fact, if we wish to remain in a mathematical framework, it is sufficient to say that evolution is a continuous homomorphism of the observables whose infinitesimal generator is the Hamiltonian.

So much for the notion of object. So far, what we have said can be applied equally to individual objects or to ensembles of statistical objects. But if we wish to move on to the notion of state, which describes mathematically the knowledge we have of physical systems, then we need to know what information regarding the system is available.

9.1.2 The Notion of State

The information we have regarding a system is irremediably probabilistic—in quantum mechanics we cannot have it another way. And therefore if it is probabilistic, quantum mechanics forces us to describe not individual systems but ensemble of statistical objects. A statistical ensemble E is an ensemble of systems that although identically prepared can produce different outcomes when we experiment on them. If we want to consider an individual system, the only option is to suppose that it is an element of a virtual ensemble. But we can also consider that individual systems belong to real ensembles. For example, when we consider a measurement, the measurement is a real ensemble E of many runs. We will also be required to introduce sub ensembles E_{sub}, by extracting from ensemble E certain runs—in the same way, when tossing a coin, we can extract different groups of draws; depending on the sub ensemble, the proportion of heads or tails will be arbitrary.

Ultimately, the information will be provided in the form of expectation values, and as a result of fluctuations, correlations or, if we consider individual systems, autocorrelations, namely the correlations of a system at a given time with itself at another time. This autocorrelation is also a probabilistic object. I must specify that "probability" is to be understood for states in the Bayesian sense, namely characterizing our knowledge of the system; it can also be thought of as a frequency for a series of experiments providing different outcomes with given probabilities.

9.1.3 The State of a System

The state of a system is represented mathematically as an application of the algebra of observables on real numbers, application that provides the expectation value of each observable O. This correspondence summarizes all the information we are likely to extract from the system, or more exactly from the statistical ensemble to which the system belongs. This correspondence must be linear, real, standardized, and positive. We can show that this allows the construction of a Hilbert space which the observables act on. Once this space is built, the state can be represented by a density operator D, the expectation value being obtained by tracing the product of D and O in this Hilbert space [2].

This correspondence allows us to evaluate variances, which are obtained as the expectation values of squares, and probabilities, which are obtained as the expectation values of projections. However, the density operator itself is not a usual probability since it is a mathematical object that allows the calculation of all sorts of expectation values, even of non-commuting objects, which is not taken into account in probability theory as it is usually taught. The real probabilities emerge during the measurement process.

I will now place myself in the Schrödinger picture, equivalent to the Liouville-von Neumann picture, but where states, and not observables, evolve, which is mathematically equivalent in the special case where we do not consider autocorrelations but only quantities at a single point in time.

All things considered, the probabilistic interpretation seems very similar to classical statistical mechanics, which is nothing more than a correspondence between classical observables (namely elements of a commutative algebra that behave like ordinary random variables) and their expectation values. But here, everything is irreducibly probabilistic: we cannot generally obtain certainties, as is illustrated by Heseinberg's inequality: we are not allowed to think of the values the observables could take. The values (and the ordinary probabilities) that we will assign to the observables will emerge from a measurement, as the outcome of an inference of the observation of a macroscopic pointer towards a property of the microscopic system. The latter is governed by laws that we can apprehend only through mathematical means, and our concrete observations require macroscopic instruments.

9.1.4 Attribution of a State to a System in a Given Situation

The attribution of a state to a system in a given situation can be achieved either by complete preparation with filtration, or—much more frequently when dealing with large objects—by partial preparation. The latter case is analogous to the attribution of a probability law to a given system: it is necessary to make an unbiased choice, which can rest on certain symmetry criteria. In quantum mechanics, it is unitary symmetry that makes all representations of the observables equivalent. This criterion generalizes Laplace's method, which consists in saying that if multiple events, all on the same plane, are possible, then they are equiprobable. In quantum mechanics, this is translated into the fact that the density operator is proportional to the unit in the absence of any indication. If certain expectation values are given, they provide constraints on the density operator, and we can show, by using the maximal entropy criterion, that the determination of the least biased density operator among those that satisfy these constraints results from Laplace's principle.

Catherine Pépin. What is the difference between a complete preparation and a partial preparation? Can you give an example of a complete preparation?

Roger Balian. In a complete preparation we take, for example, a system whose fundamental is non-degenerate and empty it of its energy as much as possible. It is then in its fundamental state. It could also be a spin that we polarize so that it is pure. As for an incomplete preparation, I could give the example of the initial state of an apparatus during a measurement. It being macroscopic, the only way to describe it is probabilistic. If the apparatus is an object that is brought to a certain temperature, we use the maximal entropy criterion, which results in assigning a state characterized by a canonical distribution.

9.1.5 The Status of Pure States

So far in my presentation, there is no conceptual difference between a pure and a mixed state, as in all instances it is an association between the observables and their expectation values. The only difference is that pure states are maximal states within the space of different states. They are only special cases. But the other states present a very troublesome property, having no equivalent in ordinary probability theory: if we start with a statistical mixture and decompose it, we can mathematically write this decomposition as:

$$D = \Sigma_k \, p_k \, |\psi_k\rangle\langle\psi_k|.$$

This equation seems easy to interpret: the statistical ensemble considered, represented by a mixture D, is none other than the reunion of statistical sub ensembles described by the pure state $|\psi_k\rangle$ with a proportion p_k. But this interpretation, which would be natural in classical probability theory, is totally fallacious and wrong in quantum mechanics because the decomposition is not unique. It is a quantum ambiguity that I will come back to later on.

Hervé Zwirn. The $|\psi_k\rangle$ are pure states, are they not?

Roger Balian. Yes. Let us take the trivial example of a non-polarized spin. We could equally say that this non-polarized spin describes a population of spins that are polarized in equal proportions following Oz or $-Oz$. But we could also say that they are polarized in any direction, with the same probability. The second description seems more reasonable, but it is no more or less reasonable than the first description, as the first one exists. More generally, we can decompose D mathematically in the sum $\Sigma_k \, p_k \, D_k$, where D_k appears to describe a sub ensemble, but here again the decomposition can take place in an infinite number of ways and this interpretation is fallacious.

As multiple decompositions of the same density operator exist, if we take an individual system from the system described by D, we need to state to which particular sub ensemble it belongs. However, the density operators we assign to it are completely different from one decomposition to another and even incompatible since they can lead to different predictions for the same system. It is impossible to decompose a real statistical ensemble, described by D and constituting true real systems, into real statistical sub ensembles in the absence of information other than the knowledge of D. Sub ensembles k or k' that are defined by a decomposition $\Sigma_k \, p_k \, D_k$ or by another decomposition $\Sigma_{k'} \, p'_{k'} \, D'_{k'}$ are virtual objects that have meaning only mathematically and cannot strictly have a physical meaning. Indeed, we cannot say that a system belongs to a sub ensemble k where it would be in state D_k and at the same time to a sub ensemble k' where it would be in state $D_{k'}$—since these two states are incompatible. This point, essential for the analysis of a measurement process, will be the subject of the last part of my presentation.

If we wish to give meaning to a decomposition of this type, we must consider not only the complete statistical ensemble to which the system in question belongs but also all the possible sub ensembles that it could contain. And we must try to obtain additional information on these real sub ensembles.

This is all I wanted to say on interpretation. I will continue my presentation within this framework to try to describe a measurement.

9.2 The Curie-Weiss Model

The model we have considered and resolved in great detail is made up first of all of the microscopic system S we want to test, a spin ½ for which we want to measure the component s_z along direction *Oz*. We use an apparatus A = M + B that describes, within a rather realistic model, a magnetic dot, namely a macroscopic object whose magnetization is likely to be oriented along *Oz*, either in the positive or negative direction. Within it we can distinguish the magnet M, assembled in *N* spins ½, which functions as a macroscopic pointer—the sign of its magnetization $m = (1/N)\ \Sigma_n\ \sigma_z^{(n)}$ will emerge from the measurement process. These *N* spins interact through Ising coupling $-½JNm^2$ in the *Oz* direction. Initially M is in the paramagnetic metastable state at temperature $T(\langle m\rangle = 0)$. Direct coupling $-Ngs_zm$ of the system S under examination and the pointer M will trigger a transition towards one or the other of the two stable ferromagnetic states, of magnetization $\langle m\rangle = +m_F$ or $-m_F$. But in itself, this magnetic dot would not be dynamic. That is why the apparatus A has a second part, a thermal phonon bath B at temperature *T*. This temperature is inferior to that at transition *J*; the bath imposes its temperature on M and the heat transfer from M to B leads M towards a ferromagnetic equilibrium. The phonon/spin coupling is characterized by a weak coupling constant γ between M and B. One of the advantages of this model is that it allows us to find a solution for multiple values of the parameters, and to see for example in which case the process will be a measurement and in which case it will fail.

We have tested this model from every angle. I will restrict myself here to the simplest elements. The variable that functions as a pointer is the magnetization sign $\langle m\rangle$ following direction *Oz* of the magnetic dot. The dynamics are Hamiltonian—the most standard formulation of quantum mechanics—with the three couplings M-M, S-M and M-B defined above. The Hamiltonian also has a part associated with the bath phonons. The density operator D(*t*) under study describes the global (isolated) system S + A. The notations we use for the marginal density operators are *r* for S, R for A, R_M for M, R_B for B, and *D* for S + M. The initial state D(0) is factorized in the tensor product $r(0) \times R_M(0) \times R_B(0)$, where *r*(0) describes the state of S we want to test, $R_M(0)$ the metastable paramagnetic state of M and $R_B(0)$ the canonical equilibrium state of the bath at temperature *T*. In order to determine D(*t*) at any time, we apply the Liouville–von Neumann equation $i\hbar \mathrm{D}(t)/\mathrm{d}t = [H, \mathrm{D}(t)]$, which is, in statistical mechanics, the equivalent of Schrödinger's equation (We have to place ourselves within the framework of statistical mechanics since the apparatus is necessarily a macroscopic object: it cannot be described within the framework of pure states, as it cannot be prepared in such a pure state).

The conservation of the measured observable s_z allows us to decompose the density matrix D into 4 blocks: two diagonal ones associated respectively with the values 1 and −1 of s_z and two off-diagonal ones. These 4 blocks evolve independently since $[H, s_z] = 0$. Given that the coupling between the magnet and the bath is weak, it is treated at the lowest order of perturbation theory. The magnet, whose invariance $m \leftrightarrow -m$ breaks down at $T < J$, will have a tendency to reach one or the other of its ferromagnetic equilibrium states, written ↑ or ↓, under the effect of its interactions with S; conversely S is perturbed by this interaction. Contrary to what happens in a usual phase transition, the states ↑ and ↓of M will both contribute to the final state $D(t_f)$ of S + A.

Bernard d'Espagnat. I believe you explain in your paper that spontaneously, apparatus M would take position, but that is doesn't know quite where to go and that it would take a very long time. It is the spin that makes it do it.

Roger Balian. Not only would it be long, but it could equally go up or down. The dynamics of an invariance that spontaneously breaks are slow. Here, it is the spin that triggers the breaking process, so that it is faster, as when in the presence of a small magnetic field g or $-g$.

In the mechanism of evolution, the interaction between the bath and the magnet allows an energy transfer from M to B that is necessary for the phase transition (the paramagnetic energy, which is greater than the ferromagnetic energy, needs to go somewhere). On the other hand, it is the interaction between system S and magnet M that acts as a trigger, therefore that chooses—correctly, we hope—the outcome ↑ for $s_z = 1$, ↓ for $s_z = -1$ and the probabilistic result predicted by Born's rule for an arbitrary state $r(0)$. If the parameters are chosen badly, the trigger goes wrong. But today, I place myself in the position where all goes well. At the same time as system S triggers the transition, magnet M necessarily perturbs S.

9.2.1 Required Conditions for a Repeated Experiment on Identically Prepared Systems to Be a Measurement

First of all, a well-defined outcome must be obtained by the pointer for each individual process: the "measurement problem" consists of understanding this. Then, a total correlation must be established between what the pointer will say and the value of the spin after measurement; this correlation appears in the diagonal blocks of D. In addition, the von Neumann reduction, defined in the strong sense, must take place: each individual process must provide a well-defined result $D_\uparrow$ or $D_\downarrow$; this implies that no element must remain in the off-diagonal blocks of the global density matrix $D(t_f)$, but also in the final density matrices associated with all measurement sub ensembles. Finally, we need to demonstrate Born's rule, which links the probability $p_\uparrow(p_\downarrow)$ that pointer M gives the indication ↑(↓) to the initial state $r(0)$ of S.

Catherine Pépin. In von Neumann's reduction, is it required that there be no off-diagonal states in the system and all subsystems?

Roger Balian. In all sub ensembles, not subsystems! There should be no off-diagonal elements in the density matrix of the system made up of S + A—and that not only for the ensemble of all measurements, but also for all measurement sub ensembles that we will see come out at the end.

Catherine Pépin. Why?

Roger Balian. I will come back to it later on, but I will anticipate by expressing mathematically the properties I have just listed. For each individual measurement that has produced the outcome $\uparrow(\downarrow)$, the state of S + A at time t_f must be represented, according the laws mentioned above, by the reduced density operator $D_{\uparrow(\downarrow)} = r_{\uparrow(\downarrow)} \times R_{\uparrow(\downarrow)}$, where $r_{\uparrow(\downarrow)} = |\uparrow(\downarrow)\rangle\langle\uparrow(\downarrow)|$ represents the state of S for which $s_z = 1(-1)$ and where $R_{\uparrow(\downarrow)}$ represents the ferromagnetic equilibrium state of A with the magnetization m_F $(-m_F)$. According to Born's rule, the proportion of results $\uparrow(\downarrow)$ is $p_{\uparrow(\downarrow)}$, so that $D(t_f)$ must be equal to $D(t_f) = p_\uparrow D_\uparrow + p_\downarrow D_\downarrow$. But in addition, to all measurement sub ensembles having produced outcomes $\uparrow(\downarrow)$ in proportions $q_{\uparrow(\downarrow)}$ must be associated a density operator of the form $D_{sub}(t_f) = q_\uparrow D_\uparrow + q_\downarrow D_\downarrow$. We will call this property the "hierarchical structure of sub ensembles". It is the strongest we could impose within the framework of the statistical interpretation, which allows the description of individual systems only when they are considered as elements of statistical ensembles. Here we consider all sub ensembles to which an individual measurement run can belong.
In addition we must establish this condition for the global systems S°A and not for single systems S.

9.3 The Ensemble of All Possible Measurements

This ensemble E is simply characterized by the initial state D(0) as the tensor product of the initial state of the object with the initial paramagnetic state of the apparatus. We want to demonstrate, as I have just said, that the final state of ensemble E of all these measurements has the form $D(t_f) = p_\uparrow D_\uparrow + p_\downarrow D_\downarrow$, where p are the diagonal elements of $r(0)$. We therefore want the final state to not contain off-diagonal blocks, and for the diagonal blocks to correlate the spin and the apparatus. It is necessary, but far from sufficient—I will come back to this point.

The aim is therefore to resolve a problem of statistical mechanics that is well-formulated but rather difficult as numerous successive events occur. I will go through each of these events step by step.

9.3.1 Truncation

The first event is truncation, namely the progressive disappearance of off-diagonal blocks. All off-diagonal blocks tend towards zero. This process occurs during a short period of time $\hbar/g(2\ N)^{1/2}$ because the number of degrees of freedom of the pointer (N) is very large, and because it is principally an action of a large apparatus on a small system. It is not a decoherence in the strict sense of the term. What matters is the coupling (g) of the spin and the apparatus, and not the temperature of the bath (playing here the role of environment). In fact, numerous oscillations occur in the mechanism and add up. It is the phase shift between the different possible oscillations that occur randomly that, in the end, destroy these off-diagonal elements.

We can equally say that it is a random precession of the spin around the field according to Oz produced by M that makes the transverse components of the spin disappear, but also its correlations with M. As in the irreversibility of a Boltzmann gas when pairs of particles collide successively, there occurs a chain of correlations between S and an increasing number of spins of M, which create and destroy themselves over time. It is this chain of correlations that makes the off-diagonal elements disappear.

Bernard d'Espagnat. In classical mechanics, there would naturally be the paradox of irreversibility.

Roger Balian. It is the same thing. For example, we can demonstrate that the sum of the squared modules of the matrix elements of the off-diagonal blocks remains constant. But in reality, when we say it tends towards zero, it is in a weak sense: each module, and not the sum of the modules squared, tends towards zero. The number of matrix elements is so great that these elements dissolve and practically disappear. We do not think we will ever find them again. However, that is not strictly true because in the chosen model, there is in reality a recurrence—which constitutes the second stage.

9.3.2 Absence of Recurrence

A recurrence occurs at time $\pi\hbar/2\,g$ when only the interaction between S and M exists, but it can be eliminated by two possible mechanisms. The first is active during the couplings of the measured spin S with the different spins of pointer M that are not quite equal to g. Dispersion of these couplings is sufficient to make the recurrence time considerable, comparable to the age of the universe. The second mechanism is caused by coupling with bath B, which can also trigger a definitive relaxation, even more so than in the first mechanism.

9.3.3 Registration

The third stage of the process—you can see it is complicated!—is registration, namely what happens in the diagonal blocks of D(t) (The dynamics of the four diagonal blocks are independent as the measured variable s_z is conservative). We define registration as the creation of correlations between the sign of spin s_z and the magnetization of M. These correlations are established as each of the two diagonal blocks proceeds towards ferromagnetic equilibrium, with $m \rightarrow m_F$ for the first, and $m \rightarrow -m_F$ for the second. Relaxation is triggered by the coupling with spin S, which plays the role of a magnetic field equal to g within a block, and $-g$ within the other. It is slow, on a timescale in the order of $\hbar/J\gamma$, since it requires an energy transfer towards the bath as a result of the weak coupling γ between M and B. Registration length is much longer than that of truncation. The denominator is $\gamma \ll 1$ rather than $N \gg 1$. If these conditions of validity on the model coefficients are not met, the process fails and cannot result in the necessary outcome in the final state.

9.4 How to Move onto Individual Processes: Measurement Sub Ensembles

9.4.1 The Measurement Problem in the Statistical Interpretation

Following the comment I made earlier, quantum mechanics does not allow the description of an individual system: we must consider it as an element of one (or more) statistical ensemble, be it real or virtual—unlike what happens in classical physics, where a law of probability can be interpreted as an evaluation of the likelihood of each possible event in a given system or as a frequency of appearance of an event within an ensemble. All we have established so far is the relation $D(t_f) = p_\uparrow D_\uparrow + p_\downarrow D_\downarrow$ concerning ensemble E. It is tempting to interpret this result by saying that a proportion $p_\uparrow$ of measurement runs constitutes a sub ensemble $E_\uparrow$ for which S + A arrives at state $D_\uparrow$, and a proportion $p_\downarrow$ constitutes the complementary sub ensemble $E_\downarrow$ arriving at state $D_\downarrow$. But this interpretation is fallacious because of the ambiguity I mentioned above. The final state of S + A for ensemble E can in effect decompose itself in infinite ways as the form $D(t_f) = \Sigma_k \, p'_k \, D_{k'}$. We are then tempted to associate with this decomposition of $D(t_f)$ a decomposition of the -measurement ensemble E into sub ensembles $E_{k'}$ where each corresponds to state $D_{k'}$ of S + A. But an individual run cannot, for example, belong to the two sub ensembles $E_\downarrow$ and $E_{k'}$ at the same time; as in this case its properties would be described both by states $D_\uparrow$ and $D_{k'}$. And yet we can easily show that these two states can provide contradictory predictions for an observable of S + A, as a consequence of the quantum ambiguity of the decomposition of $D(t_f)$.

In particular, if an individual process arrives at a final state of S + A represented by the state $D_{k'}$ with elements in the off-diagonal blocks, the presence of these elements would imply (as a result of the positivity of $D_{k'}$) a presence of elements situated in the two diagonal blocks simultaneously. Then neither the pointer nor s_z could take on a well-defined value.

We must demonstrate that each individual process always leads to either state $D_\uparrow$ or state $D_\downarrow$, so that the pointer provides a well-defined outcome each time, which can be read and which signals (for our ideal measurement) that s_z takes on the corresponding value, either 1 or −1, at the end of the process. However, quantum mechanics, following the minimalist statistical interpretation I have described, deals only with statistical ensembles. If we want to have access to an individual object or process, the only way to achieve this is to start by considering all real sub ensembles it can belong to.

However the sole knowledge of $D(t_f)$ only allows the construction of mathematical decompositions that have no physical significance. Among these decompositions are necessarily those that correspond to real measurement runs. We cannot identify these, but our strategy will consist in considering all sub ensembles E_{sub} of E that are mathematically possible and in establishing for these general properties that we will then use for the real sub ensembles, the only ones that have any meaning.

Catherine Pépin. What brings you to make the hypothesis that each individual process leads to $D_\uparrow$ or $D_\downarrow$? Why is this necessary?

Roger Balian. When we make a measurement, each measurement must provide a definite outcome.

Catherine Pépin. Yes, but we could look only at the ensemble.

Roger Balian. No. The ensemble E of all real measurements is not enough, unlike what would happen for a classical ensemble. For example, we must demonstrate that coefficient $p_\uparrow$, which is only a mathematical coefficient within $D(t_f)$, is the proportion of processes for which we obtain ↑. We therefore need to say certain things on the real realized processes—whereas statistical quantum mechanics provides the density operator $D(t_f)$ and nothing else: it does not allow to separately give a meaning to its two terms $p_\uparrow D_\uparrow$ and $p_\downarrow D_\downarrow$.

We must demonstrate that the decompositions $\Sigma_k\, p'_k\, D_{k'}$ exist mathematically on paper but not in reality. This demonstration will be based on the properties of the apparatus, a macroscopic object.

I specify here that we use the word "truncation" for the disappearance of off-diagonal blocks for the entire ensemble E, and the word "reduction" in a much stronger sense, namely the attribution for each individual process of the final state $D_\uparrow$ or $D_\downarrow$. As we want to consider sub ensembles of processes, this reduction implies that the final state of S + A for all sub ensembles E_{sub} has a form $D_{sub}(t_f) = q_\uparrow D_\uparrow + q_\downarrow D_\downarrow$ with coefficients $q_\uparrow$ and $q_\downarrow$ that depend on this sub ensemble.

The density operators that I now speak of are associated with sub ensembles of real measurements, which are included in all mathematically conceivable sub ensembles, and we want to show in particular that all lose their off-diagonal blocks.

Catherine Pépin. How do we choose the basis?

Roger Balian. It is that which diagonalizes s_z.

Catherine Pépin. Once there is truncation of all real sub ensembles, do we have to work within the right basis?

Roger Balian. I will try to demonstrate there is truncation of all sub ensembles, real or not, and that truncation is imposed by the dynamics. You are anticipating my next point.

9.4.2 The Hierarchical Structure of Sub Ensembles

What we want to demonstrate now is that the density operator $D_{sub}(t)$, describing S + A for all real sub ensembles E_{sub} of measurements, arrives at the final time t_f of the process at a form $q_\uparrow D_\uparrow + q_\downarrow D_\downarrow$. In addition, we want the coefficients $q_\uparrow$ and $q_\downarrow$ obtained in this way, which depend on sub ensemble E_{sub}, to be additive when we put two disjointed sub ensembles together (given the weighting by the number of elements of each sub ensemble). In particular additivity imposes a coherence of coefficients between states associated with nested sub ensembles. We will call "hierarchical structure" these properties of sub ensembles, the universality of constitutive blocks $D_\uparrow$ and $D_\downarrow$, and coefficient additivity.

The Liouville-von Neumann equation allows the study of the evolution of the density operator $D_{sub}(t)$ for any sub ensemble E_{sub} of E, but we do not know the initial conditions of $D_{sub}(t_{split})$ at time t_{split} from which we choose to follow this evolution. All that we know is that E_{sub} b is a sub ensemble of E, for which we know state $D(t)$ so that $D_{sub}(t_{split})$ must be one of the elements of one of the possible decompositions $D(t_{split}) = \Sigma_k p'_k D_{k'}$ of $D(t_{split})$. Among all mathematically allowed decompositions of $D(t_{split})$ are physical decompositions associated with the selection from E of a sub ensemble E_{sub} of real measurement runs; but we do not know how to identify these. Our strategy will therefore consist of taking as the initial condition any that is mathematically allowed, and follow the evolution of a virtual sub ensemble whose initial state $D_{sub}(t_{split})$ arises from an arbitrary decomposition of $D(t_{split})$. If we manage to show that the dynamics of S + A lead $D_{sub}(t_f)$ to the desired form $q_\uparrow D_\uparrow + q_\downarrow D_\downarrow$, this result will obviously be established for real sub ensembles, the only ones that interest us physically.

Hervé Zwirn. Do you arrive at a final state $D(t_f)$ of the form $p_\uparrow D_\uparrow + p_\downarrow D_\downarrow$?

Roger Balian. Yes. This will arise from the hierarchical structure that we will show for all $D_{sub}(t_f) = q_\uparrow D_\uparrow + q_\downarrow D_\downarrow$. In fact $D(t_f) = p_\uparrow D_\uparrow + p_\downarrow D_\downarrow$ is already established and is now our starting point. This result only concerns the statistical ensemble E of all measurements. It has the necessary properties, but the existence of this form is not sufficient to ensure that there will not be anomalies, different from what we expect, for the sub ensembles of the process.

Hervé Zwirn. To make a parallel with the usual mechanism of decoherence, would you agree that this stage is the one we arrive at in the end when we make a partial trace over the environment?

Roger Balian. I cannot make a partial trace, because that would be cheating totally. The state is that of S + A.

Hervé Zwirn. But do you agree that by another route, without a partial trace, we arrive at a similar result?

Roger Balian. It is a lot stronger, for on the one had the result concerns sub ensembles of the process, and on the other hand because decoherence will not occur—I am getting ahead of myself—on the system but on the apparatus. It is fundamental. And that is the reason why decoherence is so effective.

Franck Laloë. You are going along same lines as what Wojciech Zurek calls the privileged basis of the measuring apparatus.

Roger Balian. If you want, yes. But there is another fundamental point: decoherence will take place on the apparatus because there is a broken invariance, namely two possible macroscopic states. This is the original point we arrived at after many months—after an unsuccessful first attempt.

Bernard d'Espagnat. I believe, if I may try to answer Hervé Zwirn's question, and for having read your article, that you show that for all sub ensembles, all goes well. Then, at the end of your demonstration, you arrive at the sum $p_{\uparrow}\mathrm{D}_{\uparrow} + p_{\downarrow}\mathrm{D}_{\downarrow}$ where there is no ambiguity left, after having eliminated all other decompositions $\Sigma_k p'_k \mathrm{D}_{k'}$.

Roger Balian. Yes, but in addition we establish the same form $q_{\uparrow}\mathrm{D}_{\uparrow} + q_{\downarrow}\mathrm{D}_{\downarrow}$ for all sub ensembles.

Bernard d'Espagnat. In this case, you will need to make another leap …

Roger Balian. … yes, there is still a small leap to make.

Bernard d'Espagnat. For the time being, you are not there yet.

Roger Balian. Exactly! I am coming to it. As I have said the number of stages is considerable.

9.4.3 Sub Ensemble Relaxation

The stage we are at consists of looking at the dynamics of all possible sub ensembles, real or hypothetical, that are compatible with one of the possible decompositions of $\mathrm{D}(t_{\mathrm{split}})$. These dynamics are examined after decoupling has taken place between the spin and the apparatus. We therefore suppose that we place ourselves at the very end, so that t_{split} is close enough to t_{f} for $\mathrm{D}(t_{\mathrm{split}}) = \mathrm{D}(t_{\mathrm{f}}) = p_{\uparrow}\mathrm{D}_{\uparrow} + p_{\downarrow}\mathrm{D}_{\downarrow}$.

This choice of t_{split} is essential as the form of $\mathrm{D}(t_{\mathrm{split}})$ imposes on all possible $\mathrm{D}_{\mathrm{sub}}(t_{\mathrm{split}})$, as initial states, a strong constraint (coming from the positivity of the density operators and from the fact that m only takes values close to m_{F} or $-m_{\mathrm{F}}$), which will allow the dynamics to arrive for $\mathrm{D}_{\mathrm{sub}}(t_{\mathrm{f}})$ at the desired final form.

The net result of these dynamics will be the following: all sub ensembles arrive at the end, after a very brief period of time, towards a structure of the same type as that of the entire ensemble—where weights $p_\uparrow$ and $p_\downarrow$ are those of Born—but with different weights $q_\uparrow$ and $q_\downarrow$ for each sub ensemble, these weights following the same hierarchical structure as in classical probabilistic theory. The complete ensemble of measurements E will therefore have the desired hierarchical structure where all its sub ensembles follow this law and have the particular form in which there is no ambiguity left.

9.4.4 Solution Within the Curie-Weiss Model

To demonstrate what I have just described within the Curie-Weiss model, we would need to modify slightly the Hamiltonian and add to it complementary interactions that are internal to the apparatus, conserving its energy, which are active after it has already relaxed towards a mix of ferromagnetic states. The absence of energy transfer and the large size of the pointer will allow relaxation time to be brief. To study the dynamics without too many technical difficulties, we start by eliminating the bath B by treating its weak coupling γ in the lowest order of perturbation theory.

First of all, we must build all the conceivable initial density operators $D_{sub}(t_{split})$ of S + M associated with an arbitrary sub ensemble E_{sub} of E. Each one must arise from a decomposition of the state $D(t_{split}) = D(t_f) = p_\uparrow D_\uparrow + p_\downarrow D_\downarrow$ of S + M associated with the entire ensemble E, where $D_{\uparrow(\downarrow)} = r_{\uparrow(\downarrow)} \times R_{\uparrow(\downarrow)}$; $r_{\uparrow(\downarrow)} = |\uparrow(\downarrow)\rangle\langle\uparrow(\downarrow)|$ where the ket $|\uparrow\rangle(|\downarrow\rangle)$ of S designates the specific state of s_z associated with the specific value 1 (−1), and where $R_{\uparrow(\downarrow)}$ is the ferromagnetic equilibrium state of M so that $m = (1/N)\Sigma_n\sigma_z^{(n)}$ is equal to $m_F(-m_F)$ (save $1/\sqrt{N}$). To simplify the discussion here, we will compare (although this is not essential) this equilibrium to a microcanonical equilibrium, so that $m = \pm m_F$. The state $R_{\uparrow(\downarrow)}$ is therefore a sum of projectors on kets $|m_F, \eta\rangle$ ($|-m_F, \eta\rangle$) of M, which designate the specific states of the spin operators $\sigma_z^{(n)}$ so that $(1/N)\Sigma_n\sigma_z^{(n)}$ is equal to m_F ($-m_F$). Index η takes on a number of values equal to G, a degeneration of the specific values $m = \pm m_F$ of $m = (1/N)\Sigma_n\sigma_z^{(n)}$ that is like an exponential of N in size. We therefore have $R_{\uparrow(\downarrow)} = (1/G)\Sigma_\eta| \pm m_F,\eta\rangle\langle\pm m_F, \eta|$.

Based on the fact that the density operator associated with a sub ensemble of E complementary to E_{sub} is positive, we show that all density operators that describe sub ensembles of E can be built like weighted sums of projectors on pure states of the form

$$|\Psi(t\text{split})\rangle = \Sigma\eta\ U\uparrow\eta|\uparrow\rangle\times|m\text{F},\ \eta\rangle+\Sigma\eta\ U\downarrow\eta|\downarrow\rangle\times|-m\text{F},\ \eta\rangle, \qquad (9.1)$$

where $U_{\uparrow\eta}$ and $U_{\downarrow\eta}$ are complex random normalized coefficients. This particular form of possible states, where only the magnetization states $\pm m_F$ of M correlated with the states $\uparrow(\downarrow)\rangle$ of S occur, will be essential for the desired relaxation; it results from a choice of t_{split} such that S + M has already reached the state $D(t_{split}) = D(t_f)$ for the entire ensemble E.

9.4.5 Dynamics of the Density Operator for Any Subensemble

We must now study the dynamics of the sub ensemble described by the density operator $D_{\mathrm{sub}}(t)$ and arising from $|\Psi(t_{\mathrm{split}})\rangle\langle|\Psi(t_{\mathrm{split}})|$. We have carried out this study within two models of standard quantum statistical mechanics, where the Hamiltonian of the apparatus causes the transitions $|m_{\mathrm{F}}, \eta\rangle \leftrightarrow |m_{\mathrm{F}}, \eta'\rangle$ or $|-m_{\mathrm{F}}, \eta\rangle \leftrightarrow |-m_{\mathrm{F}}, \eta'\rangle$. The magnetization m is therefore conserved by this interaction. In the first case, the evolution concerns only M; in the second case, the interaction of M with B causes a collisional-type relaxation with successive flip-flops of spins $\sigma_z^{(n)}$. The interactions are characterized by a parameter Δ, which measures the loss of degeneracy of states $|\pm m_{\mathrm{F}}, \eta\rangle$, and which is inferior to $J/\sqrt{N}$. The system S remains a spectator. The large number N of spins of M allows the evolution to be irreversible.

The result established first for the pure state below and thereafter for all sub ensembles is as follows:

$$D_{\mathrm{sub}}(t_{\mathrm{split}} + t') = f(t')\, D_{\mathrm{sub}}(t_{\mathrm{split}}) + [1 - f(t')][q_\uparrow D_\uparrow + q_\downarrow D_\downarrow], \tag{9.2}$$

where $q_{\uparrow(\downarrow)} = \mathrm{Tr}\, D_{\mathrm{sub}}(t_{\mathrm{split}})\, r_{\uparrow(\downarrow)}$. The function $f(t')$ decreases from 1 to 0 on a short timescale, of the order $\hbar/\Delta$, the length of the relaxation time of the sub ensembles. Any density operator of a sub ensemble therefore tends towards an incoherent mix of projectors $r_{\uparrow(\downarrow)} \times R_{\uparrow(\downarrow)}$ correlating S and M—whatever the starting point.

Two things occur simultaneously in the process: on the one hand the coherent terms disappear, hence a truncation in the strong sense occurs; on the other hand, the populations of states $|m_{\mathrm{F}}, \eta\rangle$ equalize themselves, as do those of states $|-m_{\mathrm{F}}, \eta\rangle$, and a ferromagnetic equilibrium establishes itself inside each of the two groups of states $|\pm m_{\mathrm{F}}, \eta\rangle$ that are possible for the pointer. The dynamics are the same, decoherence and the move towards the equilibrium of M occur on the same timescale. It is therefore a special form of decoherence, associated with invariance breaking and relaxation towards an equilibrium for one phase or the other. This process concerns only the apparatus and takes place only at the end of the measurement. It is because we are at the end of the measurement and because we have managed to separate within the large density operator $\mathrm{D}(t_{\mathrm{split}})$ a part $_\uparrow$ and a part$_\downarrow$ that in the end this mechanism allows us to arrive at what we are looking for.

9.4.6 Result

The above result is valid for any mathematically conceivable sub ensemble, therefore for any real sub ensemble of the measurement process. The dynamics therefore lead to the hierarchical form of density operators of all the sub ensembles, as required by physics. Quantum ambiguity disappears. This disappearance takes place as a result of the relaxation dynamics applied to the sub ensembles. One small step remains to be taken.

Bernard d'Espagant. This disappearance of the ambiguity seems to me to be indeed a remarkable result in itself, and is essential for the coherence of the proof. As you say, a small step remains to be taken, where, of course, we await your explanations.

Roger Balian. Imagine all sub ensembles of possible real measurements, extracted from E. They have ended up in states of S + A that have the hierarchical structure $D_{sub}(t_f) = q_\uparrow D_\uparrow + q_\downarrow D_\downarrow$ with the coefficients $q_\uparrow(_\downarrow)$ dependent on the sub ensemble. The additive property of these coefficients when we regroup these disjointed sub ensembles is exactly the same as that of probabilities. Besides, in the interpretation of probabilities as frequencies, this property is taken by mathematicians as a definition of the probabilities. It is therefore natural, once the quantum ambiguity has been resolved by the establishment of the universal form $D_{sub}(t_f) = q_\uparrow D_\uparrow + q_\downarrow D_\downarrow$, to interpret $q_\uparrow(_\downarrow)$ as the proportion of individual measurement runs having produced the outcome $\pm m_F$. There appears in this way a structure of classical probabilities (probabilities in the sense of relative frequencies) within the measurement ensemble. We also give to S + A, for each individual measurement run a density operator $D_\uparrow$ or $D_\downarrow$; thereby maintaining a quantum structure. The ensemble $E_\uparrow$ described by $D_\uparrow$ is a sub ensemble of E characterized by a reading of outcome $+m_F$: the reduction is the result of a selection process. The absence of ambiguity makes the interpretation of the weights $q_{\uparrow(\downarrow)}$ as proportions, and the interpretation of Born numbers $p_{\uparrow(\downarrow)}$ as classical probabilities, reasonable; in the sense of frequency.

Bernard d'Espagnat. It seems reasonable for a human being.

Roger Balian. Yes, we are human beings.

Bernard d'Espagnat. A pure spirit…

Roger Balian. … no, precisely, the statistical interpretation is only about human beings.

Bernard d'Espagnat. That is it. It only deals with what is operationally accessible to human beings. That is an important point that you have clearly suggested from the beginning of your presentation. And besides, it seems to me that it is only a small leap to make from putting, as you do, the observables "on the side of" reality and posing as "real", by definition, only the observable. It is clear in any case that the interpretation in question only deals with knowledge that any human being could have relative to what he is capable of perceiving with the help of his instruments.

Roger Balian. If we place ourselves within a pure quantum framework, this knowledge is always probabilistic. Once the particularities of quantum mechanics, ambiguity in particular, disappear, there is no reason not to admit that our ordinary logic forbids us from saying something concerning the individual processes—which is not allowed with the statistical interpretation. For good reason as, with this interpretation, there is no place for individual processes. We have extrapolated a little bit to individual processes, but this seems legitimate to me since this is based on a clear mathematical structure, the hierarchical structure with random coefficients $q_{\uparrow(\downarrow)}$.

9.5 Conclusions

Once the dynamics have allowed us to eliminate quantum ambiguity by ensuring $D_{sub}(t_f) = q_\uparrow D_\uparrow + q_\downarrow D_\downarrow$ for any measurement sub ensemble E_{sub}, we can say that each sub ensemble E_{sub} corresponds to a situation where $q_{\uparrow(\downarrow)}$ can be interpreted as the ordinary proportion of individual runs, and where $D_{\uparrow(\downarrow)}$ is attributed to the final state of each of these runs. The outcome of each individual process belongs to one or the other sub ensemble $E_{\uparrow(\downarrow)}$, characterized by the indication $\pm\, m_F$ of the pointer.

We obtain in this way a well-defined outcome $D_{\uparrow(\downarrow)}$ for each sub ensemble $E_{\uparrow(\downarrow)}$, and for the corresponding individual outcomes, for the pointer ($\pm m_F$) as for the system ($r_{\uparrow(\downarrow)}$), through the correlations established between S and A by the dynamics, and which have been conserved in the relaxation of the sub ensembles. If we select a certain outcome on the pointer, this constitutes a preparation of system S towards a given state $r_{\uparrow(\downarrow)}$.

At the start, we have a density operator D(0) for the entire ensemble; the dynamics of S + A give for this entire ensemble E the state $D(t_f)$. The last step is selection, allowed by the unicity of the outcome of each run, itself guaranteed theoretically by relaxation of the sub ensembles and by their hierarchical structure. If we select a certain outcome for the pointer, for example $+m_F$, this selection allows us to go from density operator $D(t_f)$ of S + A for the entire ensemble to $D_\uparrow$ which concerns sub ensemble $E_\uparrow$, and equally from $r(0)$ to $r_\uparrow$ if we consider only the evolution of S. If we call this passage a reduction, it is the result of selection and not evolution. The observer has selected some information and, in so doing, has changed the laws of probability.

The rest is trivial. Born's $p_{\uparrow(\downarrow)}$ is interpreted as probability in the sense of frequency (the number of elements in each sub ensemble). The repetition of a measurement after selecting an outcome produces this outcome once again. Furthermore, imperfect processes have allowed us to discuss what happens when a process is not a measurement.

We observe in this way an emergence of classical concepts: well-defined individual systems (which is not the case in quantum mechanics), ordinary probabilities and ordinary correlations—as a result of the passage from the microscopic to the macroscopic (because the apparatus is macroscopic). This requires approximations, exactly as for the irreversibility paradox. All this is true not in a mathematical sense but with a probability almost equal to one on a reasonable timescale. And all is based on the dynamics, which play an absolutely essential role.

These dynamics were studied principally in the Schrödinger picture: we have made the state of S + A evolve using the Liouville-von Neumann equation, but we could very well have decided to look at the evolution of the observables of the system and of the apparatus in the Heisenberg picture, which sheds a new light on the measurement problem. In the latter, it is the observables that evolve. But, if the algebra is non-commutative at the start, its non-commutative parts are transferred towards completely inaccessible observables, which correspond to a very large number of spins, with very complex degrees of freedom, of the phonon bath. If we eliminate from the algebra those things that we do not observe and that will never again have any visible effect within a reasonable future, then only commutative observables remain. From there, we have reasons to think that we are in classical physical statistics.

Bernard d'Espagnat. Thank you so much for this excellent presentation. We will now start the discussion.

9.6 Discussion

Hervé Zwirn. I place your conclusion alongside the one from our last session on decoherence. Once the algebra of observables becomes commutative because we remove all inaccessible observables, we are in the same situation as in the theory of decoherence: it is "*for all practical purposes*".

Roger Balian. Yes, except that we have trawled through all the literature on existing models; it appears that each time "decoherence" is mentioned, there is in fact no complete study of the Hamiltonian dynamics of the system and the environment, and no timescale, even if decoherence has brought about some progress. Decoherence works here only because we make it act on the apparatus and at the end of the measurement, and furthermore, as we have seen, it is a particular type of decoherence associated with a relaxation towards a thermodynamic equilibrium of the pointer. This study of sub ensembles has allowed us to resolve the ambiguity and demonstrate the unicity of the outcome of each individual process. We must note that the programme consisting of using the Heisenberg picture rather than the Schrödinger picture has not yet been realized, even though it could be conceptually interesting.

Hervé Zwirn. You provide the technical proof that we can resolve the ambiguity, but the philosophical conclusion that we can draw from it is nevertheless similar to what we usually conclude in decoherence, since we do not succeed at completely eliminating the fact—principally due to unitarity—that we need make the decision at some point to put aside what is inaccessible.

Roger Balian. Of course. That is exactly Boltzmann's (or Leibovitz's) approach, which consists of solving the irreversibility paradox by putting aside the elements that, in any case, are totally inaccessible. It is the same philosophy. We are not doing mathematics but physics and a result that is true at 99.99999999% is good enough. It is statistical quantum mechanics and not quantum mechanics in its pure state. Indeed in the latter we cannot come out of pure states. We cannot therefore understand bifurcations.

Franck Laloë. On the condition that we postulate that observing two outcomes at once is inaccessible to man. Is this what you postulate?

Roger Balian. No. It is inaccessible to man to observe correlations between the measured object and a very large number of spins of the measuring apparatus.

Franck Laloë. You concluded your last stage by considering that it was not unreasonable to suppose that classical probabilities are reduced to a single one.

Roger Balian. What we are doing is simply, by selecting the outcome, going from an ensemble E characterized by probabilities $p_{\uparrow(\downarrow)}$ to a sub ensemble $E_{\uparrow(\downarrow)}$ where only one possibility exists. It is not only reasonable but trivial.

Franck Laloë. It is essential.

Bernard d'Espagnat. We find this in decoherence theory. It is one of the difficulties we encounter, and that is the unicity of reality.

Roger Balian. Is it a difficulty? I don't see any difficulty provided there exists a hierarchical structure of all real measurement ensembles. For each sub ensemble of real processes, we obtain $D_{sub}(t_f) = q_\uparrow D_\uparrow + q_\downarrow D_\downarrow$ and this structure eventually gives a direction to $D_\uparrow$ and $D_\downarrow$, direction which up to that point did not exist as there is no seed of these two possible outcomes in the initial state D(0).

Bernard d'Espagnat. What you indicate half-way through your presentation is very true: these $p_{\uparrow(\downarrow)}$ are mathematical measurements.

Roger Balian. At the start they are only mathematical measurements, and take on a meaning of physical probabilities only at the end.

Bernard d'Espagnat. We make a measurement. And normally, a measurement only has one answer. It cannot have two answers that are incoherent with one another.

Roger Balian. Of course.

Bernard d'Espagnat. Therefore from the moment we make a measurement, we must have only one answer.

Roger Balian. It is the case here. It is demonstrated since, if we make only one measurement, it will belong to the hierarchical structure of sub ensembles that does not allow ambiguity.

Franck Laloë. You have not demonstrated this.

Roger Balian. Yes we have. We have eliminated the ambiguity, then interpreted the general structure $D_{sub}(t_f) = q_\uparrow D_\uparrow + q_\downarrow D_\downarrow$ which has then given a direction to $p_{\uparrow(\downarrow)}$.

Bernard d'Espagnat. I find myself in complete agreement with you when you state that statistical interpretations only deal with knowledge that human beings can have. Therefore I believe I can reformulate your thoughts without betraying them in a way that, probably as a result of my own mental habits, appears clearer to me. What you are saying seems to me to be completely in line with what us physicists have (for the most part) now come to realize: namely that what we—as human beings—can reasonably ask of physics is not to describe reality as some pure spirit would see it but, less ambitiously, to answer those questions we ask ourselves regarding what we observe in such and such circumstances. Which makes me think that your answer to Franck Laloë is that, ultimately, under this conception of physics and in the light of your results, you have nothing left to prove since a question can only have one answer, or at least not two answers that are incompatible with one another.

Roger Balian. It is precisely the fact that we demonstrate that because any real sub ensemble has the right property $D_{sub}(t_f) = q_{\uparrow}D_{\uparrow} + q_{\downarrow}D_{\downarrow}$, the elements of this decomposition start having a significance for the individual constituents of the sub ensembles.

Bernard d'Espagnat. Yes, a significance for human beings, because once again, human beings ask questions regarding what they will see and because each question can only have one answer at a time. Whereas a pure spirit that knows mathematics but has never heard of solid objects would not be able to deduce that $p_{\uparrow(\downarrow)}$ are probabilities.

Roger Balian. It would not be able to deduce anything since we have chosen the statistical interpretation of quantum mechanics where nothing else is said of the properties of ensembles and sub ensembles. Since quantum mechanics only concerns statistical ensembles and sub ensembles, there are no individual objects in this interpretation. If we want to consider individual objects, we must extrapolate slightly. If this extrapolation can be done logically, which is the case by taking ensemble and sub ensembles of all sizes, and if there is no ambiguity (all D'_k disappear), this means that objects will emerge from the dynamics.

Franck Laloë. It seems to me that only the right basis emerges. But you cannot show that as you restrict…

Roger Balian. … I cannot demonstrate that among all real sub ensembles there is one for which a single q equals 1.

Franck Laloë. That is what I was going to say. We would need to demonstrate that as the sub ensemble becomes smaller, one probability tends towards 1 and all the others tend towards 0. This is the problem that everyone has stumbled over for decades.

Roger Balian. No. People stumble over the ambiguity, one way or another.

Franck Laloë. The basis ambiguity?

Roger Balian. The ambiguity of the decomposition of the global density operator. Indeed, practically everyone who has resolved a measurement problem studies the dynamics of the process starting with the initial state and arriving at a final state $D(t_f) = p_\uparrow D_\uparrow + p_\downarrow D_\downarrow$, associated with all possible measurements, then makes a leap by interpreting this result. Here, we complete this type of result by demonstrating that for any sub ensemble, no matter how small, the hierarchical structure $D_{sub}(t_f) = q_\uparrow D_\uparrow + q_\downarrow D_\downarrow$ is reached without ambiguity.

Franck Laloë. You somewhat underestimate Zurek in my opinion. His models show well that a basis is favoured. You do it from a change of basis…

Roger Balian. … from the dynamics.

Franck Laloë. He does it from the contact of the measuring apparatus with the outside. But the result is close, nonetheless.

Roger Balian. The result is close. We would need to look at these models, but none are really satisfactory. Ambiguity is resolved by decoherence, but there is no registration, and resolving the ambiguity supposes that registration as defined above was carried out earlier.

Franck Laloë. The basis we arrive at is unique. Is this what you are saying?

Roger Balian. Not the basis exactly, but the basis in relation to the measured object. It is the blocks that are unique. The basis in relation to the measured object emerges only as a result of decoherence on the apparatus… which is rather amusing.

Franck Laloë. That is clear. It is also the case with Zurek.

Jean Petitot. What is particularly interesting in your model is that it calls upon the theory of magnetic phase transitions. You have insisted a lot—and you have just come back to it—on the role it plays in your demonstration. You call upon critical phenomena within the measuring apparatus. For you, is it simply a special case of something much more general, or do you think that the notion of criticality really needs to be introduced?

Roger Balian. “Critical”, no. There is no critical fluctuation in all this, on the contrary, since the average field is almost exactly at equilibrium in our model. There is invariance breaking, with a transition temperature. It is simply a paradigm that is behind the fundamental idea according to which the pointer can arrive at multiple different stable states with no a priori bias with regards to these different states.

Jean Petitot. Here you say much more than that. There are for example necessarily phenomena of symmetry breaking.

Roger Balian. There is necessarily breaking.

Jean Petitot. A thermometer can indicate multiple temperatures without there being symmetry breaking.

Roger Balian. Indeed, but we cannot use a thermometer for making a quantum measurement.

Jean Petitot. Precisely. This is question I am asking you.

Roger Balian. The pointer needs to be a macroscopic object such that it arrives at multiple different stable macroscopic states, under the same external conditions, through dynamics from the same initial metastable state.

Jean Petitot. This means that underlying this, there needs to be dynamically some bifurcations.

Roger Balian. Yes. The possibility of bifurcation can exist for phase transitions, be there needs to be ergodicity breaking.

Jean Petitot. This may be a form of axiom that would need to be introduced—and which, in general, is not introduced.

Roger Balian. Implicitly, it is necessarily introduced, since a pointer must be able to arrive towards multiple stable states under the same circumstances.

Jean Petitot. No.

Roger Balian. From the moment we say that an apparatus can produce different outcomes…

Jean Petitot. … there is not necessarily symmetry breaking underlying this.

Roger Balian. There is not necessarily symmetry breaking, but dynamic ergodicity breaking.

Jean Petitot. I agree.

Roger Balian. It is necessary and everyone has this in mind, since those who talk about pointers think the pointer can produce different outcomes, and therefore that there are multiple equilibrium states. These states do not necessarily correspond to a broken invariance, but they are necessary.

Jean Petitot. I agree with you. I thought this was one of the original aspects of your work, but you seem to say it is completely commonplace.

Roger Balian. It is not that original.

Jean Petitot. As a mathematician, or at least as someone who knows bifurcation theory well, I cannot claim that all the physics papers dealing with the problem of measurement that I read call upon dynamic bifurcation theory… unless I am not reading the right papers.

Roger Balian. Unfortunately, a large part of the literature focuses on decoherence and forgets, in the measuring process, the registration by the apparatus. And yet in a measurement, there is necessarily a large object that is an apparatus, with multiple stable states. People forget this fact.

Jean Petitot. It is a key fact.

Roger Balian. It is often missed in a certain number of papers.

Jean Petitot. Am I mistaken by saying that not only do you stress this fact, but you also consider it to be a fundamental fact, constituent of the problem of measurement?

Roger Balian. Of course, it is a fundamental fact in a measurement. But obviously we are not the only ones saying this. This is why I was pointing out that it was not original.

Catherine Pépin. I would like to come back to the notion of ergodicity that you have just mentioned. Does this dynamic trajectory cover all possible configurations?

Roger Balian. I chose the word ergodicity badly.

Catherine Pépin. I have never seen this elsewhere in your theory. You still make the hypothesis that a dynamic trajectory determines the state we observe.

Roger Balian. Yes; but I have badly chosen the word ergodicity. It is a purely quantum dynamic process. What I wanted to express by ergodicity breaking is that in this process the transitions from a macroscopic equilibrium state to another have no chance of occurring within a reasonable timescale.

Catherine Pépin. There are states that we observe and others that we do not observe. In this regard, it is similar to the point Mr. d'Espagnat made.

Roger Balian. In the evolution of this model, we have to start from an initial paramagnetic state, namely one where there is nearly the same number of spin-up and down spin-down. The interaction with S favours flips in a certain direction for each diagonal block. If it favours upward flips, for example, coupling with the bath will allow most spins to orient themselves upwards and we will end up with a configuration $+m_F$. Starting from a pure state, we will have something very random in the evolution, with a trajectory. In statistical mean, we will have a Fokker-Planck type probability evolution. In the end, magnetization will be $+m_F$, give or take $1/\sqrt{N}$, with a purge of everything along the way.

Catherine Pépin. You cover therefore only one part of the states.

Roger Balian. It is the dynamics that do this, and which practically forbid the distancing of m from $+m_F$.

Catherine Pépin. My question was more philosophical, in a way, and tried to link Mr. d'Espagnat's comments that it is more the work of a human being than a pure spirit. I understand the dynamics mean that you do not explore all possible states of S + A. They choose the final outcome with a certain trajectory that does not necessarily explore everything—but this partial exploration is sufficient. It reminds me of Ludwig Boltzmann's ergodic hypothesis.

Roger Balian. I used the word "ergodic", which is wrong. In fact, there is nothing ergodic in all this. It is quantum, therefore linear. I should have used the term "erratic". There is nonetheless some decoherence within, in a way. But it is a very special type of decoherence.

Catherine Pépin. I must admit I have not quite understood the distinction between human and pure spirit. How would a pure spirit see things differently from a human being? Could you please clarify?

Roger Balian. May I translate what you were saying, Mr. d'Espagnat, in the way I understood it? You used the term "human" meaning a quest for affirmations concerning individual objects, whereas quantum mechanics is "inhuman" in that it does not say very much about these objects. It is us, as human beings, who want to attribute these things to individual objects. Is that it?

Bernard d'Espagnat. Yes, it is one aspect of the way I see things, but this one is more "radical". I consider that not only does standard quantum mechanics (without hidden variables, etc., and where quantum statistical mechanics derives) says nothing about individual objects, but in order to make it say something about these objects we need to "force it", in a way, with approximations like 0.9999999 = 1, justified by the fact that human ability is limited. We need, for example, to deliberately ignore the existence—even though it is obvious—of quantities that are in theory measurable, and whose measurement, according to quantum mechanics itself, would jeopardize the notion of object in itself (for example, that of a well-localized object) … but that the universe's too brief duration or similar "contingencies" would preclude us from measuring. This contrasts with the point of view of classical physics because according to it, not only do objects exist in themselves completely independently from human knowledge, but the role of science, and that of physics in particular (I am not including statistical mechanics) is precisely to lift the veil on appearances and describe them as they are. I think we are in agreement when we consider that standard quantum mechanics (without any hidden variables), as well as classical *statistical* mechanics, is not capable of doing this. I would say it is not even its goal, which would be more like accounting for the fact that we have the *impression* that we see things and individual objects. I reckon that is also what you have shown us, in a different way.

Roger Balian. No, not at all!

Bernard d'Espagnat. What I want to say is that—as you stated before—like Boltzmann, you cast aside de facto what is inaccessible. That is why I do not think you achieve—no more than him either!—knowledge of reality as it *is*.

Roger Balian. I do not agree with you. I come back to the interpretation of quantum mechanics. There is reality. For my part, I believe in reality.

Bernard d'Espagnat. Certainly. So do I. But is reality actually knowable as it really is?

Roger Balian. Microscopic reality is not totally knowable. I agree completely.

Bernard d'Espagnat. Ah! Per se, or will we know it one day?

Roger Balian. Within the framework of current quantum mechanics, it is not totally knowable and we cannot answer this question. The goal of physics is to say as much as possible, which is achieved through what we call "states"—meaning "what we know probabilistically of a statistical ensemble of systems that have the same starting properties". However, this does not preclude the existence of individual systems.

Bernard d'Espagnat. It does not preclude their existence but nothing proves they exist independently from us. The least we can say is that, unlike classical physics, quantum physics does not provide a conceptual framework that would lead us to believe in their existence!

Roger Balian. For my part, I believe in it. It is a question of belief. It seems to me unthinkable to not believe in it.

Bernard d'Espagnat. For you, are macroscopic systems real per se?

Roger Balian. They are as real as microscopic systems. But they are as difficult to know as quantum systems. In addition, they are much more difficult to know than microscopic systems. And it is even worse than that! Within the framework of statistical mechanics, I do not know the detailed structure of a sheet of paper. And my lack of knowledge is much greater than the lack of knowledge I have…

Bernard d'Espagnat. … ultimately, nothing is knowable!

Catherine Pépin. What do we mean by "knowable"?

Roger Balian. Mathematically describable and perfectly predictable.

Bernard d'Espagnat. What are mathematically describable are phenomena; etymologically "what appears". Experience and observation provide us with "appearances, (which are) the same for all of us". Schrödinger—the physicist, *our* Schrödinger!—has much insisted on this point. We cannot prove that our sensations and reality are isomorphic. In order to prove this, we would need to be able to compare them to each other. Yet he points out (after many philosophers) that we know our sensations, but not a reality that is supposedly distinct from these.

Roger Balian. They are not isomorphic. It is an image.

Bertrand Saint-Sernin. I am sorry to be so ignorant and to not have been able to follow your presentation. This leads me to ask two questions. The first is the following. You have spoken about statistical mechanics. Part of Boltzmann's work consists of the *Populäre Schriften*. And Schrödinger also wrote some popular works. All things considered, could what you have presented be made accessible, at least in part, either to laypersons not specialized in quantum mechanics, or to scientists from other fields—who must be as lost as I am?

The second thing that has touched and moved me was to find almost word for word passages from *Timaeus*. On page 54D, Timaeus says that we must make judgement calls since certain observations we can make on nature do not allow us to choose irrefutably one direction over another. In some instances, there is a sort of insufficiency of empirical contributions. Then, further on, he says that if we had the possibility of knowing singular processes instead of seeing things as a whole, this would deepen our understanding of nature. I was under the impression that in the experiment you presented, you cited this obstacle in passing from a statistical view to a perception or conception of singular events—even if we cannot know them completely. Here again, you are absolutely faithful to what Plato said in *Timaeus* on page 57C, I think.

Roger Balian. I have never read *Timaeus*, unfortunately—not even page 57C! But I believe it is exactly that, with one addition: I quite like the idea that the most fundamental things in science (the impossibility to go faster than the speed of light, the impossibility to go back in time, to create energy, to obtain work with a single heat source, etc.) are limits imposed on us by nature. It seems to me precisely that quantum mechanics places limits to what is knowable. We can try to move forwards, but hidden underneath are what we are analysing as the non-commutation of observables. We have managed to find a mathematical tool that works well with this microscopic physics, but we have no intuition of it. And the fact that we have no intuition of it means that when we try to go down at that scale, which is far removed from our everyday experiences, we find ourselves stuck in our descent—whereas in the passage you cited, there was still perhaps a possibility to go down even further.

Bertrand Saint-Sernin. Plato said "only for God and a few of his friends".

Roger Balian. Yes! That is exactly what I think.

To come back to your first question, it is really very difficult to get across, not only to philosophers but also to scientists other than physicists or chemists, this idea of a hidden non-commutative algebra, this idea of an imposed barrier, or even the idea itself of quantum probabilities (which are not ordinary probabilities—mathematicians themselves find that hard to accept). We manage to master them well mathematically. We manage to link this type of mathematics with reality. But we do not manage to link this with our macroscopic way of thinking. It is the reason why the question you ask unfortunately has a negative answer.

Catherine Pépin. Do you have the same definition as Mr. Balian of what is knowable, Mr. Saint-Sernin? Being neither a mathematician nor a physicist, do you consider that what is knowable is what is mathematically describable and predictable? Or do you have another definition?

Bertrand Saint-Sernin. What is knowable is not necessarily predictable. I am convinced that probabilities play an absolutely fundamental role. One thing that struck me was that even my probabilistic colleagues experience great psychological difficulties in admitting the importance of probabilities in life. Acknowledging probabilities when they do not affect us is one thing. But acknowledging them when they affect us in everyday life is another thing altogether, and is much more difficult. It is something that was developed by Cournot, a 19th century mathematician who was also a great philosopher, and whose beginnings were very strange. He left school at 15, was a solicitor's clerk for four years before going to Besançon where he prepared *Normale* sciences. He was awarded first prize, but the school closed down the following year. He became tutor to the sons of Gouvion-Saint-Cyr, then professor of analysis and school administrator.

Roger Balian. Is it the same Cournot who was a pioneer in economics?

Bertrand Saint-Sernin. Yes, it is he! And he has written much about probabilities.

Roger Balian. I have always been surprised to see what we have made Laplace say concerning probabilities. Indeed, at the beginning of the *Traité des Probabilités*, he takes great care to say that absolute determinism (a notion we still credit him for) is unthinkable, which is why we must use probabilities. This phrase is often lifted out of context when in fact it was an introduction to the *Traité des Probabilités* whose goal was to promote the need for probabilities. It is odd.

Olivier Rey. He sums up what was thought at the time in order to propose something else.

Roger Balian. We speak of "Laplacian determinism" when in fact it is the exact opposite.

Bertrand Saint-Sernin. Indeed.

Catherine Pépin. From a personal point of view, I think the same thing applies to Descartes.

Bernard d'Espagnat. This discussion is fascinating but unfortunately it is getting late. We must close the session. Thank you.

References

1. Armen E. Allahverdyan, Roger Balian, Theo M. Nieuwenhuizen, «Understanding quantum measurement from the solution of dynamical models», *Physics Reports*, 2012, p. 1–187. Available on the IPhT website http://ipht.cea.fr.
2. François David gives a lecture on the use C*-algebras as the starting point for quantum mechanics; he explains why these types of algebra come into play and in particular why these algebras act on complex numbers (they are the only ones to have separability).

Author Biography

Roger Balian is a researcher at the *Institut de Physique Théorique* at Saclay and member of the *Académie des Sciences*. He has also been professor at the *École Polytechnique*, director of the Summer school of theoretical physics at Les Houches and president of the *Société Française de Physique*. His work has often used the methods of statistical physics, and has even had some impact in mathematics. His research included such wide-ranging topics as the prediction of phase B of superfluid helium 3, the discovery of a strong uncertainty principle, the relationship between waves and trajectories, transition phases, the Casimir effect, the distribution of galaxies, the theory of quantum measurement. He has published a treatise on statistical physics in two volumes, *From Microphysics to Macrophysics* (Springer Verlag, 2006).

Chapter 10
Loop Quantum Gravity

Carlo Rovelli

Bernard d'Espagnat. We welcome today Carlo Rovelli, who has kindly accepted to speak to us about his book *Quantum Gravity* [1]. Carlo has already participated several times in our discussions; consequently we are familiar with his very novel ideas on quantum mechanics. We are delighted to have the opportunity to hear him speak about his ideas, not unrelated to the former, on cosmology. So without further ado, I give him the floor.

10.1 What Is Loop Quantum Gravity?

Carlo Rovelli. Thank you for giving me the opportunity to speak about quantum gravity. I do not believe you were expecting on my part a mass of technical and mathematical details on the subject. I will speak instead of the conceptual manner in which we can elaborate a quantum space-time theory.

I had the opportunity, alongside Matteo Smerlak [2], to present to you the relational interpretation of quantum mechanics. The relational interpretation of quantum mechanics and the quantum space-time theory are two related subjects. Speaking today about quantum gravity will allow me to clarify the many connections between the two themes.

First of all, the points I made on the relational interpretation of quantum mechanics were prompted in part by what we have learned about the physical world with general relativity, starting with Einstein in 1915 and with the interpretational efforts that followed. Secondly, several questions came up during my presentation on relational quantum mechanics. How should we think of this theory in

C. Rovelli (✉)
University of Aix-Marseille, Marseille, France

B. d'Espagnat and H. Zwirn (eds.), *The Quantum World*,
The Frontiers Collection, DOI 10.1007/978-3-319-55420-4_10

space-time? How should we think of it within the context of quantum field theory or in quantum space-time? My presentation today on quantum space-time theory will allow us to come back to these questions, the aim being to provide you with a physical image and show you how relational quantum mechanics, relativity and quantization of space-time interact.

10.1.1 Relational Quantum Mechanics

After a long gestation period, with the works of Max Planck and Albert Einstein in 1905 (the discovery of the "granularity" of the electromagnetic field), then those of the Copenhagen School, quantum mechanics was born in 1927 with on the one hand the works of Heisenberg and on the other those of Erwin Schrödinger—two independent approaches from which Paul Dirac will build the general formal structure of the theory.

For Heisenberg, quantum mechanics speaks of "number tables"; for Schrödinger, of a "wave in space" that quickly became a wave in configuration space. In a sense, these two ways of apprehending quantum mechanics still exist today. Their difference does not stem from a dichotomy between a realist interpretation and an operational interpretation. A realist interpretation is possible for either of them, and so is an operational interpretation.

I would like to say a few words about Heisenberg, as we must start with him in order to understand what is being done on quantum relativity and the relational interpretation. Heisenberg has told that the idea for his theory came one night when he was out in a park in Copenhagen. Everything was dark except near a few streetlights. Heisenberg noted the presence of another person walking, who disappeared and reappeared depending on whether he was close or not to a streetlight. Heisenberg knows that the person walking does not actually disappear when he is in the dark. He can even imagine him walking even if he cannot see him. For all that, he wonders if he could have the certainty to think the same thing for very small objects: is it certain that an electron we see intermittently when it interacts with other physical systems continues to exist when we no longer see it? The number tables only describe the appearance of the electron at certain locations, when it is interaction with something else such as light.

Hence my suggestion that it is perhaps easier to describe reality if we abandon a part of classical realism to concentrate on the description of the possible interactions of one system with another. That is the central idea behind relational quantum mechanics: we do not speak of the world in space but only of quantum events that occur through interactions: "*The best description of reality is the way things can affect one another*".

It is better to describe reality in terms of interactions rather than objects, thereby concentrating on a process and not on entities. Processes are facts, meaning they are what happens. Reality is the ensemble of these facts. Furthermore, each process is delimited by boundaries. The states describe what happens at the boundaries of these processes—i.e. how what happens there interacts with the outside.

Catherine Pépin. Can you define what the boundary of a process is?

Carlo Rovelli. I will come back to it in more detail when I will speak about quantum gravity. But to give an example, at CERN [3] a process is a set of particles that enter the collision and come out of it, plus the region of space-time where the collision takes place. In quantum mechanics, it is the transition from the initial state to the final state. Or still, it is the way in which a process interacts, affects or influences something else on the outside.

Furthermore, a specific characteristic of quantum mechanics is *discreteness*—I do not know the French equivalent.

Bernard d'Espagnat. There is no French equivalent. In the relevant sense of the term, implied for instance in the expression "discrete spectrum", there is—unfortunately!—no noun derived from the adjective "discrete".

Hervé Zwirn. Perhaps we should use the term "discrétitude"?

Carlo Rovelli. I sometimes get the impression that when we speak of quantum mechanics, we underestimate the notion of discreteness, even though it is a central aspect of quantum mechanics. Indeed, quantum mechanics describes discrete spectra, i.e. things that occur in a discrete way. Admittedly, this does not explain everything.

Bernard d'Espagnat. Discrete spectra are not really things that occur.

Carlo Rovelli. Indeed. However, the discrete aspect is still universal and concerns interactions in the following sense: what is discrete is the action. The action is the measure of the volume of regions in phase space. Quantum mechanics is characterized by the fact that we cannot precisely localize a system in phase space. An interaction allows us to distinguish between a finite number of orthogonal states, of different states. In classical theory, the number of possibilities is infinite. By contrast, in quantum theory, this number is finite. Phase space encodes what we know of the system. If the system is in a certain region of phase space, we can determine how many different states it can be in. One of the characteristics of quantum mechanics is that this information is always limited, always finite. If the information is understood in terms of the number of different states something can be in (Shannon), quantum mechanics provides a discrete structure of the information that physical systems can have.

This discrete structure is fundamental—and yet, it is seldom considered. That is one of the reasons why I consider that a realist interpretation of the wave function leads us on the wrong path. A quantum particle has *less* information that a classical particle. We cannot imagine that reality is characterized by more numbers than in its classical description, when the information is less. The wave function tells us less than a point in phase space, and not more.

10.1.2 General Relativity

In 1915, Einstein had the extraordinary idea that the gravitational field was the same thing as Newton's space-time. He reinterpreted the strange entity that is Newtonian space—which will then become space-time—as the description of a physical field when we discard all dynamic aspects. It is a radical simplification of the world. Instead of having a space with fields (and particles) within this space, we simply have fields—the gravitational field is only one of them—that, in a way, live on top of each other. That is, to my mind, at the core of Einstein's physics. Everything derives from this. Einstein's equations are the equations of these fields.

From that point on, the notion of localization, i.e. the meaning of the question "where?", changed radically. Before Einstein, a particle was thought to be localized within space. After the theory of general relativity, space as the container of the world no longer existed and there were only fields. Things could only be localized in relation to one another. General relativity shows that positions can only be expressed in relation to a field. A position is therefore defined for each dynamic entity only in relation to others. It is the relational aspect of general relativity: localization in space and time is relative to things, one in relation to the other.

10.1.3 Quantum Field Theory

In quantum field theory, fields behave on a small scale as if they were sets of particles (e.g. photons). This structure is described mathematically by Hilbert space (Fock space) with a set (algebra) of operators and rules on the transition probabilities (Feynman rules). These rules show how states evolve in time, with links between an initial interaction and a final interaction. The states describe the position of particles (field quanta) in relation to space (or their momentum).

10.1.4 Quantum Gravity

How do we carry out quantum theory not of a field in space, but of space itself? From a mathematical point of view, the granularity characterizing the electromagnetic field is the same as that of a gravitational field. In Hilbert space, which characterizes the quantum states of the electromagnetic field, the operators have a discrete spectrum. It is the same for the gravitational field: we can build a Hilbert space with operators with a discrete spectrum. These operators are described by the gravitational field—which is, according to Einstein, the metric of space-time, i.e. the quantities that determine the lengths, areas and volumes. In quantum theory, the operators correspond to lengths, areas and volumes. We can write them with operator algebra, and we can thus study the spectrum. It appears that the spectrum of quantities, volumes in particular, is discrete. There is therefore a set of discrete volumes, exactly like the energy of photons. The volume of fragments of space is discrete. The image is therefore that of a set of quanta of space.

The loop quantum gravitation theory defines a Hilbert space, an operator algebra and transition amplitudes. The Hilbert space has states which do not have a position in relation to a space (or momentum). Indeed, no position exists in relation to which quanta of space can be localized, because they are themselves space. That being, there are a number of quanta of space, with associated volumes and information according to which the quanta are close to one another. The state space is a space of quanta in their continuity relation with one another. Operators correspond to the metric (dimensions, areas, etc.) and there are transition amplitudes.

Bernard d'Espagnat. We were expecting you tell us that, in the same way that photons are electromagnetic field quanta, by quantifying the gravitational field we find particles called gravitons. However, this is not what you are saying.

Carlo Rovelli. No. We were expecting a discrete structure, but in a different sense.

The basic idea dates back to the 1930s. At the time when Heisenberg wrote about uncertainty relations, Lev Landau made the mistake of deducing from it that because of quantum mechanics, we cannot measure a component of the electric field at a point. Niels Bohr instantly recognized that it was a mistake. He published a well-known article with Léon Rosenfeld [4] on field measurability where he showed that Heisenberg uncertainty relations do not preclude, in principle, measuring an electric field at a point of space-time. It seemed at the time that the matter was closed. But not so!

A young colleague of Landau redid Bohr's analysis for a gravitational field. He realized that it no longer worked: it is not possible to measure a gravitational field at a point of space-time. Put simply: to measure a very small thing requires condensing the position to a very small region—which results in considerable fluctuation of the momentum, thus a lot of energy. If there is a lot of energy, there needs to be a lot of mass, thus space-time is curved and a horizon appears. That is general relativity. When the horizon becomes larger than the dimension we want to observe, we are lost: the region we want to observe falls into its own black hole. It generates a limit to measurability on a small scale.

The basic idea underlying quantum gravity is that there is a minimal length below which we cannot go. We find this minimal length in the quantification of discrete spectra of volume. It has nothing to do with gravitons. Indeed, we do not take this into account with gravitons. We forget about the non-linear structure of general relativity. Gravitons are an approximation on a large scale, like phonons. We expect gravitational waves to be quantified like photons, but the graviton approach is not good for small distances, meaning at very high energy levels or at very short wavelengths. This notion is suitable only for long distances. At short wavelengths, at high energy levels, the discrete structure of space at the Planck length is not captured by the notion of graviton.

Hervé Zwirn. What would be the equivalent? A Planck mass in a Planck volume [5] that would generate a black hole beyond whose horizon we cannot go?

Carlo Rovelli. Yes. It is a problem of algebra. It is the point where the horizon becomes greater than the delta x of what we want to measure.

Jean Petitot. In addition, gravitons are conceptually paradoxical. Indeed, they are associated with a defined field in classical space-time.

Carlo Rovelli. Indeed. To be able to include them, we need to forget about general relativity, meaning the fact that the gravity field is the same thing as space-time.

A Hilbert space is determined by a random graph. Operators are defined in Hilbert space and give a geometrical interpretation to the graph.

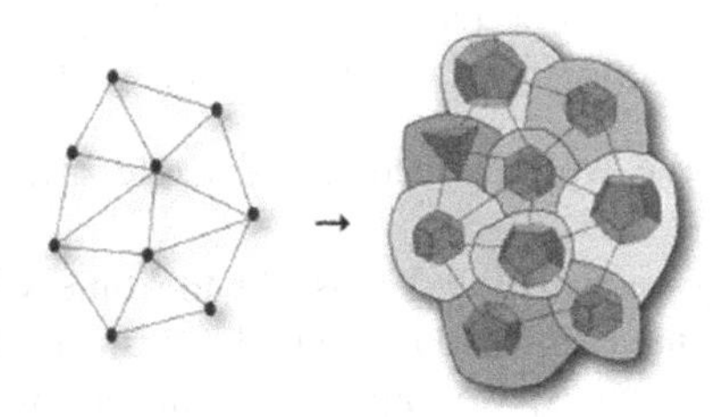

The states occupy this random graph, each node being a "quantum of space". The quantum gravity theory shows very precisely that even though the operators are associated with the volumes, areas and angles, they do not commute among themselves. Since all the quantities required for defining the geometry do not commute, the geometry remains fuzzy. The quanta of space have specific discrete values, but remain fuzzy. The crucial thing to remember is that gravitational field quanta are not *in* space, but form *in themselves* the space.

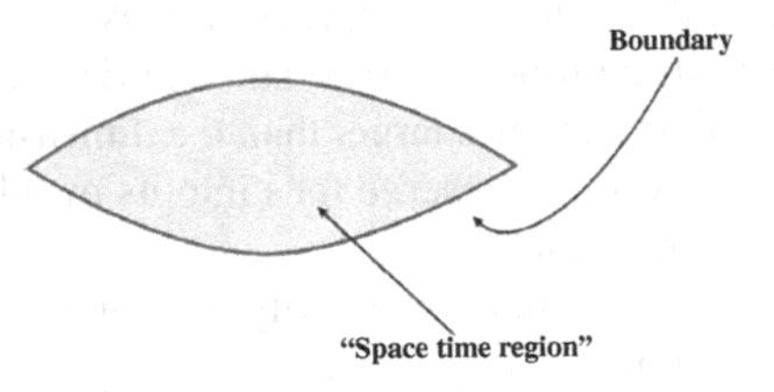

Space time region

I now come to the transition amplitudes. The transition amplitudes show how to go from an initial state to a final state. We can imagine a space-time region with a boundary, the states being on the boundary of this space-time region.

How do we calculate transition probabilities? Where is space? Where is time? The main idea is that to describe for instance a collision of particles, we need to include in the description the space-time region where the collision takes place. To calculate what happens, we need to know the field state at the boundary of this region—including the gravitational field state, which is the boundary geometry, and thus the metric, thus the distances and time intervals between the entering and exiting particles. There is no need to say when and where the collision takes place, because the distance between the past and the future is precisely the state at the boundary of the gravity field. All we need to know are the transition amplitudes between the initial state and the final state. There is no need for any other spatial or temporal information.

Catherine Pépin. Is this the idea according to which states are at the boundaries?

Carlo Rovelli. Yes. That is the reason why states are not localized and the theory does not specify any space or time variable. All we need to know is that the gravitational field is at the boundary. The gravitational field is itself the information of the distances and the time lapse.

Hervé Zwirn. The boundary is a state of the gravitational field, which contains in itself space and time.

Carlo Rovelli. Exactly.

Hervé Zwirn. A transition between two states is therefore a transition between two states of space and time.

Carlo Rovelli. Yes.

Hervé Zwirn. So, how can we interpret a transition probability over time? How can this be represented intuitively?

Carlo Rovelli. Two interactions are possible, with an associated amplitude (a complex number). We can therefore compare two probabilities—one more likely and one less likely.

Hervé Zwirn. There is therefore a probability to go from time T0 to time T1, but also from time T2 to time—T2?

Carlo Rovelli. Yes, but comparing probabilities is only possible for comparable things. Furthermore, it is interesting to note that time is part of the field, and so is itself a variable.

Hervé Zwirn. What happens if we only consider gravitational field theory, with nothing else added?

Carlo Rovelli. The same thing.

Hervé Zwirn. Let us imagine that there is only the gravitational field, without any particle or anything else. What is the intuitive interpretation of these transition probabilities from one state to another?

Carlo Rovelli. It is the same thing, because the gravitational field is complex. Imagine a boundary state, with time elapsing and the passage of two gravity wavelets that interact. In principle, we do not need matter in this theory. Even if, obviously, it is simpler when there is some!

In summary, the presence of discrete structures corresponds to the existence of quanta of space. Probability corresponds to the transition amplitudes between configurations within the gravitational field. Processes themselves are regions of space. States are on the boundaries and are exclusively regions of space.

One point remains unclear, but is to my mind the most interesting aspect. I would like to mention it here. On the one hand, we have the idea that quantum mechanics speaks about interactions. It shows the way systems influence each other, interact with each other. On the other hand, we have a relationism that is typical of general relativity, for which the localization of things is relational. Ultimately, it is in the vein of Descartes or Aristotle: for Descartes, there is no space-time but only a relative position of things. And there is clearly a link between the two: between the relational aspect of quantum mechanics and the relational aspect of general relativity. Indeed, things can interact if they are in contact. And at the same time, what does being in contact mean other than to interact? How can we see that two things are side by side other than by seeing an interaction between them?

Ultimately, to interact is the same as being side by side in space-time. The structure of systems that interact with one another is very closely linked to that of space-time. These ideas are not yet very clear!

10.1.5 Conclusion

Physics is becoming simpler in a way. For Newton, there were particles that moved for a certain length of time in space. Faraday and Maxwell introduced fields: matter was not only made up of particles, but of particles and fields. Faraday wrote some beautiful passages where he showed that fields were real. Then, with special relativity, we had to describe space and time in a more concise way, as space-time. Later, quantum mechanics, particularly quantum field theory, showed that we could describe particles like excitations of quanta of field. We therefore ended up with quantum fields in space-time. However, with general relativity, we realized that space-time was ultimately just a field: a gravitational field. Ultimately, with

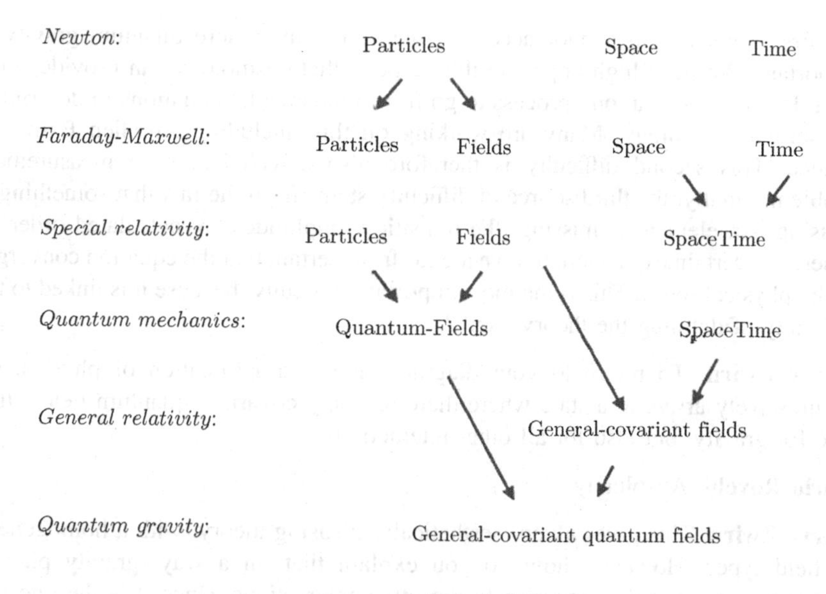

Sequence of physical theories

quantum gravity, we end up with a single entity and nothing else: a quantum field that does not occur in space but in itself.

Contemporary fundamental physics tells us that best ontology for describing nature given what we know is to consider that an object is a quantum field with general covariance, not occurring in space but as part of a set of quantum fields existing on each other, and that quanta weave the structure of space itself.

This is the general framework, admittedly not always very clear, I believe we must work into try to bring together quantum mechanics and general relativity. To finish, I would like to remind you that a theory becomes true once its predictions are verified. In this case, we are not there yet.

Bernard d'Espagnat. Thank you very much.

10.2 Discussion

Catherine Pépin. What difficulties does this theory encounter to become predictable? Does one difficulty stem from the high number of variables? Or are these simply technical difficulties?

Carlo Rovelli. I can see three sources of difficulty. The first is that the equations may be very elegant, but it is very hard to use them to perform calculations. That is a technical difficulty. The second source of difficulty, perhaps more important than the first, comes from a poor access to measurements where quantum gravity is important. We have high hopes, in this respect, that cosmology can provide some data. However, it is a long process to go from fundamental equations to describing primordial cosmology. Many are working on this, including Aurélien Barrau in France. This second difficulty is therefore also a technical or a measurement problem. Finally, the third source of difficulty stems from the fact that something is missing. An element is missing: the transition amplitudes are calculated order by order, in a certain expansion, but we are far from certain that the equation converges in the physical sense. This is the most important difficulty, because it is linked to the difficulty of defining the theory itself.

Hervé Zwirn. To return to your diagram on the simplification of physics, we progressively arrive at a state where there are only covariant quantum fields. It is true for gravity, but also for all other interactions.

Carlo Rovelli. Absolutely.

Hervé Zwirn. This gives a very aesthetically pleasing theory, with a homogeneity of field types. However, how do you explain that, in a way, gravity plays a somewhat special role compared to the other interactions, since it is the one that gives rise to space-time in which all the other interactions are described? Why is there not a complete symmetry between all the fields? Is it due to a different spin? Is there an explanation or is this a stupid question?

Carlo Rovelli. Your question is not stupid at all, but it is as relevant to classical mechanics as it is to quantum mechanics. Once we have general relativity, everything becomes covariant. There are only covariant fields. They are all on the same level. Consequently, the question is to know in what way the gravity field differs from the others. I do not have any convincing answer. That being, the world we see is described well by a certain action, in which the gravitational field interacts with everything. The theory remains coherent if we remove certain elements, except the gravitational field. In Einstein's action, there is no electromagnetic field. However in Maxwell's field action, there is the metric. It is this universal interaction with everything that differs.

Hervé Zwirn. The gravitational field really does play a rather particular role.

Carlo Rovelli. Yes, even though I am tempted to downplay this role. I think we are far from having understood everything.

Bernard d'Espagnat. This is fascinating. In the description of physics you have just given us, what must we make of our classical representation? In the way I understand quantum mechanics, without hidden variables, etc., I consider that the latter does not tell us how things are, but describes how we see them. From your point of view, it is not quite the same since you say it describes "things", systems, how other systems see them. In either case, there is a relative aspect. But in my view of things, we can tell ourselves that since quantum physics is the truth, classicality is only an appearance that works very well for anything ordinary but important to us. Since it is only an appearance, did it also function nearer to the Big Bang? Perhaps at that time, the universe was in a quantum state and we cannot interpret this state *now* as having been quasi-classical *then*. This question has been examined and I believe that certain theories consider that it seems we can interpret the state of the universe as having been quasi-classical, even in times far removed. Does your theory say anything about this?

Carlo Rovelli. There is an ongoing reflection on this point within loop quantum gravity theory. In the Hilbert space of this theory, there are many possible states. Some have semi-classical characteristics. These are the coherent states. We know they can describe things well, like in quantum mechanics. Besides, the formal structure is that of quantum mechanics. Is the semi-classical approximation the right one for describing what happens near the Big Bang? I do not think so. Indeed, once the energy density reaches the order of the Planck energy density, fluctuations of the metric are such that we can no longer speak of a semi-classical approximation. In principle, we can still use the theory of states and transitions, but there is no trajectory. Many calculations have been performed, with contradictory results: with a certain approximation (by taking a homogenous universe and by observing the evolution of its scale factor), we are able to reconstruct a near-classical metric. I am personally not convinced by these calculations and think there is still much left to understand.

In short, I do not know how to answer your question! My intuition tells me that space-time becomes completely non-classical during the Big Bang.

Bernard d'Espagnat. According to your intuition, we can therefore not represent what happened more or less at the Big Bang. The images we try to have of it, and which are published in newspapers, do not hold up.

Carlo Rovelli. Indeed.

Alexis de Saint-Ours. In your book, you explain that your work focuses as much on quantum gravity as it does on time. Indeed, part of your research consists in trying to show how in a generally covariant formalism, we work with a physics that has no time. Could you remind us why you think that at the Planck scale, there are no good clocks? Is that still your position today?

Besides, this vision of a physics without time in a generally covariant formalism is relatively well-accepted by the quantum community, except by Lee Smolin [6] who keeps on claiming that time is real. What is your opinion on his position and on this subject?

Carlo Rovelli. In my book, I have rewritten classical mechanics in a language where we do not need to consider the notion of time, before showing that this language is necessary for dealing with general relativity and even more so with quantum gravity. In transition probabilities, there is no "T" or Schrödinger equation. Where is time? Time is in the boundary state. The claim that "at the Planck scale, there are no good clocks" is a matter of intuition, but I think it is correct. If we forget about gravity, we can describe a physical system and, next to it, a clock that measures time—with no interaction between the two. We can thus use this clock to characterize different time points. Once we have gravity, masses change the metric, thus the way time passes. In other words, the clock is influenced by gravity. It must therefore be described at the same time as the rest of the gravitational field. If the gravitational field is weak and classical, we can forget about it. However, this is not possible if it is strong and quantum. To put in place this way of seeing classical and quantum mechanics without time has been a long journey. Ultimately, I did a type of collage. Indeed, Jean-Marie Souriau [7] already had a formalism that allowed it. The ideas were there. I reprised the existing formulas. In any case, I believe I was sincere in my approach.

As for Lee Smolin, he is a very good friend and I think highly of him—but I think he is wrong! It seems to me that he confuses different subjects. First of all, the absence of time does not imply that there is no local time. It simply means that we cannot take an external variable T for the whole universe. I am not saying that there is no time, but that there is not a time with all the characteristics. Furthermore, I think Lee Smolin is under the influence of Roberto Unger. This philosopher, who teaches at Harvard, promotes the idea that everything evolves over time. He considers for instance that the laws of chemistry, which are scientific laws, did not exist prior to the formation of atoms, or that the laws of biology did not exist prior to the origin of life. That is all very well, but it has nothing to do with a description at the Planck scale. Of course, we live in time. Our knowledge is in time. Our knowledge is limited. But that has nothing to do with what we apprehend within the fundamental physical structure. I think there is confusion of style.

Alexei Grinbaum. Listening to your presentation, it occurred to me that the theory of quantum gravity resembled dangerously Alfred North Whitehead's [8] process philosophy. It seems that it would be wise to take a few precautions to distance ourselves from this vision, because in my opinion it is not at the right level of abstraction. We must clearly distinguish your use of words like "process", a word that also struck Whitehead, and your philosophical point of view, from his.

Bertrand Saint-Sernin. Or at least, I would say, Whitehead's followers. Einstein used to say that Whitehead's work on the theory of general relativity was very accurate, but seemed to him excessively complicated.

Alexei Grinbaum. It is not a masterpiece, that is for sure.

Alexis de Saint-Ours. The term process can also make us think of Gilbert Simondon.

Alexei Grinbaum. In this precise case, I do not think so.

Jean Petitot. Quantum mechanics is far removed from Simondon!

Alexei Grinbaum. In any case, some of the terms you use are dangerously labelled as Whiteheadian.

Carlo Rovelli. My initial considerations were vague, but I take note of your comment. I do not have a precise way of describing them. I simply wanted to highlight that I was not speaking of objects but of processes.

Alexei Grinbaum. Whereas words such as "relation", "relational" or "interaction" characterize relational quantum mechanics, words like "process" carry this label.

Bertrand Saint-Sernin. With Whitehead, the word that predominates is relation. Indeed, in his view, scientific progress consists of substituting independent disciplines with inter-related disciplines.

Alexei Grinbaum. Indeed. That being, the use of the word "relation" goes far beyond Whitehead, whereas the term "process" is quite Whiteheadian.

Bertrand Saint-Sernin. Or at least bears the stamp of Whitehead's followers.

Hervé Zwirn. You speak of a theory where time is not a parameter at the start, but emerges as a trajectory of the metric. It is a way of defining time, which is not an initial variable. In string theories, certain ways of performing symmetry breaking lead to structures in which there is no time. At least, that is how it is often presented. I have never really understood what that meant and I would like to know how you link loop quantum gravity and its equivalent in string theories, i.e. symmetry breaking, where the final geometrical structures generate models of the universe where time does not feature at all. Is there a link between the two? How is this opposition understood mathematically? What does a universe without time mean?

Carlo Rovelli. String theory is a perturbative theory of a given space-time, with things in common with what I have just presented. I am thinking in particular of the structure of dimensions which is the same. There is a fundamental structure. In both cases, it is not a field theory in the usual sense. However, the theory I have just presented is a total theory which is not the case for string theory. But I do not always find my way in this great world of possibilities. It is like a fish that we never manage to catch! String theory lives on the hope that there is a unified whole. But no one is capable of formulating it. It involves more mathematical calculations, but less clarity with regards to the conceptual structure on which it is based.

Hervé Zwirn. Is there in loop quantum gravity theory an equivalent to the countless possibilities of the types of universes that can coexist in string theory? Let us imagine that string theory is true and that the many different universes that it is capable of describing exist, as some would claim—in a realist sense. How does loop quantum gravity adapt to this situation?

Carlo Rovelli. Let us take quantum chromodynamics. It is a coherent theory of strong interactions. It is a very nice theory. But it is a closed theory. We can build a quantum theory of strong interactions while forgetting that there are weak electron interactions or gravity, for example. It is therefore not necessary to have a unified theory of everything to understand strong interactions. In the same way, I think it is not necessary to have a unified theory of everything to understand gravitational interactions. Far from me the idea that quantum gravity theory is the ultimate knowledge we can have of the world. However, I think that we are not capable of making predictions and that we do not have the conceptual structure capable of bringing together what Einstein discovered on the absence of a fixed space-time and quantum mechanics. What I present is a possible solution—although perhaps only a part of a greater whole.

Hervé Zwirn. A future theory could therefore be a sort of conjunction between string theory and quantum gravity as a structure. However, we often tend to oppose them.

Carlo Rovelli. They could, indeed. That being, I think string theory is going in the wrong direction. It is a very complex intellectual conception. But others will say the opposite!

Mathieu Guillermin. The previous comment regarding Lee Smolin brings a question to mind. I am thinking of those attempts by Robert Griffiths [9] and others in trying to consider that the fact that decoherence takes place is not specifically linked to human observation, but to interactions between systems, which allows us to grant an independent existence to entities described by theory, even if we cannot do so solely from experiments. Does your approach contain this type of decoherence phenomena which would allow us to make this leap? Or does this type of constraint not exist?

Carlo Rovelli. The discovery of decoherence is very interesting and enlightening. It is very useful for explaining why certain things do not take place or are not seen. For all that, it does not explain the mystery of quantum mechanics. What the relational interpretation contributes is the possibility of freeing ourselves from the idea that we need a particular observer to give sense to quantum mechanics. In Bohr's original interpretation, there was something special: the observer—"special" having a variety of meanings. Relational quantum mechanics is an attempt to reformulate the interpretation of quantum mechanics without anything special, the cost of this approach being precisely the relational aspect, the interaction between systems.

I consider that the relational interpretation is useful for doing quantum gravity. All interpretations of quantum mechanics are easily attacked and not easily defendable. However, for quantum gravity theory, the problem is clear and simple: can we construct a coherent theory with quantum mechanics, one that is not in contradiction with what we know of the world, which would have general relativity as classical limit and which would be finite? This problem is well-posed mathematically. It differs from the question "Is it the right description of the world?" To know if it is the right description of the world, we would need to carry out measurements to see if these predictions are verified. It is a preliminary step for constructing a coherent theory.

After more than forty years of research in quantum gravity, we have very few theories that are coherent and finite. String theory, for instance, is not capable of describing our world in four dimensions. A number of questions remain open.

Jean Petitot. You have spoken very little of the formalisms, but you have still explained the position of principle which consists of saying that we must quantify geometry, that we must go deep into what geometry is and apply a quantum approach to its fundamental concepts. We must quantify space and all the geometrical concepts. I believe you are fundamentally right. For example, you explain how structures of a simplicial type, which allow the sticking together of fragments of space into a larger space, can be quantified and you insist on the fact that intervening operators do not commute. Where are you now, without going into any detail, in this project of quantification of the concepts of Riemannian geometry? Have you managed to more or less carry out this quantification or are you still far from it?

Carlo Rovelli. We have practically finished, although it is always possible to go further. The structure is clear and simple: there is a Hilbert space and an operator algebra with Riemannian geometry as the classical limit. Everything is well-defined.

Jean Petitot. It should nonetheless involve some non-commutative geometry.

Carlo Rovelli. It is the case. We observe non-commutative structures. Many articles have broached this subject.

Jean Petitot. Do you find things like cyclic cohomology, for example?

Carlo Rovelli. No. For all that is topology, we do not. As I have said, the question remains open. However, the structure is well-defined. It is the structure of a representation of an algebra system, from which we understand how coherent states represent Riemannian geometry.

Bernard d'Espagnat. Thank you again, dear Carlo, for such an informative presentation on the main guiding principles that matter to us.

References

1. Cambridge University Press, 2007.
2. See session VII, "The relational interpretation in quantum mechanics and the EPR paradox".
3. European organization for nuclear research, known as the *Conseil Européen pour la Recherche Nucléaire*.
4. Niels Bohr & Léon Rosenfeld, "Zur Frage der Messbarkeit der elektromagnetischen Feldgrössen", Kgl. Danske Videnskabernes Selskab Mat.-Fys, 12(8), 1933 ("Field and Charge Measurements in Quantum Electrodynamics", Physical Review, 78, 1950, 794–798).
5. The Planck mass is the mass where the Schwarzschild radius is equal to the Planck length of $1{,}6.10^{-33}$ centimeter, below which it becomes a black hole.
6. Physicist who contributed alongside Carlo Rovelli to the birth of this theory.
7. Mathematician who also worked on general relativity.
8. British philosopher, logician and mathematician.
9. Physicist who also worked with Roland Omnès and Murray Gell-Mann on decoherence.

Titles in this Series

Quantum Mechanics and Gravity
By Mendel Sachs

Quantum-Classical Correspondence
Dynamical Quantization and the Classical Limit
By Dr. A.O. Bolivar

Knowledge and the World: Challenges Beyond the Science Wars
Ed. by M. Carrier, J. Roggenhofer, G. Küppers and P. Blanchard

Quantum-Classical Analogies
By Daniela Dragoman and Mircea Dragoman

Life—As a Matter of Fat
The Emerging Science of Lipidomics
By Ole G. Mouritsen

Quo Vadis Quantum Mechanics?
Ed. by Avshalom C. Elitzur, Shahar Dolev and Nancy Kolenda

Information and Its Role in Nature
By Juan G. Roederer

Extreme Events in Nature and Society
Ed. by Sergio Albeverio, Volker Jentsch and Holger Kantz

The Thermodynamic Machinery of Life
By Michal Kurzynski

Weak Links
The Universal Key to the Stability of Networks and Complex Systems
By Csermely Peter

The Emerging Physics of Consciousness
Ed. by Jack A. Tuszynski

B. d'Espagnat and H. Zwirn (eds.), *The Quantum World*,
The Frontiers Collection, DOI 10.1007/978-3-319-55420-4

Quantum Mechanics at the Crossroads
New Perspectives from History, Philosophy and Physics
Ed. by James Evans and Alan S. Thorndike

How Should Humanity Steer the Future
Ed. by Anthony Aguirre, Brendan Foster and Zeeya Merali

Mind, Matter and the Implicate Order
By Paavo T.I. Pylkkanen

Particle Metaphysics
A Critical Account of Subatomic Reality
By Brigitte Falkenburg

The Physical Basis of the Direction of Time
By H. Dieter Zeh

Asymmetry: The Foundation of Information
By Scott J. Muller

Decoherence and the Quantum-To-Classical Transition
By Maximilian A. Schlosshauer

The Nonlinear Universe
Chaos, Emergence, Life
By Alwyn C. Scott

Quantum Superposition
Counterintuitive Consequences of Coherence, Entanglement, and Interference
By Mark P. Silverman

Symmetry Rules
How Science and Nature are Founded on Symmetry
By Joseph Rosen

Mind, Matter and Quantum Mechanics
By Henry P. Stapp

Entanglement, Information, and the Interpretation of Quantum Mechanics
By Gregg Jaeger

Relativity and the Nature of Spacetime
By Vesselin Petkov

The Biological Evolution of Religious Mind and Behavior
Ed. by Eckart Voland and Wulf Schiefenhovel

Homo Novus-A Human without Illusions
Ed. by Ulrich J. Frey, Charlotte Stormer and Kai P. Willfiihr

From the Web to the Grid and Beyond
Computing Paradigms Driven by High-Energy Physics
Ed. by Rene Brun, Federico Carminati and Giuliana Galli Carminati

The Language Phenomenon
Human Communication from Milliseconds to Millennia
Ed. by P.-M. Binder and K. Smith

The Dual Nature of Life
By Gennadiy Zhegunov

Natural Fabrications
By William Seager

Ultimate Horizons
By Helmut Satz

Physics, Nature and Society
By Joaquín Marro

Extraterrestrial Altruism
Ed. by Douglas A. Vakoch

The Beginning and the End
By Clément Vidal

A Brief History of String Theory
By Dean Rickles

Singularity Hypotheses
Ed. by Amnon H. Eden, James H. Moor, Johnny H. Søraker and Eric Steinhart

Why More Is Different
Philosophical Issues in Condensed Matter Physics and Complex Systems
Ed. by Brigitte Falkenburg and Margaret Morrison

Questioning the Foundations of Physics
Ed. by Anthony Aguirre, Brendan Foster and Zeeya Merali

It From Bit or Bit From It?
Ed. by Anthony Aguirre, Brendan Foster and Zeeya Merali

Trick or Truth?
Ed. by Anthony Aguirre, Brendan Foster and Zeeya Merali

The Challenge of Chance
Ed. By Klaas Landsman

Quantum (Un)Speakables II
Half a Century of Bell's Theorem
Ed. by Reinhold Bertlmann, Anton Zeilinger

Energy, Complexity and Wealth Maximization
Ed. by Robert Ayres

Ancestors, Territoriality And Gods
By Ina Wunn

Space,Time and the Limits of Human Understanding
By Sham Wuppuluri

Information and Interaction
Ed. by Durham

The Technological Singularity
Ed. by Victor Callaghan

Printed in the United States
By Bookmasters